Studies in Economic Design

Economic Design comprises the creative art and science of inventing, analyzing and testing economic as well as social and political institutions and mechanisms aimed at achieving individual objectives and social goals. The accumulated traditions and wealth of knowledge in normative and positive economics and the strategic analysis of Game Theory are applied with novel ideas in the creative tasks of designing and assembling diverse legal-economic instruments. These include constitutions and other assignments of rights, mechanisms for allocation or regulation, tax and incentive schemes, contract forms, voting and other choice aggregation procedures, markets, auctions, organizational forms such as partnerships and networks, together with supporting membership and other property rights, and information systems, including computational aspects.

The series was initially started in 2002 and with its relaunch in 2017 seeks to incorporate recent developments in the field and highlight topics for future research in Economic Design.

More information about this series at http://www.springer.com/series/4734

Walter Trockel
Editor

Social Design

Essays in Memory of Leonid Hurwicz

Springer

Editor
Walter Trockel
Center for Mathematical Economics (IMW)
Bielefeld University
Bielefeld, Germany

ISSN 2510-3970 ISSN 2510-3989 (electronic)
Studies in Economic Design
ISBN 978-3-319-93808-0 ISBN 978-3-319-93809-7 (eBook)
https://doi.org/10.1007/978-3-319-93809-7

This Springer imprint is published by the registered company Springer Nature Switzerland AG
The registered company address is: Gewerbestrasse 11, 6330 Cham, Switzerland

Leonid Hurwicz and Walter Trockel at Bielefeld University on December 4th, 2004

Foreword

Take the 1969 paper in the *American Economic Review* and the 1972 chapter in the *Decision and Organization* book as two critical moments in the broadcasting of Leo Hurwicz's message to the economic profession (they were earlier ones, but to many young researchers in those days, like myself, this is where we discovered his work) and consider their intellectual context. The field then called "Mathematical Economics" is booming and only a few years away from bursting into the main stream of our profession.

This is happening around three main themes.

First, the successful formalization of the competitive equilibrium by Arrow and Debreu (not to forget Allais and McKenzie) provides rock-solid logical foundations to the central concept of economic analysis, going much beyond Walras' ideas by elucidating precisely which mathematical assumptions, on preferences and feasible transactions, ensure its existence and stability, or not.

Second, von Neumann and Morgenstern's early intuition that the behavior of economic agents in the small, e.g., face-to-face, is logically related, perhaps even isomorphic, to that of players in strategic games like chess or poker is slowly developing into the next paradigm of economic modeling, through the pioneer work of Shubik (*Strategy and Market Structure*, 1959), Debreu and Scarf (*Equivalence of the Core and Competitive Equilibrium*, 1963), Shapley and Shubik (*Market Games*, 1969, and *Assignment Games*, 1971), and Gale and Shapley (*Stable Matching*, 1962: still a mathematical curiosity at this point).

Third, the spectacular entrance of the axiomatic methodology in the social sciences, in Arrow's Social Choice Model (1951), states the first genuine mechanism design question in the language of economics. It promptly initiates much research on the meaning and design of "good" voting rules; a particularly challenging conjecture is the nonexistence of reasonable and strongly incentive-compatible (strategyproof) voting rules for more than two candidates (Dummett and Farquharson, 1961). It is about to be proved by Gibbard (1973) and Satterthwaite (1974).

Thus, the key modeling tools for economics as we know it today are already present in these critical two decades of the 1950s and 1960s. But they are mostly developing, so to speak, in parallel universes. To my mind, Leo Hurwicz's fundamental contribution was to set for his fellow economists a goal of a higher order, a "theory of everything" that has been the mechanism design program ever since: we need to understand what high-level goals can be implemented in possibly yet to be built economic institutions, once we take into account informational constraints and strategic incentives (what Myerson aptly calls adverse selection and moral hazard constraints).

This volume offers a contemporary snapshot of Leo Hurwicz's influence, captured in the work of a small set of scholars who may or may not have been in personal contact with him, but all acknowledge their intellectual debt to him by contributing to this Festschrift.

For the reader unfamiliar with Hurwicz's specific contributions, I recommend Postlewaite and Schmeidler' effective review of his approach to the implementation problem and the difficulties still ahead. Then Myerson explains Hurwicz's own inspiration in the historical debate on the feasibility of socialism and goes on to develop a rich interpretation of his 1998 "But who will guard the guardians?" paper. Finally, the reprinted 1995 chapter by Hurwicz, Maskin, and Postlewaite is not a comment on his work but the real thing.

Classic implementation theory, the one that starts in a straight line from Hurwicz to Maskin and beyond, accounts, appropriately, for one-third of the volume, authored by some of the main actors in the field then and now. Edelman and Weymark discuss an extension of Rochet's theorem about implementation by cash transfers. Dutta explains how the small twist of endowing agents with a costless preference for honesty turns the theory upside down. Peleg and Peters find new relations between implementation in strong equilibrium and elimination voting procedures. D'Aspremont and Cremer explain why implementation in the Bayesian context may or may not align incentives and full efficiency. Roy, Sadhukhan, and Sen extend and qualify the results of Barbera, Sonnenschein, and Zhou on the strategyproof formation of a committee when the voting rule is random. Barbera, Berga, and Moreno explore the incentives of experts in Condorcet's celebrated jury problem and uncover relevant properties of the information structure. Finally, Hammond, in the only nontechnical paper of this batch, reinterprets mechanism design a la Hurwicz as a way to avoid "institutional externalities."

The next third is a bouquet of mainstream research in economic theory. Ledyard discusses an important informational aspect of the cap and trade mechanism for the provision of a public bad. Kannai and Raimondo offer a very general existence result for financial markets. Marschak and Wei show that in the simplest format of the principal agent problem, the welfare consequences of an improvement of the monitoring technology are not straightforward. Thomson deploys the axiomatic methodology to evaluate a simple rationing rule between two agents. La Mura shows that correlating strategies by quantum bits can improve cooperation in auctions. And two papers report on experiments testing new mechanisms to solve traditional social dilemmas: Van Essen and Walker for the public good provision problem and Saijo for the prisoners' dilemma.

Finally, three papers suggest new questions for the mechanism design of tomorrow. Demange discusses common misinterpretations of the algorithms that surround us in the age of Internet and derives new challenges for the designer. Vega Redondo proposes network design as a new research direction: e.g., the proactive design of financial networks can insure banks optimally against both small frequent shocks and large unfrequent ones. Chiu and Koeppl explain why blockchain systems are not immune to tampering, a difficulty that mechanism design should work to alleviate.

Glasgow, UK
May 2018

Hervé Moulin

Contents

In Lieu of an Introduction: How I Remember Leonid Hurwicz

Walter Trockel

1 On This Volume

All essays in this volume have been contributed by co-laureates, co-authors, colleagues, or former students of Leonid Hurwicz or by contributors to fields he had founded, fundamentally contributed to, or just successfully worked on. I am very grateful to all of them, a "star spangled list of contributors" as Robert Aumann had remarked in a recent letter to me, that they by their accepting my invitation and by their valuable contributions have made this exciting volume in memory of Leonid Hurwicz possible.

The timing for this book project, that with the permanent invaluable editorial support of Martina Bihn was started in 2017 and finished with the publication of the volume in 2019, has been selected in order to simultaneously commemorate three anniversaries: the birth of Leonid Hurwicz on August 21, 1917, the award of the Prize in Economic Science in Memory of Alfred Nobel to him on October 15, 2007, and the day of his death on June 24, 2008.

My motivation for trying to realize this present book resulted from my admiration for Leo Hurwicz as a great person and an outstanding scientist. It was my intention to contribute to increase his own and his work's popularity, in particular among young people in our profession.

I feel that the abstracts of these essays make a topical introduction unnecessary. Instead, I want to focus on the commemoration of Leo. Because information about most aspects of his adventurous life and his outstanding scientific work is publicly available, I will just describe how I remember some of my encounters with Leo and with his work.

W. Trockel (✉)
Center for Mathematical Economics (IMW), Bielefeld University, Bielefeld, Germany
e-mail: walter.trockel@uni-bielefeld.de

W. Trockel (ed.), *Social Design*, Studies in Economic Design,
https://doi.org/10.1007/978-3-319-93809-7_1

2 Leo Hurwicz

The first time I had met Leo Hurwicz was in 1981 when, during a stay in Berkeley, I was invited for a seminar at the University of Minnesota. Leo came to my talk—about 15 minutes late. After the talk when I introduced myself to him he revealed to my surprise that he knew my dissertation from 1974. In the following years, I met him several times at conferences. I became interested in mechanisms and implementation, read several of his articles, and was impressed, in particular, by his conceptual clarity. It turned out that we shared many interests, and I enjoyed the opportunities to listen to and learn from him, whether he talked on economic theory, politics, music, or ancient civilizations, and became influenced by him. During the Conference on Economic Design at Istanbul's Boğaziçi University in 2000, Leo had suggested to me to take the next opportunity to visit Ḫattuša, the ancient capital of the Hittites, in Anatolia. When I was visiting Bilkent University in 2001 I followed his advice.

On one evening in 2002, during the next Conference on Economic Design at the NYU, Leo and I found ourselves to be the two last persons still in the hall of the conference building. He proposed to join for dinner and asked, whether I would mind to have that close to his hotel. Our dinner and intensive conversation till late in the night became the starting point for friendly personal relations during the years to follow.

When I started with the organization of the 12th European Workshop on General Equilibrium Theory at Bielefeld University in December 2003, I thought about enriching the usual program by adding a plenary lecture and asked Leo whether he would accept an invitation to give such a lecture at the workshop. He accepted the invitation and promised to come, even though I could not guarantee the financing of a business class ticket. At that time Leo was 86 years old! After his arrival in Bielefeld, he suggested to me to choose as the time for his lecture the last session before lunch. "That gives me an incentive to finish in time" he remarked with a smile. The workshop was a success and so were Leo's presence and presentation. The plenary lecture became institutionalized as *Debreu Lecture* at the next and for every following European Workshop.

Many young researchers were impressed and attracted by Leo's personality and behavior. One young woman from a regional newspaper who had interviewed Leo asked him, impressed by his vitality, for an advice how one could reach an age of 90 years. Leo smiled at her and answered: "First you have to make sure that you reach the 89 and then you behave for one year very carefully."

One year later, on December 3, 2004, Leo Hurwicz was awarded the academic degree of a *Doctor Rerum Politicarum Honoris Causa* by the Faculty of Economics and Management of Bielefeld University. In my *laudatio* I also stressed the sufferings of Leo and his family from the Germans twice in his lifetime. In his following acceptance speech, Leo remarked that in 1939 he never would have expected that something so good could come to him from Germany.

Two days later when he left Bielefeld, I asked him why I had not observed any bitterness in his conversations with the young German students. His answer was:

"There was no bitterness." Leo Hurwicz had impressed everybody by his warm-hearted behavior and his charisma.

We stayed in contact in the following years but did not meet until 2007. When coming home from a one month stay in California at end of March 2007, I found a letter with an invitation to participate in the workshop *Perspectives on Leo Hurwicz—A Celebration of 90 Years* at the University of Minnesota on April 14, 2007. In Minneapolis I found Leo lined with stress from the regular dialysis but mentally crystal clear. During that *Celebration* several conversations centered on a theme that somehow was wafting through the room: Why not a Nobel Prize for Leo? Exactly half a year later he had been awarded, together with Eric Maskin and Roger Myerson, the Sverige's Riksbank Prize in Economic Sciences in Memory of Alfred Nobel 2007 *for having laid the foundations of mechanism design theory*. Leo Hurwicz said at a news conference: "There were times when other people said I was on the short list, but as time passed and nothing happened, I didn't expect the recognition would come because people who were familiar with my work were slowly dying off, ..."

So, the *recognition had come*, and there is good reason to predict that the set of *people familiar with his work* will not die off. In fact, Leo Hurwicz's memorable work is encompassing much more than mechanism theory. The impressive spectrum of his relevant scientific contributions had been recognized in his Bielefeld honorary doctoral document that in addition to his "foundation and economic application of mechanism design theory" lists his "*groundbreaking work on the allocation of resources, his fundamental contributions to the analysis of the stability of competitive equilibria, his pioneering work on decision theory, and the theory of organization and of enterprises, the enrichment of economic theory by innovative mathematical methods and important results in Econometrics and Optimization theory, his early recognition of the importance of game theory for social sciences and his innovative ideas on genuine implementation and the design of effective institutions*."

Sometimes Nobel prizes had been awarded for outstanding past achievements even when these had turned out to be not so fruitful and promising anymore. This is definitely not the case with the theory of implementation and mechanism design whose methods have recently been used in other fields like operations research, computer science, engineering, artificial intelligence, medicine, and law.

The scientific work of Leonid Hurwicz had deeply impressed me. But, also his conceptual clarity made him a role model. He never became tired to emphasize the obvious, when necessary, like the difference between games and game forms. I have learned from him in my own publications to insist on the differences between solutions and social choice rules, between noncooperative support and implementation and to see the abundance of possibilities to factorize payoff functions into outcome functions and utility functions in contrast to the problem of finding one such factorization that would be consistent with a given outcome space (cf. Trockel 2000, 2002a, b; Haake and Trockel 2010).

Yet, what I admired most was Leo's personality, and his just "being a 'Mensch.'"

3 Enforcement and Genuine Implementation

The broad relevance of the Theory of Mechanism Design for so many facets of human society under various aspects together with the late Leo Hurwicz's interest in the roles of implementation and enforcement in institutional modeling, what he described by the term *genuine implementation*, has motivated the title *Social Design* for this volume.

When I am observing the world with wars and refugees everywhere, with more and more autocrats and enormously increasing ideological "-isms" and pressure on democratic structures, with growing pollution, racism, violence, selfishness, greed, and disdaining science and facts, with big companies behaving unethically, or often even illegally, developing high-tech cheating technologies, I feel that the only way out to a more humane society may come from intelligent social design and effective enforcement of suitable institutions as indispensable tools for almost all aspects of organizing sustainable cooperative and fair structures on our planet.

It is this conviction that drives me to direct the readers' attentions in particular toward the late work of Leo Hurwicz, that concentrated on *institutional modeling*, as, for instance, in Hurwicz (1987, 1993, 1994, 1996, 1998). Though the remaining pages will be a little more technical, I will forgo a coherent detailed discussion and instead refer to the excellent essay of Myerson (2009) that is reprinted in this volume.

In Hurwicz (1998)—a revised version of which is his Nobel speech *But who will guard the Guardians* published in Hurwicz (2008)—the problem of *genuine implementation* is addressed and the question is asked "Are Nash equilibria self-enforcing?" Then Hurwicz hints at the fact that in Nash equilibria there may be advantageous unilateral deviations for players to physically feasible but illegal strategies. That insight leads him first to the question of how to *enforce* legal strategies and then to the distinction between a given "legal game" and a bigger "true game," in which the legal game is embedded. Hurwicz (1998, p. 9) introduces two notions of "successful enforcement": "To say that the *legal game rules are being successfully enforced* means that the outcomes of the true game ensure that illegal strategies are less attractive than legal strategies. A strong formulation of successful enforcement might require that for every player, every illegal strategy is *dominated* by (that is, is less attractive than) some legal strategy. A weak domination would require only that a player at least be no worse off by staying within the law. However, this may be asking too much: if everyone else is acting illegally, a player may not find it possible to remain law-abiding. It seems, therefore, more reasonable to adopt a somewhat *weaker concept of successful enforcement* of the rules of a given mechanism ..."

This weaker concept defines successful enforcement of the *legal game* G in the *true game* H by the postulate that the sets of Nash equilibria of G and of H coincide and are nonempty. Myerson (2009) suggests a weakening of successful enforcement of the l*egal game* G in the *true game* H by requiring the set of strategy profiles of G to be *c*losed *u*nder *r*ational *b*ehavior, making it a *CURB set*, as defined by Basu and Weibull (1991), in the set of strategy profiles of the game H. None of these two weaker definitions does imply the validity of the other one.

Hurwicz (1998) concludes his section on *Successful enforcement and implementation* by answering his previous question: "*A reason why Nash equilibria cannot be considered self-implementing is that the assumption of the effectiveness of the outcome function h (.) hides the need for institutional arrangements typically required to accomplish this*."

When it comes to *institutional design* the distinction between the legal and the true game is only the starting point. If players of a higher institutional layer determine and control the rules of the true game on a lower level by acting as the *guardians* for these rules, we are confronted with Hurwicz's question: "*But who will guard the guardians?*"

In his section *Back to Juvenal* Hurwicz (1998) comes to the conclusion that "*either there is no way to guard the guardians*" or one needs also "*guardians of the second order*" for guarding these guardians. So Hurwicz continues (1998, p. 11): "*But then, if those are also subject to corruption, guardians of the third order are also necessary, and so on. This conjures the image of an infinite regress of guardians, with the guardian of order k needed to guard the guardian of order k,*[1] *with k = 2, 3 ... ad infinitum. If an infinity of guardians is not usually available, this seems to preclude the possibility of enforcement*." Nevertheless, Hurwicz argues, based on "*casual empiricism*", that in many situations rules are in fact implemented or enforced, "*even if not perfectly*" and continues: "*Somewhere at a finite end in the chain of guardians, there may be guardians (individual or collective) who are in sympathy with the rule (game-form) that makes certain behavior illegal, e.g., whose ethical standards rule out corrupt behavior, and who have the ability (through power, financial assets, personal charisma or status, combined with the population's respect for it), as well as the inclination to act so as to discourage improper behavior of the guardians of lower order. In such a situation the rule is likely to be successfully enforced. Well-functioning societies try to choose judges and rulers from among such individuals*".

In several of his articles these special guardians are called *intervenors*. Hurwicz (1993, p. 59) remarks: "Beyond modeling issues there is an interesting substantive question. To what extent is the presence of intervenors essential for the achievement of social goals when enforcement is necessary?"

Hurwicz (1998, pp. 12, 13) provides a (partial) answer when he writes: "But we do not have to rely on the presence of intervenors. There are other structures conducive to successful enforcement." Then he suggests finitely many layers, say *k*, of game forms where the citizens playing the legal game of the first layer are simultaneously the guardians of *k*-th order controlling the game form on the last layer. He concludes: "This type of structure is also closely related (but not identical with) the notion of separation of powers."

[1]This *k* is a typo and has in fact to be $k - 1$ (my correction).

Myerson (2009) described and discussed in detail Hurwicz's ideas and indicated further potential developments. Unfortunately, there appears to be, apart from Myerson (2004, 2008, 2009), not much theoretical literature concerning enforcement, genuine implementation, and their relation to a fundamental theory of institutions. There is, however, the recent article by Myerson and Weibull (2015) where they combine their CURB set approach to successful enforcement with Schelling's (1960) *focal points*.

Another promising new line of research has opened up recently relating Hurwicz's model of successful enforcement to the notion of a *Social System* of Debreu (1952). This concept, also known in the literature as *abstract economy* or *generalized game*, turns out to be closely related to the Hurwicz approach based on the distinction between legal and illegal strategies (cf. Trockel and Haake 2017).

I would be very happy to see this volume attract many young researchers to the theory of mechanism design and convince them of its general importance and, in particular, its relevance for a *Hurwicz Program* of developing a complete fundamental theory of institutions and their genuine implementation.

References

Basu, K., & Weibull, J. (1991). Strategy subsets closed under rational behavior. *Economics Letters, 36*, 141–146.

Debreu, G. (1952). A social equilibrium existence theorem. *Proceedings of the National Academy of Sciences (USA), 38*, 886–893.

Haake, C.-J., & Trockel, W. (2010). On Maskin monotonicity of solution based social choice rules. *Review of Economic Design, 14*, 17–25.

Hurwicz, L. (1987). New institutions: The design perspective. *American Journal of Agricultural Economics, 69*, 395–402.

Hurwicz, L. (1993). Implementation and enforcement in institutional modeling. In Barnett, Hinich, & Schofield (Eds.), *Political economy: Institutions, competition and representation*. New York: Cambridge University Press.

Hurwicz, L. (1994). Economic design, adjustment processes, mechanisms, and institutions. *Review of Economic Design, 1*, 1–14.

Hurwicz, L. (1996). Institutions as families of game forms. *Japanese Economic Review, 47*, 113–132.

Hurwicz, L. (1998). *But who will guard the guardians*. University of Minnesota paper. http://www.econ.umn.edu/workingpapers/hurwicz_guardians.pdf, revised for Nobel Lecture in Hurwicz (2008).

Hurwicz, L. (2008). But who will guard the guardians? *American Economic Review, 98*, 577–585.

Myerson, R. B. (2004). Justice, institutions, and multiple equilibria. *Chicago Journal of International Law, 5*, 91–107.

Myerson, R. B. (2008). The autocrat's credibility problem and foundations of the constitutional state. *American Political Science Review, 102*(1), 125–139.

Myerson, R. B. (2009). Fundamental theory of institutions: A lecture in honor of Leo Huwicz. *Review of Economic Design, 13*, 59–75.

Myerson, R., & Weibull, J. (2015). Tenable strategy blocks and settled equilibria. *Econometrica, 83*, 943–967.

Schelling, T. C. (1960). *Strategy of conflict*. Cambridge, MA: Harvard University Press.

Trockel, W. (2000). Implementations of the Nash solution based on its Walrasian characterization. *Economic Theory, 16*, 277–294.

Trockel, W. (2002a). Integrating the Nash program into mechanism theory. *Review of Economic Design, 7*, 27–43.

Trockel, W. (2002b). A universal meta bargaining implementation of the Nash solution. *Social Choice and Welfare, 19*, 581–586.

Trockel, W., & Haake, C.-J. (2017). *Thoughts on social design* (Working Paper 577). Center of Mathematical Economics (IMW), Bielefeld University.

Part I
Institution Design

Technical Change and the Decentralization Penalty

Thomas Marschak and Dong Wei

1 Introduction

Does the case for decentralizing a firm get stronger or weaker when the production technology used by one or more of its divisions improves? Consider the Organizer of the firm, who seeks a good balance between the cost of the divisions' efforts and the revenue which those efforts yield. One way to achieve a good balance may be intrusive but perfect monitoring and policing, which fully reveals the chosen efforts and guarantees that they are those the Organizer prefers.

Perfect monitoring/policing may be very costly. A better mode of organizing might be "decentralization," where the divisions are totally autonomous, though their choices may be influenced by appropriate rewards and penalties. In the decentralized mode that we shall study there is a Principal who treats each division as an Agent. Each Agent freely chooses her effort and bears the effort's cost. The Principal observes the realized revenue and rewards the Agents. Each Agent's reward is a function of revenue, and her net earnings are her reward minus the cost of her chosen effort. The reward functions the Principal chooses are acceptable to the Agents and are preferred by the Principal to other possible reward functions that are also acceptable to the Agents. The Principal pockets the residual revenue which is left over after the rewards have been paid. When an Agent's technology improves, the cost of a given effort drops.

The Organizer compares the decentralized Principal/Agents mode with perfect monitoring/policing. Many production technologies rapidly improve, but at the

T. Marschak (✉)
Walter A. Haas School of Business, University of California, Berkeley, CA, USA
e-mail: marschak@berkeley.edu

D. Wei
Department of Economics, University of California, Berkeley, CA, USA

W. Trockel (ed.), *Social Design*, Studies in Economic Design,
https://doi.org/10.1007/978-3-319-93809-7_2

same time the costs of perfect monitoring may rapidly drop as well, because of dramatic advances in monitoring techniques. So, the relative merit of the two modes requires regular reassessment. We shall let the Organizer take a "welfare" point of view in comparing the two modes. The Organizer's focus is the firm's surplus: the revenue earned by the divisions' efforts minus the cost of those efforts. Perfect monitoring/policing guarantees maximal surplus. The Decentralization Penalty is the welfare loss due to decentralizing. It is the gap between maximal surplus and the surplus achieved in the decentralized Principal/Agents mode.

Our central question is whether the Decentralization Penalty grows or shrinks when a technical advance lowers Agents' effort costs. If the Penalty substantially grows, then perfect monitoring may now be worth what it costs. (We will not explicitly model the cost of monitoring). If the Penalty shrinks, then perfect monitoring becomes less attractive even if the monitoring techniques have advanced. Our central question is tricky for the following reason: when the Agents' technology improves, maximal surplus rises (under weak assumptions). Maximal surplus is a "moving target." Decentralized surplus also rises, under reasonable assumptions. But, that does *not* mean, in general, that as technology improves, the rising decentralized surplus gets *closer* to the moving surplus target. Our question appears to be very rarely asked in the abundant Principal/Agent literature. The cost of an Agent's effort appears in many papers and so does the welfare loss due to Agents' second-best choices. But, the effect of cost improvement on welfare loss seems to be widely neglected.

2 The Model

We shall study a highly simplified model. There is a single effort variable x, chosen from a set $\Sigma \subseteq \mathbb{R}^+$ of possible positive efforts. The set Σ may be finite or it may be a continuum. There is no uncertainty about the consequences of a given effort. The effort x generates a positive *revenue* $R(x)$, where R is strictly increasing. The effort x costs $t \cdot C(x)$, where C is positive and strictly increasing. A drop in t occurs when technology improves (or there is a fall in the price of the inputs which effort requires). For a given t, we consider the *surplus at the effort* x, denote $\tilde{W}(x, t)$. Thus,

$$\tilde{W}(x, t) = R(x) - t \cdot C(x).$$

In the centralized mode, perfect monitoring/policing guarantees that effort is "first-best": it maximizes surplus. In the decentralized mode, there is no direct monitoring. Instead, there is a self-interested Principal and a single self-interested Agent who freely chooses $x \in \Sigma$ and bears the cost $t \cdot C(x)$. The functions R and C, and the technology parameter t, are known to both parties. The Principal observes the revenue $R(x)$. Since R is strictly increasing, that observation also reveals the

Agent's chosen x. The Principal rewards the Agent, using a reward which is a function of the observed revenue. We study an extremely simple reward scheme, namely linear revenue sharing. The Principal pays the Agent a *share* $r \in (0, 1]$ of the revenue. So, if the Agent chooses the effort x, she earns $rR(x) - t \cdot C(x)$ and the net amount received by the Principal is the residual $(1 - r) \cdot R(x)$. We will assume that for every (r, t) there is an effort $x \in \Sigma$ such that the Agent's gain $rR(t) - tC(x)$ is nonnegative, and this is sufficient for the Agent to be willing to participate. The Agent chooses to exert the effort $\hat{x}(r, t)$, the smallest maximizer of $rR(x) - t \cdot C(x)$ on the set Σ. We denote *the surplus when the share is r* by $W(r, t)$. So,

$$W(r, t) \equiv \tilde{W}(\hat{x}(r, t), t) = R(\hat{x}(r, t)) - t \cdot C(\hat{x}(r, t)).$$

Note that if $r = 1$, then the Agent's effort choice $\hat{x}(1, t)$ is surplus-maximizing. Thus,

$$W(1, t) = \tilde{W}(\hat{x}(1, t), t) \textit{ is the largest possible surplus.}$$

In the centralized mode, perfect monitoring/policing insures that $W(1, t)$ is achieved.

In our study of the Decentralization Penalty, we consider two cases. In the *exogenous* case, the reward share is determined outside the model. It might, for example, be the result of previous bargaining between Principal and Agent, or it might be prescribed by law. In the *endogenous case*, the Principal considers all the shares in the open interval $(0, 1)$ and chooses a share which maximizes $(1 - r) \cdot R(\hat{x}(r, t))$, the residual when the Agent uses the best-effort function $\hat{x}$ in responding to a given share. We let $r^*(t)$ denote the maximizer which the Principal chooses. So, in the endogenous case the Agent's effort is $\hat{x}(r^*(t), t)$ and surplus is $\tilde{W}(\hat{x}(r^*(t), t), t) = W(r^*(t), t)$.

3 The Main Results

The "moving target" remark that we made above suggests that the effect of a drop in t on the Decentralization Penalty is subtle. On the other hand, it is hard to imagine a model simpler than ours. So, one might hope that in our simple model there are simple conditions on Σ, R, C under which the Penalty rises (falls) when t drops. It turns out, however, that even in our model there is a striking diversity of results. There are simple examples where the Penalty rises and simple examples where it falls. That is the case in both the exogenous and endogenous settings.

There are, however, basic results that do not directly concern the Penalty and hold for all examples, whether the effort set is finite or a continuum, and whether or not the functions R, C are differentiable. Exogenous-case basic results are given in Theorem 1 and endogenous-case basic results in Theorem 2. The main findings of Theorem 1 are that the Agent never works less hard when the share rises and when technology improves (t drops); that surplus cannot fall when t drops and maximal ("first-best") surplus must rise; that a drop in t is never bad news for the Principal, must be good news for the Agent, and is never bad news from the welfare point of view. A final finding is that a rise in the share r is never bad from the welfare point of view and is good if and only if the Agent's effort changes. Thus, if r is determined through Principal/Agent bargaining, then it is in the "social" interest to strengthen the bargaining power of the Agent (who prefers larger values of r).

There are fewer basic results for the more difficult endogenous case, where the share is $r^*(t)$, a maximizer of $(1 - r) \cdot R(\hat{x}(r, t))$ on the interval $(0, 1)$. Theorem 2 finds that when t drops, there cannot be a fall in the ratio $\frac{r^*(t)}{t}$ or in the Agent's effort $\hat{x}(r^*(t), t)$. A drop in t, moreover, can never be bad news for the Principal and must be good news from the welfare point of view.

In the examples which follow, we obtain very diverse results about the Decentralization Penalty. To bring some order to this diversity, we divide examples (Σ, R, C) into classes. To do so, we consider the effect of a drop in t on the example's best effort $\hat{x}$ and on the example's endogenous-case best share r^*. A higher share stimulates the Agent to work harder, but the strength of the stimulus depends on t. Consider any pair of shares (r_L, r_H), where $0 < r_L < r_H < 1$, and suppose that t drops. In some examples, the effort increase $\hat{x}(r_H, t) - \hat{x}(r_L, t)$ rises and in other examples it falls. In some examples, moreover, the Principal's best share r^* rises when t drops, and in other examples it falls. So, we have four classes of examples.

For each class, we examine the *effort gap*—the amount by which decentralized effort falls short of the "first-best" effort. The effort gap is $\hat{x}(1, t) - \hat{x}(r, t)$ when r is exogenous and it is $\hat{x}(1, t) - \hat{x}(r^*(t), t)$ in the endogenous case. The effect of a drop in t on the effort gap is again tricky—just as it was for the Decentralization Penalty, or "surplus gap." Under broad conditions, both terms of the gap rise when t drops, but the gap itself may rise or fall.

The effect of a drop in t on the effort gap is interesting in itself. There are classes of examples, moreover, in which the effort gap *tracks* the surplus gap (the Decentralization Penalty): when t drops, the two gaps move in the same direction. Imagine that first-best effort $\hat{x}(1, t)$ has been studied for many triples (R, C, t). Then for an impending new technology t, first-best effort is already known but the welfare effects of decentralizing remain to be discovered. If we indeed have tracking, then it suffices to observe the Agent's work to see whether, with the new technology, her effort has moved closer to first-best effort or further away from it. In the former case, we know—if we indeed have tracking—that the new technology has shrunk the Decentralization Penalty, so it has made perfect monitoring/policing less attractive. In the latter case, it has increased the Penalty.

Our first result about tracking is Theorem 4, which requires R and C to be thrice differentiable. The theorem concerns the exogenous case. It considers the effectiveness of a share increase in stimulating higher effort and classifies examples according to the change in effectiveness when t drops, i.e., the sign of the cross partial $\hat{x}_{rt}(r,t)$. It finds that we indeed have tracking, provided that the following monotonicity condition holds: either $\hat{x}_{rt}(r,t) > 0$ at all (r,t) or $\hat{x}_{rt}(r,t) < 0$ at all (r,t). The theorem has a Corollary which directly addresses our central question for the exogenous case. It finds conditions on the signs of C'', R'', C''', R''' under which the Decentralization Penalty rises when technology improves (t drops) and conditions under which the Penalty falls.

As one would expect, the tracking question is more difficult in the endogenous case. Theorem 5 again classifies examples with regard to the effect of technical improvement on effectiveness (the sign of $\hat{x}_{rt}(r,t)$) but also classifies them with regard to the effect of technical improvement on the Principal's endogenous-case "generosity," i.e., the sign of the derivative $r^{*'}(t)$. It finds that we have tracking if either of the following conditions hold: (i) for every possible (r,t), $\hat{x}_{rt}(r,t) < 0$ and $r^{*'}(t) \geq 0$; and (ii) for every possible (r,t), $\hat{x}_{rt}(r,t) > 0$ and $r^{*'}(t) < 0$. There is now no Corollary, analogous to Theorem 4's Corollary, in which the effect of technical improvement on the Penalty is related to the signs of the second and third derivatives of R and C.

Theorem 6 shows that we cannot have an example where marginal revenue is declining ($R'' < 0$), and a drop in t (weakly) increases both effectiveness and the Principal's generosity (i.e., at every possible (r,t) we have both $\hat{x}_{rt}(r,t) \geq 0$ and $r^{*'}(t) \geq 0$).

Finally, Theorem 7 addresses bargaining between Principal and Agent over the share r. For a given t, consider the curve which shows the Principal's gain $(1-r) \cdot R(\hat{x}(r,t))$ as a function of $r \in (0,1)$. Suppose the curve is single-peaked, i.e., the gain rises to a peak at the principal's preferred share $r^*(t)$ and then falls (perhaps after an interval where it is flat). In the interval $(0, r^*(t))$, where the gain curve rises, both parties favor higher r. (That follows from the results in Theorem 1). The negotiation set—where one party prefers higher r and the other lower r—is the interval $(r^*(t), 1)$. The final result in Theorem 1 showed that welfare increases when the exogenous share increases. So—informally speaking, it is in the "social" interest for the Agent's bargaining strength to be high. Moreover, if r^* is increasing (decreasing) in t, then the negotiation interval $(r^*(t), 1)$ shrinks (widens) when technology improves. To conclude that this shrinkage (widening) raises or lowers the share on which the bargainers finally agree would require a precise model of the bargaining procedure. In any case, Theorem 7 provides conditions on R and C under which the Principal's gain curve is indeed single-peaked.

Plan of the Remainder of the Paper In Sect. 5, we examine six examples where the set of possible efforts and the set of possible values of t are not finite and calculus methods can be used. The examples will illustrate the theorems we informally sketched above. The examples, together with the preceding sketch of the main results, provide an extensive preview of the theorems. It then remains to state each of them formally. In Sect. 6, we formally state the exogenous-case Theorem 1 and the endogenous-case Theorem 2 (which do not require differentiability) and we comment on the proof techniques. Section 7 states two exogenous-case theorems which require differentiability. The second of these is Theorem 4 which concerns tracking and has a Corollary that relates directly to our central question—when does a drop in t increase (decrease) the Decentralization Penalty? Section 8 presents two endogenous-case theorems which again require differentiability: Theorem 5, which again concerns tracking, and Theorem 6, which concerns the case where marginal revenue is decreasing. Section 9 presents Theorem 7, about bargaining and the shape of the Principal's gain curve. Section 10 sketches several of the many possible extensions and modifications of our model. An Appendix contains portions of the proofs. The complete proofs are given in Liang, Marschak, and Wei (2017), abbreviated henceforth as LMW.

4 Related Literature

A great many Principal/Agent papers, starting with the earliest ones, use a framework that allows the Agent's effort to have a cost. The Agent has a utility function on her actions and rewards. In many papers, Agent utility for the action a and the reward y takes the form $V(y) - g(a)$. Among the early papers where this occurs are Holmstrom (1979, 1982) and Grossman and Hart (1983). The action a might be effort and $g(a)$ could be its cost. Welfare loss also appears very early in the literature. Ross (1973), for example, finds conditions under which the solution to the Principal's problem maximizes welfare (as measured by the sum of Agent's utility and Principal's utility) and notes that these conditions are very strong. But, Principal/Agent papers whose main concern is the relation between effort cost and welfare loss are scarce.

The Principal/Agent papers closest to ours are Balmaceda et al. (2016) and Nasri et al. (2015). They study an Agent who has m possible efforts. Each effort has a cost, which the Agent pays. There are n possible revenues. For a given effort, the probability of each of the possible revenues is common knowledge, but the revenue actually realized only becomes known after the Agent's effort choice has been made. The Principal announces a vector of n nonnegative wages. For each of the n possible revenues, the vector specifies a wage received by the Agent when that revenue is realized. Both Principal and Agent are risk-neutral. Surplus for a given effort equals expected revenue minus the effort's cost. The socially preferred effort maximizes surplus. In the decentralized (Principal/Agent) mode, on the other hand, surplus is not maximal. Instead, it is the surplus when the effort is the one the Principal chooses to induce. The papers study a fraction. Its numerator is maximal surplus and its denominator is "worst-case" Principal/Agent surplus. (When the Principal is indifferent between several efforts, the denominator of the fraction selects the one that is socially worst). The fraction is a measure of the welfare loss due to decentralizing. It is shown, under standard assumptions on the probabilities and on the possible (revenue,effort-cost) pairs, that the ratio cannot exceed m, the number of efforts. That upper bound does not depend on the effort costs, so the papers are silent on the effect of a drop in those costs on welfare loss.

Note that welfare loss is also defined as a fraction in a larger literature, initially developed by computer scientists. Typically, the object of study is a game. The fraction studied is often called "the price of anarchy." Its numerator is the payoff sum in the "socially worst" equilibrium of the game. Its denominator—attainable when the players cooperate—is the largest possible payoff sum.[1] In our setting, it is natural to use the surplus gap rather than a ratio in defining the Penalty (welfare loss) due to decentralizing. Perfect monitoring (centralization) would eliminate the gap, but in reality it would be expensive. If its cost exceeds the gap then decentralization is the preferred mode.

If we allow more than one Agent, then parts of the large literature on the design of organizations become relevant. The designer has a goal, say surplus (profit) maximization, and can choose between a structure where a single member commands the choices made by all the others, and a structure where everyone is autonomous. The latter structure might be modeled as a game. A rather small piece of the design literature studies the communication and computation costs of each structure and the trade-off between those costs and some measure of gross performance (e.g., gross expected surplus, before the costs are subtracted). The problem is far more complex than the one we consider here and the results remain scarce and specialized.[2]

Finally, it seems appropriate to mention a paper co-authored by Leo Hurwicz, whom this volume honors (Hurwicz and Shapiro, 1978). Here, the Principal is a landlord and the Agent is a sharecropper who chooses how hard to work. The landlord knows neither the sharecropper's utility function nor her production function but does observe the revenue that the sharecropper's labor has achieved. The landlord rewards the sharecropper with a share of the revenue. It is shown that a fifty/fifty split is preferred by the Principal. If the reward function is required to be linear, then that split is also socially optimal, where "optimal" means that the

[1] A variety of social situations are studied from this point of view. One of them concerns optimal versus "selfish" routing in transportation (Roughgarden, 2005). Others are found in Nissan et al. (2007). Many of these studies develop bounds on the price of anarchy. Several of them (e.g., Balbaieff et al. 2009) consider a Principal/Agent setting.

[2] Surveys of the design literature with communication and computation costs are found in Garicano and Prat (2013) and Marschak (2006). A model in which revenue is shared by a group of game-playing Agents is studied in Courtney and Marschak (2009). Each player chooses effort and bears its cost. Equilibria of the game are compared with the welfare-maximizing efforts. The paper finds conditions under which the welfare loss drops (rises) when effort costs shift down.

largest possible social "regret" is minimized. One could study the welfare effect of lowering the Agent's cost for every effort, but the paper does not do so.

5 Examples

In each example, we specify a triple (Σ, R, C) and we also specify a set Γ of possible pairs (r, t). The set Γ has the property that for each of its pairs (r, t): (1) $0 < r < 1$; and (2) there exists a positive effort $\hat{x}(r, t)$ which maximizes $R(x) - tC(x)$ on Σ. It will be convenient to use the symbol $\tilde{\Gamma}$ for the set of possible values of t. Thus,

$$\tilde{\Gamma} \equiv \{t \ : \ (r, t) \in \Gamma \text{ for some } r\}.$$

In discussing our example, we will use the terms *exogenous tracking* and *(exogenous) "opposite directions"*. Here is the definition:

Definition 1 An example (R, C, Γ, Σ), with R and C thrice differentiable, has the *exogenous tracking (exogenous "opposite directions")* property if

$$\frac{d}{dt}\left[\hat{x}(1, t) - \hat{x}(r, t)\right] \cdot \frac{d}{dt}\left[W(1, t) - W(r, t)\right] > 0 \ (< 0) \text{ at all } (r, t) \in \Gamma.$$

Each of our examples will be an *interior* example. In such an example, first-order conditions suffice to identify both $\hat{x}(r, t)$ and the Principal's endogenous-case share $r^*(t)$.

Definition 2 An example (Σ, R, C, Γ) is Interior if

- $\Sigma \subseteq \mathbb{R}^+$, and $\Gamma \subseteq \mathbb{R}^{2^+}$ are open sets.
- R, C are thrice differentiable on Σ and $R' > 0, C' > 0$.
- There exists a twice differentiable function $\hat{x} \ : \ (0, 1] \times \tilde{\Gamma} \to \Sigma$ such that for $r \in (0, 1]$, $\hat{x}(r, t)$ satisfies the first-order condition $0 = rR'(x) - tC'(x)$ and is the unique maximizer of $rR(x) - tC(x)$ on Σ.
- For every $t \in \tilde{\Gamma}$, there exists a share $r^*(t) \in (0, 1)$ which satisfies the first-order condition $0 = \frac{d}{dr}\left[(1 - r) \cdot R(\hat{x}(r, t))\right]$ and is the unique maximizer of $(1 - r) \cdot R(\hat{x}(r, t))$ on $(0, 1)$.

In discussing an interior example, we use the terms *endogenous tracking* and *endogenous "opposite directions."* The definitions are analogous to Definition 1.

Definition 3 An interior example (R, C, Γ, Σ) has the *endogenous tracking (endogenous "opposite directions")* property if

$$\frac{d}{dt}\left[\hat{x}(1, t) - \hat{x}(r^*(t), t)\right] \cdot \frac{d}{dt}\left[W(1, t) - W(r^*(t), t)\right] > 0 \ (< 0) \text{ at all } t \in \tilde{\Gamma}$$

$$= \{t \ : \ (r, t) \in \Gamma \text{ for some } r\}.$$

Example 1: A Classic Monopoly

For brevity, we shall call this the Classic example. The firm is a monopolist and the effort x is product quantity. Price is $A - Bx$, where $A > 0$, $B > 0$, so revenue is $R(x) = Ax - Bx^2$. Cost is $t \cdot C(x) = tx$. Marginal revenue becomes negative at $x = \frac{A}{2B}$. To keep price and marginal revenue positive, our set of possible efforts will be

$$\Sigma = \left(0, \frac{A}{2B}\right).$$

In the decentralized mode, the monopolist acts as a Principal, lets an Agent choose quantity, and announces a share $r \in (0, 1)$. We consider the following set Γ of possible pairs (r, t):

$$\Gamma \equiv \{(r, t) \ : \ 0 < r < 1; \ 0 < t < Ar\}.$$

Thus, the set of possible values of t is $\tilde{\Gamma} = (0, A)$. If $(r, t) \in \Gamma$, the Agent's best quantity is

$$\hat{x}(r, t) = \frac{A}{2B} - \frac{t}{2Br}.$$

That belongs to Σ and is the unique maximizer of the Agent's net gain $R(x) - t \cdot C(x)$.

Note that our Γ in this example is the interior of a triangle. In a diagram with r on the horizontal axis and t on the vertical axis, the triangle has vertices at the points $(0, 0)$, $(1, 0)$, and $(1, A)$. In other examples, Γ might be a rectangle, as in Example 3 below. In still other examples, one of the boundaries of Γ might have curvature.

We now differentiate $\hat{x}$ and obtain some exogenous-case statements. The subsequent theorems will generalize them.

- $\hat{x}_r(r, t) = \frac{t}{2Br^2}$, which is positive. For a given t, increasing the share evokes more effort. It is easily shown, in Part **(a)** of Theorem 1, that in any example, finite or nonfinite, increasing the share never evokes less effort.
- $\hat{x}_t(r, t) = -\frac{1}{2Br}$, which is negative. When r is fixed and technology improves, the Agent works harder. Part **(b)** of Theorem 1 says that the Agent never works less when t drops.
- $\hat{x}_{rt}(r, t) = \frac{1}{2Br^2} > 0$. So, technology improvement (a drop in t) *diminishes* the effectiveness of a small rise in the share as a stimulus to higher effort. We use effectiveness (the sign of $\hat{x}_{rt}(r, t)$) in classifying examples. The classification will be especially important in the endogenous case.
- When the Agent uses the best effort $\hat{x}(r, t)$, he receives $r \cdot R(\hat{x}(r, t) - t \cdot \hat{x}(r, t)$. The derivative of that expression with respect to t is negative.[3] So, technology

[3]The derivative is $\hat{x}_t(r, t) \cdot [rR'(\hat{x}(r, t)) - t \cdot C'(\hat{x}(r, t))] - C(\hat{x}(r, t))$. That is negative, since $0 < r < 1$ and $\hat{x}(r, t)$ satisfies the first-order condition $0 = rR' - tC'$.

improvement is good news for the Agent. Part **(f)** of Theorem 1 uses a simple argument to show that this is always true in the exogenous case.

- We find that surplus is

$$W(r,t) = R(\hat{x}(r,t)) - t \cdot C(\hat{x}(r,t)) = \frac{1}{4B^2r^2} \cdot [(Ar - t) \cdot (BAr + Bt - 2Brt)].$$

The derivative with respect to t of the expression in square brackets is

$$-2BAr^2 - 2Bt + 4Brt.$$

Our requirement that $t < Ar$ implies that this is negative.[4] So, for a fixed $r < 1$, decentralized exogenous-case surplus rises when technology improves (t drops). Part **(g)** of Theorem 1 says that this always holds.
- For all $t \in \tilde{\Gamma}$, we have $W_t(1,t) < 0$. Maximal surplus rises when technology improves (t drops). In Part **(d)** of Theorem 1, a trivial argument shows that this always holds.
- $W_{rt}(r,t) = \dfrac{(1-r) \cdot t}{Br^3} > 0$. So, $W_{rt}(r,t)$ and $\hat{x}_{rt}(r,t)$ have the same sign. Theorem 3 shows, using a very simple argument, that whenever $\hat{x}_{rt}(r,t) > 0\ (< 0)$, we also have $\dfrac{d}{dt}[\hat{x}(1,t) - \hat{x}(r,t)] > 0\ (< 0)$. An analogous argument shows that whenever $\hat{W}_{rt}(r,t) > 0\ (< 0)$ we also have $\dfrac{d}{dt}[\hat{W}(1,t) - W(r,t)] > 0\ (< 0)$. So, in our example, the exogenous Decentralization Penalty (surplus gap) $W(1,t) - W(r,t)$ and the exogenous effort gap $\hat{x}(1,t) - \hat{x}(r,t)$ move in the same direction when technology improves, i.e., the exogenous surplus gap *tracks* the exogenous effort gap. Theorem 4 shows that this must be so as long as R and C are thrice differentiable.

We now turn to the endogenous case. In our example, we can verify that the Principal's gain $(1-r) \cdot (R(\hat{x}(r,t))$ is positive for all $r \in (0,1)$ and is concave on $(0,1)$. That implies—as Theorem 7 shows—that there is a share in $(0,1)$, denoted $r^*(t)$, which solves the first-order condition:

$$0 = \frac{d}{dr}[(1-r) \cdot R(\hat{x}(r,t)] = -R(\hat{x}(r,t)) + (1-r) \cdot R'(\hat{x}(r,t)) \cdot \hat{x}_r(r,t)$$

and maximizes the Principal's gain on the set $(0,1)$. In our example, the Principal's first-order condition turns out to be the cubic equation:

$$0 = A^2r^3 + rt^2 - 2t^2.$$

---

[4]The derivative is negative if $Ar^2 > t \cdot (2r-1)$. That is the case at $r = 0$ and at $r = 1$ (since $t < A$). At all $r \in (0,1)$, our requirement $t < Ar$ implies that $2Ar$, the derivative of the left side of the inequality with respect to r, exceeds $2t$, the derivative of the right side. So, at all $(r,t) \in \Gamma$ the inequality holds.

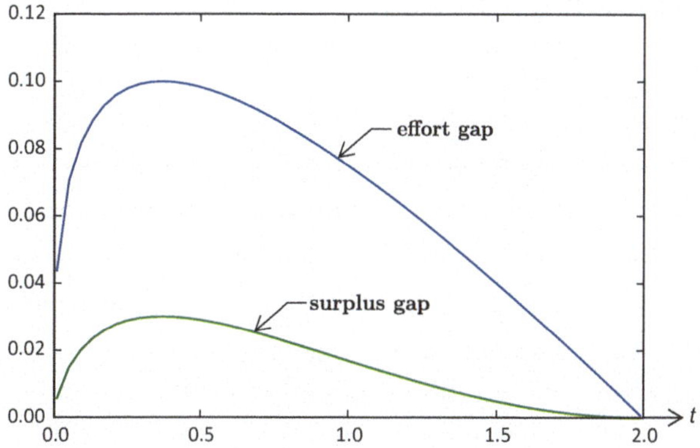

Fig. 1 The two gaps: surplus gap ($W(1,t) - W(r^*(t),t)$) and effort gap ($\hat{x}(1,t) - \hat{x}(r^*(t),t)$) for the Classic case, with $A = 2$, $B = 3$

When we graph the implicit function $r^*(t)$, we find that for the case $A = 2$, $B = 3$, r^* is increasing in t. The share-choosing Principal becomes less generous when technology improves. But without any graphing, Theorem 6 tells that r^* cannot be *decreasing* in t if $\hat{x}_{rt}(r,t) > 0$ (as in our example), and in addition $R'' < 0$, which holds in our example, since $R'' = -2B < 0$.

Next, consider the endogenous effort $\hat{x}(r^*(t),t)$. If we graph this for our example, we find that it rises when technology improves (t drops). Part **(b)** of Theorem 2 shows that this must happen, for the endogenous case, in every example, finite or nonfinite.

We now turn to the tracking question. Figure 1 shows both the Penalty (surplus gap) and the effort gap $\hat{x}(1,t) - \hat{x}(r^*(t),t)$. Figure 1 shows that when t increases each gap first rises and then falls and for each t the gaps move in the same direction, so we indeed have tracking.

But, it is *not* the case that endogenous tracking in our example is implied by the fact that we have $\hat{x}_{rt}(r,t) > 0$ and $r^{*'} < 0$. The next example has the same inequalities but it does not exhibit endogenous tracking.

Example 2: A "Cubic-Revenue" Example
In this example: $R(x) = x^3 - x^2$, $C(x) = x$ and the set of possible (r,t) pairs is the triangle $\Gamma = \{(r,t) \ : \ r \in (0,1); t \le r\}$, so the set of possible values of t is $\tilde{\Gamma} = (0,1)$. We find[5] that—just as in the Classic Monopoly example—we have $\hat{x}_{rt}(r,t) > 0$ for all (r,t) in Γ and $r^{*'}(t) < 0$ for all t in $\tilde{\Gamma}$. But when we graph the two endogenous-case gaps, we find that for t in the interval $(0.48, 0.63)$, the effort gap rises but the surplus gap falls.

[5]The details of this calculation, as well as a graph of effort gap and surplus gap, are given in LMW (2017). Details for the remaining examples are given there as well.

Example 3: A "Price-Taker" Example

The Principal takes the price of one as given and the cost function is quadratic. The set of possible efforts is $\Sigma = \mathbb{R}^+$; $R(x) = x$; $C(x) = \frac{1}{2}(x-1)^2$. The set of possible pairs (r, t) is the rectangle $\Gamma = \{(r,t) : 0 < r < 1; 0 < t < 1\}$. The Agent's exogenous-case effort choice is $\hat{x}(r,t) = \frac{r}{t} + 1$. So, $\hat{x}_{rt}(r,t) = \frac{-1}{t^2} < 0$. That contrasts with Examples 1 and 2. Surplus-maximizing (first-best) effort is $\frac{1}{t}+1$. The exogenous effort gap is $\frac{1-r}{t}$, which has a negative derivative with respect to t. Exogenous surplus is $R(\hat{x}(r,t)) - t \cdot C(\hat{x}(r,t)) = \frac{r}{t} + 1 - \frac{r^2}{2t}$ and maximal surplus is $1 + \frac{1}{2t}$. Hence, the exogenous surplus gap (the Penalty) is $\frac{1}{2t} - \frac{r}{t} + \frac{r^2}{2t}$. Its derivative with respect to t is negative, just like the derivative of effort gap. So—as Theorem 4 tells us, the exogenous surplus gap tracks the exogenous effort gap.

We now turn to the endogenous case. The unique solution to the Principal's first-order condition $0 = \frac{d}{dr}[R(\hat{x}(r,t)) - t \cdot C(\hat{x}(r,t))]$ is $r^*(t) = \frac{1-t}{2}$. So, we have $r^{*'}(t) < 0$ at every possible t. (Recall that $t < 1$). That contrasts sharply with Example 1 (Classic monopoly). Is a drop in t good news from the welfare point of view in the endogenous case? That cannot be directly answered in Example 1, where there is no closed form for welfare. But in the present example, it is easily answered. We have

$$\frac{d}{dt}\left[R(\hat{x}(r^*(t),t)) - t \cdot C(\hat{x}(r^*(t),t))\right] = [\hat{x}_r \cdot r^{*'} + \hat{x}_t] \cdot (R' - tC') - C.$$

We have $R' - tC' > 0$ (because of the first-order condition $rR' - tC' = 0$, where $0 < r < 1$). Since $\hat{x}_t < 0$ and $r^{*'} \leq 0$, we conclude that the derivative is negative, so we indeed have "good news" from the welfare point of view. Part (**d**) of Theorem 2, moreover, shows that this *must* be the case, with or without differentiability.

The endogenous Penalty (surplus gap) is $\frac{1}{4} + \frac{1}{8t} + \frac{t}{8}$. Its derivative with respect to t is $\frac{1}{8t^2} \cdot (t^2 - 1)$, which is negative, since $t < 1$. The Penalty always rises when technology improves. Note the contrast with Example 1 (Classic Monopoly), where the Penalty *drops* when technology improves, once t has dropped below a critical value. The endogenous effort gap is $\frac{1+t}{t} - \frac{r^*(t)+t}{t} = \frac{1}{2t} + \frac{1}{2}$. That also has a negative derivative. So, the endogenous effort gap tracks the endogenous surplus gap. But that is *not* implied, as we shall see, by the fact that $\hat{x}_{rt} < 0$ and $r^{*'}(t) < 0$, just as the endogenous tracking that we found in Example 1 was not implied by the fact that $\hat{x}_{rt} > 0$ and $r^{*'}(t) > 0$. *We will find a sharply different pattern in our final two examples (5 and 6); there we have endogenous tracking and that does follow from the signs of* $\hat{x}_{rt}$ *and* $r^{*'}(t)$.

Example 4: A "Cubic-cost" Example

In this example,

$$R(x) = \frac{1}{2}x^2$$

and

$$C(x) = \frac{1}{3}x^3 + \frac{a}{2}x^2 - \epsilon x,$$

where $\epsilon > 0$ and $a > 0$. The numbers a, ϵ and the set Σ of possible efforts will be chosen as we proceed. The triple (a, ϵ, Σ) will have the property that $C(x) > 0$ for all $x \in \Sigma$.

The Agent's first-order condition for given r, t is

$$rx = t \cdot (x^2 + ax - \epsilon).$$

This is solved by:

$$\hat{x}(r,t) = \frac{\sqrt{\left(a - \frac{r}{t}\right)^2 + 4\epsilon} - \left(a - \frac{r}{t}\right)}{2} > 0.$$

Our set Γ of possible (r, t) pairs will be

$$\Gamma = \left\{(r,t)\ ;\ t \in \left(\frac{1}{a}, \frac{2}{\sqrt{a^2 + 4\epsilon}}\right); 0 < r < 1\right\}.$$

Now, assume that

- $t \geq \frac{1}{a}$
- $\epsilon < \frac{3}{4}a^2$.

Then, $\dfrac{1}{a} < \dfrac{2}{\sqrt{a^2 + 4\epsilon}}$, so Γ is not empty. Moreover, $a - r/t \geq 0$ for all $r \in (0, 1)$. Under these assumptions, we can show[6] that $\hat{x}_{rt}(r, t) < 0$ at all (r, t) in Γ.

Turning to the endogenous case, we find that the Principal's chosen share $r^*(t)$ must satisfy

$$r^*(t) = 1 - \frac{t\sqrt{\left(a - \frac{r^*(t)}{t}\right)^2 + 4\epsilon}}{2}. \tag{+}$$

That allows us to show that for every t in our set $\tilde{\Gamma} = \left(\dfrac{1}{a}, \dfrac{2}{\sqrt{a^2 + 4\epsilon}}\right)$ of possible values of t: there is a unique $r^*(t)$ satisfying (+) and, moreover, $r^{*'}(t) < 0$.

[6]Once again, the details are in LMW.

Now, consider the case where $a = 1$ and $\epsilon = 0.6$. That meets our requirement $\epsilon < \frac{3}{4}a^2$. Define our set of possible efforts to be

$$\Sigma = (1, \infty].$$

Then, $\tilde{\Gamma} = (1, 1.084)$ and $C(x) > 0$ for every $x \in \Sigma$. If we graph the surplus and effort gaps, we find that the surplus gap rises in that interval but the effort gap falls. Instead of tracking, we have "opposite directions." In the Price-taker example, we also had $\hat{x}_{rt}(r, t) < 0$ and $r^{*'}(t) < 0$, but there we had tracking.

Example 5: A "Rising-Marginals" Example
In this example, marginal revenue rises but marginal cost rises faster, so there is an interior effort maximizing the Agent's gain. The set of possible efforts is $\Sigma = \mathbb{R}^+$; $R(x) = x^a$; and $C(x) = x^b$, where $0 < a < b$. The set of possible pairs (r, t) is $\Gamma = \{(r, t) : 0 < r < 1; t > 0\}$. We find that $\hat{x}(r, t) = \left(\frac{tb}{ra}\right)^{\frac{1}{a-b}}$ and $\hat{x}_{rt}(r, t) = -\frac{1}{(a-b)^2} \cdot t^{1/(a-b)-1} \cdot \left(\frac{b}{a}\right)^{1/(a-b)} \cdot r^{1/(b-a)-1}$, which is negative.

Turning to the endogenous case, we find that $r^*(t) = \frac{a}{b}$. The Principal' chosen share is independent of t. Even though we have an explicit expression for r^*, computing the derivative of endogenous effort gap (Penalty) with respect to t and the derivative of endogenous surplus gap with respect to t is cumbersome. It turns out that both are negative. So, the endogenous surplus gap tracks the endogenous effort gap. Theorem 5 will show that this follows from the fact that we have both $\hat{x}_{rt}(r, t) < 0$ and $r^{*'}(t) \leq 0$.

Example 6: An "Exploding-Marginals" Example
There remains one class, in our four-way classification of interior examples, which we have not yet illustrated. This is the class where we have both $\hat{x}_{rt}(r, t) > 0$ and $r^{*'}(t) < 0$. Theorem 6 will show us that in such an example we cannot have $R'' \leq 0$. So, our search for an example is narrowed to the case $R'' > 0$. Moreover, preliminary exercises show that a modestly increasing marginal revenue (e.g., $R'' = 1$) is not enough. Marginal revenue has to rise rapidly and marginal cost has to rise even faster. In the following example, both marginals "explode."

We have:

- $\Sigma = (0, 1)$.
- $\Gamma = \{(r, t) : 0 < r < 1; \frac{r}{t} \in (e, e^e)\}$ (e is the base of the natural logarithms).
- $R(x) = e^{x^2}$.
- $C(x) = \int_0^x \left[2e^{e^p} \cdot e^p \cdot p\right] dp$.

Since $\hat{x}_{rt}(r, t) > 0$ and $r^{*'}(t) < 0$, Part (**b**) of Theorem 5 tells us that in this example we have endogenous tracking.

A summary of the six interior examples and their relation to our theorems is provided in the table which follows.

THE EFFECT OF IMPROVED TECHNOLOGY (A DROP IN t) IN FOUR GROUPS OF INTERIOR EXAMPLES

	WHEN t DROPS, EFFECTIVENESS OF A SHARE INCREASE *FALLS*. HENCE, SO DOES THE EXOGENOUS EFFORT GAP (SEE THEOREM 3). $\hat{x}_{rt} > 0$ and hence $\frac{d}{dt}[\hat{x}(1,t) - \hat{x}(r,t)] > 0$	WHEN t DROPS, EFFECTIVENESS OF A SHARE INCREASE *RISES*. HENCE, SO DOES THE EXOGENOUS EFFORT GAP (SEE THEOREM 3). $\hat{x}_{rt} < 0$ and hence $\frac{d}{dt}[\hat{x}(1,t) - \hat{x}(r,t)] < 0$
WHEN t DROPS, PRINCIPAL BECOMES *LESS* GENEROUS OR GENEROSITY STAYS THE SAME. $r^{*'} \geq 0$	1 *SEE "CLASSIC" AND "CUBIC-REVENUE" EXAMPLES.* WE HAVE ENDOGENOUS TRACKING IN THE CLASSIC EXAMPLE BUT IN THE CUBIC-REVENUE EXAMPLE, WE HAVE "OPPOSITE DIRECTIONS" (IF THE SET OF POSSIBLE VALUES OF t IS PROPERLY CHOSEN).	2 *SEE "RISING MARGINALS" EXAMPLE.* EVERY EXAMPLE THAT LIES IN THIS BOX HAS THE TRACKING PROPERTY. *(See Theorem 5, Part (a)).*
WHEN t DROPS, PRINCIPAL BECOMES *MORE* GENEROUS. $r^{*'} < 0$	3 *SEE "EXPLODING MARGINALS" EXAMPLE.* EVERY EXAMPLE THAT LIES IN THIS BOX HAS THE ENDOGENOUS TRACKING PROPERTY. *(See Theorem 5, Part (b)). AN EXAMPLE WITH $R'' < 0$ CANNOT BE IN THIS BOX. (See Theorem 6).*	4 *SEE THE "PRICE-TAKER" EXAMPLE, WHERE WE HAVE ENDOGENOUS TRACKING AND THE "CUBIC-COST" EXAMPLE, WHERE WE HAVE "OPPOSITE DIRECTIONS."*

6 Basic Results that Do Not Require Differentiability

Theorem 1 *Let R and C be strictly increasing on Σ. Then:*

(a) $\hat{x}(r_H,t) \geq \hat{x}(r_L,t)$ *and* $R(\hat{x}(r_H,t)) - tC(\hat{x}(r_H,t)) > R(\hat{x}(r_L,t)) - tC(\hat{x}(r_L,t))$ *whenever* $(r_L,t) \in \Gamma$, $(r_H,t) \in \Gamma$, *and* $0 < r_L < r_H < 1$.
(b) $\hat{x}(r,t_L) \geq \hat{x}(r,t_H)$ *whenever* $(r,t_L) \in \Gamma$, $(r,t_H) \in \Gamma$, *and* $0 < t_L < t_H$.
(c) $\hat{x}(1,t_L) \geq \hat{x}(1,t_H)$ *whenever* $t_L, t_H \in \tilde{\Gamma}$, *and* $0 < t_L < t_H$.
(d) $W(1,t_L) > W(1,t_H)$ *whenever* $t_L, t_H \in \tilde{\Gamma}$ *and* $0 < t_L < t_H$.
(e) $(1-r) \cdot R(\hat{x}(r,t_L)) \geq (1-r) \cdot R(\hat{x}(r,t_H))$ *whenever* $(r,t_L) \in \Gamma$, $(r,t_H) \in \Gamma$, *and* $0 < t_L < t_H$.
(f) $rR(\hat{x}(r,t_L)) - t_L C(\hat{x}(r,t_L)) > rR(\hat{x}(r,t_H)) - t_H C(\hat{x}(r,t_H))$ *whenever* $(r,t_L) \in \Gamma$, $(r,t_H) \in \Gamma$, *and* $0 < t_L < t_H$.
(g) $W(r,t_L) > W(r,t_H)$ *whenever* $(r,t_L) \in \Gamma$, $(r,t_H) \in \Gamma$, *and* $0 < t_L < t_H$.
(h) $W(r_H,t) \geq W(r_L,t)$ *whenever* $(r_H,t) \in \Gamma$, $(r_L,t) \in \Gamma$, *and* $0 < r_L < r_H < 1s$. *The inequality is strict if and only if* $\hat{x}(r_H,t) \neq \hat{x}(r_L,t)$.

In proving Parts **(a), (b), (c), (d)**, we use a basic proposition from monotone comparative statics. It concerns a function of two variables with strictly increasing differences and describes the direction in which a maximizer moves when one of the variables increases.[7] For the other parts, we use the simple observation that when t drops or r rises the Agent could continue to muse the same effort as before the

[7] See, for example, Sundaram (1996).

change. To show the pattern of the first argument, the Appendix provides the proof of part (**a**)). To show the second argument, it gives the proofs of (**h**) and (**g**).[8]

Theorem 2 concerns the endogenous case.

Theorem 2 *Let R and C be strictly increasing on* Σ. *Let* $r^*(t)$ *denote a maximizer of* $(1 - r) \cdot R(\hat{x}(r, t))$ *on the interval* $(0, 1)$. *Then,*

(a) $\frac{r^*(t_L)}{t_L} \geq \frac{r^*(t_H)}{t_H}$ *whenever* $t_L, t_H \in \tilde{\Gamma}$ *and* $0 < t_L < t_H$.
(b) $\hat{x}(r^*(t_L), t_L) \geq \hat{x}(r^*(t_H), t_H)$ *whenever* $t_L, t_H \in \tilde{\Gamma}$ *and* $0 < t_L < t_H$.
(c) $(1-r^*(t_L))\cdot R(\hat{x}(r^*(t_L), t_L) \geq (1-r^*(t_H))\cdot R(\hat{x}(r^*(t_H), t_H)$ *whenever* $t_L, t_H \in \tilde{\Gamma}$ *and* $0 < t_L < t_H$.
(d) $W(r^*(t_L), t_L) > W(r^*(t_H), t_H)$ *whenever* $t_L \in \tilde{\Gamma}, t_H \in \tilde{\Gamma}$, *and* $0 < t_L < t_H$.

The Appendix provides the proofs of Parts (**a**) and (**b**). It is interesting to note that while Part (*c*) of Theorem 2 tells us that in the endogenous case technical improvement can never be bad news for the Principal, the situation is different for the Agent. We can construct examples where a drop in t leads to smaller net gain for the Agent. Informally: *in the endogenous case, the Principal is never the enemy of technical progress but the Agent might be.*

7 Two Exogenous-Case Theorems Which Require Differentiability

Theorem 3 *Let* Γ *be an open set in* $\mathbb{R}^{2^+}$. *Suppose that the functions R and C are thrice differentiable. Suppose that the following monotonicity condition is met:*

we either have

$$\hat{x}_{rt} > 0 \textit{ for all } (r, t) \in \Gamma$$

or

$$\hat{x}_{rt} < 0 \textit{ for all } (r, t) \in \Gamma.$$

Suppose, in addition, $\hat{x}_t$ *is continuous with respect to r at all points in* $(0, 1]$. *Then,* $\hat{x}_{rt}(r, t) > 0$ (< 0) *at every* $(r, t) \in \Gamma$ *if and only if*

$$\frac{d}{dt}\,[\hat{x}(1, t) - \hat{x}(r, t)] > \ (< 0) \textit{ at every } (r, t) \in \Gamma.$$

Note that the pair $(r^*(t), t)$ belongs to Γ, so the theorem applies, in particular, to $\hat{x}_{rt}(r^*(t), t)$ and the endogenous effort gap $\hat{x}(1, t) - \hat{x}(r^*(t), t)$. The proof is straightforward.

[8] As already noted, the proofs of the parts not shown are given in LMW.

The next theorem concerns exogenous tracking in Interior examples.

Theorem 4 *An interior example has the exogenous tracking property if the effort set is* $\Sigma = (0, J)$*, where* $J > 0$*, and the monotonicity condition of Theorem 2 holds (we either have* $\hat{x}_{rt} > 0$ *for all* $(r, t) \in \Gamma$ *or* $\hat{x}_{rt} < 0$ *for all* $(r, t) \in \Gamma$*).*

Straightforward calculation yields the following Corollary.

Corollary *The following hold for an interior example in which the monotonicity condition of Theorem 2 is satisfied, the effort set is* $\Sigma = (0, J)$ *(where* $J > 0$*), and* $\hat{x}_r(r, t) > 0, \hat{x}_t(r, t) < 0$ *for all* $(r, t) \in \Gamma$*:*

(i) the Decentralization Penalty (surplus gap) is decreasing in t (so the Penalty grows when technology improves) if at every effort $x \in (0, J)$ *we have* $R''(x) \geq 0, R'''(x) = C'''(x) = 0$*.*

(ii) the Decentralization Penalty (surplus gap) is increasing in t (so the Penalty shrinks when technology improves) if at every effort $x \in (0, J)$ *we have* $R''(x) < 0, C''(x) = 0, R'''(x) \leq 0$*.*

8 Endogenous-Case Results Which Require Differentiability

Theorem 5 *Consider an interior example* (Σ, Γ, R, C)*.*
(a)*. Suppose the following holds:*

for every $t \in \tilde{\Gamma}$ *we have* $r^{*'}(t) \geq 0$ *and for every* $(r, t) \in \Gamma$ *we have* $\hat{x}_{rt}(r, t) < 0$*.*

Then, we have endogenous tracking.
(b)*. Suppose the following holds:*

for every $t \in \tilde{\Gamma}$ *we have* $r^{*'}(t) < 0$ *and for every* $(r, t) \in \Gamma$ *we have* $\hat{x}_{rt}(r, t) > 0$*.*

Then, we have endogenous tracking.

The next theorem does not directly concern the two gaps. But, it implies that if marginal revenue is decreasing or constant ($R'' \leq 0$) in an interior example and the Principal has a unique best share, then the example cannot be in Box **3** of our table.

Theorem 6 *Suppose that in the interior example* (Σ, Γ, R, C) *we have:*

- $R''(x) \leq 0$ *at every* $x \in \Sigma$*.*
- $\hat{x}_{rt}(r, t) \geq 0, \hat{x}_t(r, t) < 0$ *and* $\hat{x}_r(r, t) > 0$ *at every* $(r, t) \in \Gamma$*.*
- $r^*(t)$ *is the unique maximizer of* $(1 - r) \cdot R(\hat{x}(r, t))$ *on* $(0, 1)$*,*

Then, $r^{*'}(t) \geq 0$ *for all* $t \in \Gamma$*.*

It is difficult to give a clear intuition for Theorems 4 and 5. That is a little easier for Theorem 6, which says that if marginal revenue is decreasing, and effectiveness drops when technology improves, then when technology improves, the Principal does not become more generous ($r^{*'}(t) \geq 0$), i.e., we cannot be in Box **3**. Intuitively, one might say: *when t drops, increasing the share above its previous level would damage the Principal, because the extra revenue due to extra effort has dropped (marginal revenue has declined) and at the same time the extra effort evoked by a share increase has dropped as well.*

9 Finding the Principal's Best Share for a Given t: When Is the Principal's Gain a Concave Function of the Share?

For a fixed t, consider the Principal's gain $(1-r)\cdot R(\hat{x}(r,t))$ as a function of $r \in (0,1)$. Our discussion in the "main results" section above argued that if we want to study bargaining between Principal and Agent over the share r, then it is very helpful if the graph of the gain curve is single-peaked. As long as the gain is positive at some $r \in (0,1)$, the curve is single-peaked if it is concave on $(0,1)$. The following theorem provides conditions under which the gain is indeed concave. The theorem has two parts. The first part does not require differentiability with respect to r, but the second part does. The second part says that we have concavity if marginal revenue drops ($R'' < 0$) and in addition the effectiveness of a share increase drops when the share increases ($\hat{x}_{rr} <,$).

Theorem 7

(a) *If, for a fixed t,* $R(\hat{x}(r,t))$ *is concave on* $(0,1)$, *then the Principal's gain* $(1-r)\cdot R(\hat{x}(r,t)$ *is also concave on* $(0,1)$.

(b) *Consider an interior example* (Σ,Γ,R,C) *where* $\Sigma = (0,J)$, *with* $J > 0$. *Then, R is concave on* $(0,J)$ *if for all* $x \in (0,J)$ *we have* $R''(x) < 0$, *and for all* $(r,t) \in \Gamma$ *we have* $\hat{x}_{rr}(r,t) < 0$. *If* $R''(x) < 0$, *then a sufficient condition for* $\hat{x}_{rr} < 0$ *is*

$$r \cdot R'''(x) - t \cdot C'''(x) \leq 0.$$

10 Concluding Remarks

Recall our central question: does technical improvement strengthen the case for full Agent autonomy or does it weaken it so much that perfect monitoring and policing has now become attractive? One might have reasonably hoped for a straightforward answer since our revenue-sharing Principal/Agent model is so simple. Specifically, one might have hoped that a natural condition like rising marginal cost and falling marginal revenue unambiguously implies that the Decentralization Penalty rises (or falls) when technology improves. Instead, we have found that there is no easy answer to our central question. On the other hand, we have found a rich array of other results. One of them is that in both the exogenous case and the endogenous case, an advance in technology increases welfare. Another is that an advance in technology causes the Agent to work harder. That is obvious in the exogenous case, since the Agent benefits from the advance even if she continues to use her previous effort. It is not obvious in the endogenous case.

Other interesting results for the challenging endogenous case concern the tracking question. If the effort gap always moves in the same direction as the surplus gap (the Penalty), then to see whether a technical advance has strengthened or weakened the case for autonomy, it suffices to observe (but not police) the Agent's effort before and after the advance and to compare it with first-best effort. We saw that two key properties of an example are the sign of $r^{*'}$ and the sign of $\hat{x}_{rt}$. We *must* have tracking if $r^{*'} \geq 0, \hat{x}_{rt} < 0$ or $r^{*'} < 0, \hat{x}_{rt} > 0$—if a drop in t decreases generosity (or leaves it unchanged) and increases the effectiveness of a share increase in eliciting higher effort, or the drop increases generosity and decreases effectiveness. For the other combinations of the two signs, we may have tracking but we may also have "opposite directions."

Can we obtain an easier answer to our central question if we vary or complicate the model? There are many possible variations.

We could, in particular, turn to the framework of Balmaceda et al. (2016), described in the Related Literature section above. Both Agent and Principal are risk-neutral. There are m possible efforts and n possible revenues. Consider the case $m = n = 2$. The efforts are x_L, x_H where $0 < x_L < x_H$. Letting the subscripts L, H again denote "low" and "high," the possible costs and revenues are C_L, C_H, R_L, R_H. The Agent's cost for the effort x_H (x_L) is tC_H (tC_L), where t is our technology parameter. For the effort x_L, the probability of R_H is p and $1-p$ is the probability of R_L. For the effort x_H, the probabilities are $q, 1-q$, where $q > p$. The Principal chooses a nonnegative wage pair before the Agent chooses effort. If—for a given t—the Principal wants to induce x_H, then he uses a pair (w_H^t, w_L^t), where the Agent is paid w_H^t (w_L^t) if revenue turns out to be R_H (R_L). When x_H is induced, the Agent's expected net gain is $qw_H^t + (1-q) \cdot w_L^t - tC_H$, and the Principal's expected net gain is the remainder of the expected surplus, i.e., $[qR_H + (1-q) \cdot R_L] - [qw_H^t + (1-q) \cdot w_L^t]$. Another wage pair is used to induce x_L. The chosen wage pair, among those that induce a given effort, minimizes the average wage paid by the Principal, subject to an Individual Rationality (IR) constraint (the Agent's expected net gain is nonnegative) and an Incentive Compatibility constraint (the Agent's expected net gain for the given effort is not lower than her expected net gain for the other effort).

Do we again get one of the key results in our model: is it again true that the Agent never works less when technology improves? If the Principal chooses to induce x_H^t, then will he continue to do so if t drops? It turns out[9] that if a wage pair solves the Principal's induce-x_L problem, then the IR constraint for that pair must be binding. If the IR and IC constraints are both binding for a wage pair that solves the induce-x_H problem, and if the Principal prefers to induce x_H, then a drop in t cannot reverse

[9]The details are provided in LMW.

that preference. Moreover, the Decentralization Penalty is then zero for every t. If IR is slack in the induce-x_H solution, then it remains true that a drop in t cannot reverse the Principal's preference for x_H. But now, a drop in t may raise or lower the Decentralization Penalty. A natural research path would consider all pairs (m, n) and would examine analogs of other results that we obtained in our model. One might then explore the same questions when we let the Agent be risk-averse. Does increasing risk aversion (when t is fixed) raise or lower the Penalty?

In another research path, one could change the definition of "Decentralization Penalty," so that it becomes a fraction, as in Balmaceda et al. (2016). The Penalty (in the endogenous case) would be $\frac{W(r^*(t),t)}{W(1,t)}$, rather than the surplus gap we have considered. Our central question becomes technically harder and again has no simple answer. Moreover, there are examples where some of our results about the effect of a drop in t on the Penalty are now reversed.

A third research path would let t be a random variable with common-knowledge probabilities and would let one party have better information about the true t than the other. Technical improvement lowers the expected value of t. Does it increase or decrease the Decentralization Penalty?

It was natural to start with our stripped-down model, where we already saw the unexpected challenges posed by our central question. The question of the effect of improved technology on the merits of alternative modes of organizing is well motivated but has seldom been the focus of previous research. The variations and extensions that we have noted, and numerous others, merit further attention.

Appendix

Proofs of Parts (a), (g), and (h) of Theorem 1

Proof of Part (a)

The function $r \cdot R(x) - tC(x)$, where t is fixed, displays strictly increasing differences in r, x if $r \cdot R(x)$ displays strictly increasing differences in r, x. But that is the case, since R is nondecreasing. Since, for fixed t, the effort $\hat{x}(r, t)$ maximizes $r \cdot R(x) - tC(x)$ on the effort set Σ, it is indeed the case that $\hat{x}(r_H, t) \geq \hat{x}(r_L, t)$, as (**a**) asserts. Part (**a**) also asserts that the Agent strictly prefers the higher share. That is the case since $\hat{x}(r_H, t)$ is a maximizer of $r_H \cdot R(x) - t \cdot C(x)$, so we have

$$\begin{aligned} r_H \cdot R(\hat{x}(r_H, t)) - t \cdot C(\hat{x}(r_H, t)) &\geq r_H \cdot R(\hat{x}(r_L, t)) - t \cdot C(\hat{x}(r_L, t)) \\ &> r_H \cdot R(\hat{x}(r_L, t)) - t \cdot C(\hat{x}(r_L, t)). \end{aligned}$$

Proof of Part (g)

Part **(g)** says:

$$W(r, t_L) > W(r, t_H) \text{ whenever } t_L, t_H \in \tilde{\Gamma} \text{and } 0 < t_L < t_H.$$

The effort $\hat{x}(r, t_L)$ is a maximizer of $rR(x) - t_H \cdot C(x)$. Hence,

$$r \cdot R(\hat{x}(r, t_L)) - t_L \cdot C(\hat{x}(r, t_L)) \geq r \cdot R(\hat{x}(r, t_H)) - t_L \cdot C(\hat{x}(r, t_H))$$

or

$$r \cdot [R(\hat{x}(r, t_L)) - R(\hat{x}(r, t_H))] \geq t_L \cdot [C(\hat{x}(r, t_L)) - C(\hat{x}(r, t_H))].$$

That implies—since $0 < r < 1$—that

$$R(\hat{x}(r, t_L)) - R(\hat{x}(r, t_H)) > t_L \cdot [C(\hat{x}(r, t_L)) - C(\hat{x}(r, t_H))]$$

or

$$R(\hat{x}(r, t_L)) - t_L \cdot C(\hat{x}(r, t_L)) > R(\hat{x}(r, t_H)) - t_L \cdot C(\hat{x}(r, t_H))$$

and hence (since $t_H > t_L$)

$$R(\hat{x}(r, t_L)) - t_L \cdot C(\hat{x}(r, t_L)) > R(\hat{x}(r(t_H), t_H)) - t_H \cdot C(\hat{x}(r, t_H))$$

The term on the left of the inequality is $W(r, t_L)$ and the term on the right is $W(r, t_H)$. That completes the proof of Part **(g)**.

Proof of Part (h)

When the Agent's share is r_H, he chooses an effort $\hat{x}(r_H, t)$ which satisfies

$$r_H R(\hat{x}(r_H, t)) - tC(\hat{x}(r_H, t)) \geq r_H R(\hat{x}(r_L, t)) - tC(\hat{x}(r_L, t)),$$

or equivalently

$$r_H \cdot \left[R(\hat{x}(r_H, t)) - R(\hat{x}(r_L, t))\right] \geq t \cdot \left[C(\hat{x}(r_H, t)) - C(\hat{x}(r_L, t))\right]. \quad (1)$$

Part **(a)** of Theorem 1 tells us that $\hat{x}(r_H, t) \geq \hat{x}(r_L, t)$. Since R is strictly increasing, that means that the left side of (1) is either positive or zero. First, suppose that it is positive. Then, since $r_H < 1$, (1) implies that

$$R(\hat{x}(r_H, t)) - R(\hat{x}(r_L, t)) > t \cdot \left[C(\hat{x}(r_H, t)) - C(\hat{x}(r_L, t))\right], \quad (2)$$

or equivalently

$$R(\hat{x}(r_H, t)) - t \cdot C(\hat{x}(r_H, t)) > R(\hat{x}(r_L, t)) - t \cdot C(\hat{x}(r_L, t)), \tag{3}$$

i.e.,

$$W(r_H, t) > W(r_L, t). \tag{4}$$

If $\hat{x}(r_H, t) \neq \hat{x}(r_L, t)$, then, since R is strictly increasing, the left side of (1) is indeed positive, so (4) holds. If, on the other hand, $\hat{x}(r_H, t) = \hat{x}(r_L, t)$, then both sides of (1) equal zero and (2),(3),(4) become equalities. So, as claimed, $W(r_H, t) \geq W(r_L, t)$ and the inequality is strict if and only if $\hat{x}(r_H, t) \neq \hat{x}(r_L, t)$.

Proofs of Parts (a) and (b) of Theorem 2

Proof of Part (a)

We note first that the Agent's chosen effort $\hat{x}(r, t)$ depends only on the ratio $\frac{r}{t}$, which we shall call ρ. The set of possible values of ρ is $\left(0, \frac{1}{t}\right]$. The Agent's effort is a value of x which maximizes $t \cdot (\rho R(x) - C(x))$ on the effort set Σ and is therefore a maximizer of $\rho R(x) - C(x)$. We shall use a new symbol, namely $\phi(\rho)$ to denote the Agent's chosen effort when the ratio is ρ. So, $\phi(\rho) = \hat{x}(r, t)$. The function $\rho R(x) - C(x)$ displays strictly increasing differences with respect to ρ, x. Hence, the maximizer $\phi(\rho)$ is nondecreasing in ρ, so we have

$$\phi(\rho_H) \geq \phi(\rho_L) \text{ whenever } 0 < \rho_L < \rho_H. \tag{+}$$

We can now reinterpret the Principal as the chooser of a ratio. For a given t, he chooses the ratio $\rho^*(t) = \frac{r^*(t)}{t}$, where

$$\rho^*(t) = \min\{\text{argmax}_{\rho \in (0, 1/t)} M(\rho, -t)\},$$

and

$$M(\rho, -t) = (1 - t\rho) \cdot R(\phi(\rho)) = R(\phi(\rho)) - t \cdot \rho \cdot R(\phi(\rho)).$$

The function M has strictly increasing differences in $\rho, -t$ if the function $-t \cdot \rho \cdot R(\phi(\rho))$ has strictly increasing differences in $\rho, -t$. But that is the case, since R is nondecreasing, which implies (using (+)) that $R(\phi(\ \cdot\ ))$ is also nondecreasing. Since $\rho^*(t)$ is a maximizer of $M(\rho, -t)$, we conclude that

$$\frac{r^*(t_L)}{t_L} = \rho^*(t_L) \geq \rho^*(t_H) = \frac{r^*(t_H)}{t_H} \text{ whenever } 0 < t_L < t_H,$$

as Part (**a**) asserts.

Proof of Part (b)

We use the terminology just used in the proof of Part (**a**). Since $\phi\left(\frac{r^*(t)}{t}\right) = \hat{x}(r^*(t), t)$, we have, using (+) in the proof of part (**a**), $\hat{x}(r^*(t_L), t_L) \geq \hat{x}(r^*(t_H), t_H)$, as (**b**) asserts.

References

Balbaieff, M., Feldman, M., & Nisan, N. (2009). Free riding and free labor in combinatorial agency. In M. Mavronicolas & V. G. Papadopoulou (Eds.), *SDAGT 2009: Vol. 5814. Lecture Notes in Computer Science* (pp. 109–121). Berlin: Springer.

Balmaceda, F., Balseiro, S., Correa, J., & Stier-Moses, N. (2016). Bounds on the welfare loss from moral hazard with limited liability. *Games and Economic Behavior, 95*, 137–155.

Courtney, D., & Marschak, T. (2009). Inefficiency and complementarity in sharing games. *Review of Economic Design, 13*, 7–43.

Garicano, L., & Prat, A. (2013). Organizational economics with cognitive costs. In *Advances in economics and econometrics: Theory and applications, proceedings of the tenth world congress of the econometric society*. Cambridge: Cambridge University Press.

Grossman, S., & Hart, O. (1983). An analysis of the principal-agent problem. *Econometrica, 51*, 7–45.

Holmstrom, B. (1979). Moral hazard and observability. *The Bell Journal of Economics, 10*, 74–91.

Holmstrom, B. (1982). Moral hazard in teams. *The Bell Journal of Economics, 13*, 324–340.

Hurwicz, L., & Shapiro, L. (1978). Incentive structures maximizing residual gain under incomplete information. *The Bell Journal of Economics, 9*, 180–191.

Liang, R., Marschak, T., & Wei, D. (2017). Technological improvement and the decentralization penalty in a simple principal/agent model, SSRN e-library, 2945702 [Abbreviated as LMW in the text].

Marschak, T. (2006). Organization structure. In T. Hendershott (Ed.), *Handbook of economics and information systems* (pp. 205–290). New York: Elsevier.

Nasri, M., Bastin, F., & Marcotte, P. (2015). Quantifying the social welfare loss in moral hazard models. *European Journal of Operations Research, 245*, 226–235.

Nissan,N., Roughgarden, T., Tardos, E., & Vazirani, V. (2007). *Algorithmic game theory*. Cambridge: Cambridge University Press.

Ross, S. (1973). The economic theory of agency: The principal's problem. *American Economic Review, 63*, 134–139.

Roughgarden, T. (2005). *Selfish routing and the price of anarchy*. Cambridge: MIT Press.

Sundaram, R. K. (1996). *A first course in optimization theory*. Cambridge: Cambridge University Press.

Fundamental theory of institutions: a lecture in honor of Leo Hurwicz

Roger B. Myerson

> *"The economic problem of society is not merely a problem of how to allocate 'given' resources... It is rather a problem of how to secure the best use of resources known to any of the members of society, for ends whose relative importance only these individuals know... it is a problem of the utilization of knowledge not given to anyone in its totality. This character of the fundamental problem has, I am afraid, been rather obscured than illuminated by many of the recent refinements of economic theory, particularly by many of the uses made of mathematics".*
> F. A. Hayek, "The Use of Knowledge in Society" (1945).

1 Recognizing the need for a fundamental theory of institutions

In the early twentieth century, economic theorists from left and right (Barone 1908; Lange 1938; Mises 1920; Hayek 1935) argued whether socialist reform of economic institutions was possible without loss of economic efficiency. The inconclusive nature of their debate showed that the existing framework of economic analysis was not adequate to formalize the justifications for the strongly held convictions on each side of this vital argument. To allow analytical comparison of fundamentally different forms of economic organization, a new and more general theoretical framework was needed. In an influential paper, Hayek (1945) argued that a key to this new economic theory should be the recognition that economic

---

The Hurwicz Lecture, presented at the North American Meetings of the Econometric Society, at the University of Minnesota, on June 22, 2006.
Myerson, R.B. Rev Econ Design (2009) 13: 59. https://doi.org/10.1007/s10058-008-0071-6

R. B. Myerson (✉)
Economics Department, University of Chicago, Chicago, IL, USA
e-mail: myerson@uchicago.edu; http://home.uchicago.edu/~rmyerson/research/hurwicz.pdf

W. Trockel (ed.), *Social Design*, Studies in Economic Design,
https://doi.org/10.1007/978-3-319-93809-7_3

institutions of all kinds must serve an essential function of communicating widely dispersed information about the desires and the resources of different individuals in society. From this perspective, different economic institutions should be compared as mechanisms for communication.

Hayek also alleged that the mathematical economists of his day were particularly guilty of overlooking the importance of communication in market systems. But questions about fundamental social reforms require fundamental social theory, and in a search for new fundamental theories, the abstract generality of mathematics should be particularly helpful. So the failure that Hayek perceived should not have been attributed to mathematical modeling per se, but it was evidence of a need for fundamentally new mathematical models. Among the mathematical economists who accepted this challenge from Hayek, Leo Hurwicz has long been the leader.

Over many years and decades, Leo Hurwicz has worked to show how mathematical economic models can provide a general framework for analyzing different economic institutions, like those of capitalism and socialism, as mechanisms for coordinating the individuals of society. Hurwicz (1973) noted that, in late nineteenth-century economics, the institutionalists were economists who avoided analytical modeling. Today, all this has changed, since Leo Hurwicz set the standard for mathematical economists to study institutions as coordination mechanisms.

The pivotal moment occurred when Hurwicz (1972) introduced the concept of incentive compatibility. In doing so, he took a long step beyond Hayek in advancing our ability to analyze the fundamental problems of institutions. From that point on, as Makowski and Ostroy (1993) have observed, "the issue of incentives surfaced forcefully, as if a pair of blinders had been removed." By learning to think more deeply about the nature of incentives in institutions, we have gained better insights into important social problems and policy debates. But as Hurwicz (1998) has observed, there are still basic questions in the theory of institutions that we need to understand better.

As one of many followers in this tradition, I feel privileged to have this opportunity of presenting a Hurwicz lecture. In this lecture, I want to take a broad perspective on the state of these questions and what we have learned about them. First, I will re-examine how modern analysis of incentive constraints can help us to see what was missing in the old socialist debates. Then I will follow Hurwicz (1998) in proposing an abstract general model of how institutions are defined and enforced in a broader social environment. Finally, I will consider more specific models of incentive problems in establishing the fundamental political institutions of a society. Throughout, I will suggest a shift away from Hayek's focus on communication. Although we should recognize the universal significance of informational (adverse-selection) incentive problems in all social systems, I will suggest that strategic (moral-hazard) incentive problems may be even more important for understanding the foundations of social institutions.

2 An old debate and a new theoretical framework

In a polemic against naive dreams of a socialist paradise, Mises (1920) argued that prices from a competitive market equilibrium are necessary for efficient allocation of resources. Countering this argument, Barone (1908) and Lange (1938) saw no reason why socialist managers could not be coordinated equally well by value indexes set by a socialist Ministry of Planning. Mises (1920) and Hayek (1935) expressed great skepticism about the feasibility of such central economic planning without free competitive prices, but their argument on this point remained informal, focusing largely on the intractable complexity of the resource allocation problem. It is hard to be persuasive with such arguments of intractability. After all, if the economy is too complex for our analysis, then how can we be sure that a competitive market will find an efficient solution, or that a socialist planner will not find one? For a convincing argument, they needed a simple economic model in which socialism (suitably defined) could be proven to be less efficient than capitalism.

Of course, the later twentieth century provided much evidence of capitalist economic success and socialist economic failure, but a theorist should not give up a good question simply because there seems to be evidence to answer it empirically. If our theories do not give an adequate answer, then we must continue working to develop theories that can, because one can always propose new institutional structures that do not exactly match those for which we have data. If we have no general theory about why socialism should fail, then we have no way to say that greater success could not be achieved by some new kind of socialism that is different from the socialist systems that have been tried in the past.

Economic theorists today have a strong sense of what was missing from the old debates. The old economists could model resource constraints, but not incentive constraints. Hayek and others made verbal arguments that show a basic awareness of incentive problems, but their arguments remained rhetoric without tight logical support in the absence of any general theoretical framework for analysis of incentives.

In particular, Samuelson (1954) argued that no feasible mechanism could guarantee an efficient allocation of public goods, because asking a person to pay for public goods according to his benefit creates an incentive for him to misrepresent his benefit. This remark seemed consistent with the general view that efficiency is found only in competitive private-good markets. But in trying to formalize this argument, Hurwicz (1972) found that the same incentive problems arise in the allocation of private goods, once we drop the assumptions required for perfect competition. He showed that, with finitely many individuals, no incentive-compatible mechanism can guarantee a Pareto-efficient allocation that is at least as good as autarky for all combinations of individual preferences in a broad class. Thus, the concept of incentive-compatibility was introduced.

The concept of incentive-compatibility developed rapidly after Hurwicz introduced it (Myerson 1982). We have come to understand that there are really two kinds of incentive constraints in the general social coordination problem: *informational incentive constraints* that formalize *adverse-selection* problems of gathering decentralized information, and *strategic incentive constraints* that formalize *moral-hazard* problems of controlling decentralized activity. As Hayek (1945) emphasized, economic plans must make use of decentralized information that different individuals have about their resources and desires. An individual could not be expected to honestly reveal private information that would be used against his interests, and such adverse-selection problems are formalized in economic models by informational incentive constraints. But economic plans must be implemented by decentralized actions of many different individuals, and there is a problem of getting individuals to accept appropriate guidance and direction when they have conflicting strategic incentives. An individual could not be expected to obediently refrain from opportunistic behavior that would be more rewarding to him, and such moral-hazard problems are formalized in economic models by strategic incentive constraints.

So, although the old socialist debates took place at a time when formal economic models only took account of resource constraints, we have now expanded the scope of economic analysis to take account of informational and strategic incentive constraints. If there was any validity to the intuitive arguments of Hayek and Mises, we should now be much better able to formulate them analytically in our new incentivist framework. Thus, we should ask, what is the simplest model in which we can support Mises's and Hayek's conclusions about socialism's failure?

Mises (1920) saw the essential problem arising in socialist allocation of capital, because state ownership of means of production implies the lack of any capital market. Such questions about mechanisms for allocating capital are a topic of corporate finance. Jean Tirole's *Theory of Corporate Finance* (2006) is full of models applying mechanism design to corporate finance, and we may naturally look to these models for insights into the old debate on socialism. Tirole has many models with many different features, but they are generally based on two simple models: one of moral hazard (Sect. 3.2), one of adverse selection (Sect. 6.2). Each model describes a simple world which we can transform by socialist reforms, and we can see how the efficiency of capital allocation is affected. The result may tell us something about what is truly fundamental in our models.

3 Advantages of socialism in a simple adverse-selection model

In Tirole's (2006, section 6.2) basic adverse-selection model, a manager has private information about the probability of success for a unique investment opportunity. The basic parameters of the model are $(I, A, R, p_H, p_L, \eta)$. Here I denotes the capital investment cost required for new project. The parameter A denotes the value of assets that manager can pledge to forfeit if project fails. The parameter R denotes the returns from the project if it succeeds, but the returns will be 0 if the project fails. The probability of success depends on the manager's type. If the manager's type is high then the project's probability of success in the project is p_H; but if the manager's type is low then the project's probability of success is p_L, where $p_L < p_H$. The manager knows his own type, but it is uncertain to anyone else, and the manager can lie about his type. Let η denote the probability of the manager being the high type. For simplicity here, let us assume risk neutrality and no discounting of future returns (zero interest rate). We assume that

$$p_H R > I > p_L R \quad \text{and} \quad I > A$$

so that the project is worthwhile only if the manager's type is high, but the manager does not have enough wealth to undertake the project himself.

Under socialism, there is no problem getting the manager to reveal type honestly, because he is willing to report his type honestly when we just pay him a flat wage no matter what he reports. If we want to give him strict incentives to guide social decision-making about the project, the state could pay the manager $\varepsilon(R - I)$ if the project succeeds, but make him pay εI if the project fails. For any $\varepsilon > 0$, this payment plan would give the manager a positive incentive to recommend the project only when its expected social profit is positive. Feasibility requires $\varepsilon I < A$, but for any endowment size $A > 0$, this liquidity constraint can be satisfied when $\varepsilon > 0$ is sufficiently small.

This example is interesting for Tirole (2006) because he is assuming that competition among investors in the financial market always lets the manager borrow at an interest rate such that investors get expected profit equal to zero given their information about the manager. With access to such competitive lenders, low-type managers would want to imitate high-type managers to get their favorable terms of credit. But under socialism, the monopolistic state lender can fully exploit the high-type manager, and then the low type would not want to borrow at all. So we find that socialism may actually have an advantage here, because socialism can flatten the manager's incentives to eliminate his temptation to lie about his chances of success (for other advantages and disadvantages of a monopolistic supply of credit, see Dewatripont and Maskin 1993).

4 Disadvantages of socialism in a simple moral-hazard model

In Tirole's (2006, section 3.2) basic moral-hazard model, the probability of success depends on the manager's actions (instead of the manager's hidden type). Most of the parameters here (I, A, R, p_H, p_L, B) are as in the previous model: the parameter I denotes the capital investment cost required for new project, A denotes the value of assets that manager can pledge to forfeit if project fails, and R denotes the returns from the project if it succeeds, but the returns will be 0 if the project fails. Now p_H is the probability of success if the manager behaves appropriately, but p_L is the probability of success if the manager misbehaves, where $p_L < p_H$, and B denotes the value of private benefits that the manager gets by misbehaving. We assume that

$$p_H R > I > p_L R + B \quad \text{and} \quad I > A,$$

so that the project is worthwhile only if manager behaves appropriately, but the manager cannot undertake the project alone.

As individuals should have only modest wealth under an egalitarian socialist system, let us suppose that the manager's assets are bounded by the inequality

$$A < Bp_H/(p_H - p_L).$$

In a social investment plan, let w denote the wage that will be paid to the manager if the project succeeds. Then a feasible plan must satisfy

$$p_H w - (1 - p_H)A \geq 0$$
$$p_H w - (1 - p_H)A \geq B + p_L w - (1 - p_L)A.$$

Here the first constraint is a participation constraint, that the manager should not expect to lose by participating in the project. (We are assuming that the social investment I includes a payment to the manager for the opportunity cost of his time in managing the project). The second constraint is a strategic incentive constraint, that the manager should not expect better rewards from opportunistic misbehavior. The expected social profit, to be maximized, is

$$Y = p_H(R - w) + (1 - p_H)A - I.$$

The participation constraint implies $w \geq A/p_H - A$, and the moral-hazard constraint implies $w \geq B/(p_H - p_L) - A$. So with our modest-wealth assumption, the lowest feasible wage is

$$w = B/(p_H - p_L) - A,$$

which yields expected social profit

$$Y = p_H R + A - Bp_H/(p_H - p_L) - I.$$

(Because the manager is risk neutral, we could not increase Y by adding payments to the manager when the project fails). Thus, the manager must be allowed to get a moral-hazard rent that has expected value

$$p_H w - (1 - p_H)A = Bp_H/(p_H - p_L) - A.$$

Notice that the expected social profit Y is strictly increasing in the manager's collateral A.

Now let us add the possibility that managers can be punished, and let x denote the punishment cost inflicted on manager if the project fails. Then a feasible mechanism (w, x) must satisfy the participation constraint

$$p_H w - (1 - p_H)(A + x) \geq 0,$$

and the strategic incentive constraint

$$p_H w - (1 - p_H)(A + x) \geq B + p_L w - (1 - p_L)(A + x).$$

The punishment x is not assumed to yield any social value to anyone else. So expected social profit is still

$$Y = p_H(R - w) + (1 - p_H)A - I.$$

The participation and incentive constraints now imply

$$w \geq (A + x)(1/p_H - 1) \text{ and } w \geq B/(p_H - p_L) - (A + x).$$

With modest endowments $A < Bp_H/(p_H - p_L)$, the wage cost is minimized by the punishment

$$x = Bp_H/(p_H - p_L) - A,$$

which allows the wage

$$w = B(1 - p_H)/(p_H - p_L)$$

and so yields the expected social profit

$$Y = p_H R + (1 - p_H)[A - Bp_H/(p_H - p_L)] - I.$$

Thus, punishment of failures can improve social profit. But increasing the manager's private collateral A still helps, even when punishment is allowed.

On the other hand, if there are rich agents who have assets A greater than $Bp_H/(p_H - p_L)$ then we could achieve the ideal social profit $Y = p_H R - I$, by letting the project be managed by such a rich agent for the wage $w = A(1 - p_H)/p_H$ to be paid if the project succeeds, but taking his collateral A if the project fails, with no further punishment ($x = 0$). This wage makes the participation constraint binding ($p_H w - (1 - p_H)A = 0$) and satisfies the moral-hazard constraint with $w + A \geq B/(p_H - p_L)$.

So there are two obvious ways for socialist reformers to achieve full efficiency here. First, they could allow some individuals to hold more wealth, up to $Bp_H/(p_H - p_L)$. Perhaps such favored people could be heroes of the socialist revolution (or of the Norman conquest). Second, they could drop the participation constraint and force people to become managers without compensation for punishment risks. Perhaps such disfavored people might be prisoners or enemies of the state. But either way, socialism looks rather less appealing from the perspective of this moral-hazard model, as it forces us to admit either inequality or coercion or productive inefficiency into our imagined socialist paradise. Indeed, our simple model does not do badly as a source of theoretical insights into the flaws of Soviet communism, and it formalizes some of Hayek's informal intuitive arguments: "To assume that it is possible to create conditions of full competition without making those who are responsible for the decisions pay for their mistakes seems to be pure illusion" (Hayek 1935, p. 237).

5 Comparing moral hazard and adverse selection

The comparison of these two models suggests that, when we probe the logical foundations of social institutions, moral-hazard problems may be more fundamental than adverse-selection problems. The problems of motivating hidden actions can explain why efficient institutions give individuals property rights, as owners of property are better motivated to maintain it. But property rights give people different vested interests, which can make it more difficult to motivate them to share their private information with each other. Thus, adverse selection might not be so problematic if there were no moral hazard. Socialism differs from capitalism in allowing less property rights for individuals, but moral hazard provides a fundamental economic rationale for some property rights that must apply even under socialism. So adverse-selection problems can be important under socialism, just as under capitalism.

For example, take Tirole's basic moral-hazard model with no punishment ($x = 0$), but now let us add a small probability ε that the manager is a bad type who cannot do better than the p_L probability of success (and cannot get the benefit B). With small collateral $A < p_L B/(p_H - p_L)$, such a bad manager would imitate the good type, to enjoy the positive expected benefits $(p_L B/(p_H - p_L) - A)$ from getting his project financed. So in the presence of moral hazard, the socialist system loses its ability to trivially solve informational adverse-selection problems.

On the other hand, if the uncertainty in the basic adverse-selection model were about the required investment amount I (instead of the success probability p), the socialist planner would have to allow informational rents to low-I type managers. But nobody would even try to take these rents away if the manager were a capitalist entrepreneur.

More generally, even if incentive analysis of other adverse-selection models does not reveal actual disadvantages of socialism, it can help to show that the supposed advantages of socialism may be less than its advocates would have suggested when they failed to recognize the possibility of opportunistic misrepresentations under systems other than capitalism. Analysis of mechanism design with informational incentive constraints has taught us that individuals with unique private information may have to be allowed informational rents in an efficient mechanism. But mechanism design as a conceptual framework can fit capitalist or socialist institutions, and so it can help us to see that the manager of a socialist monopoly who has private information about production costs (and can divert unaudited profits) may extract informational rents that look essentially like the profits of a monopoly in capitalism. Conversely, a capitalist monopoly's profits could be regulated away if its costs were publicly known, and it may be the monopolist's private information about costs that enables him to fend off such regulation. Thus, mechanism design teaches us that having multiple independent sources of supply may be just as important under socialism as under capitalism, which traditional market models could not show. Soviet planning may have suffered from failing to recognize such benefits of informational decentralization.

6 General theory of institutions enforced in larger games

In recent work, Hurwicz (1998) has focused on questions of how institutions are constructed and how institutional rules are enforced (see also Schotter 1981). Here strategic incentive constraints are at the heart of the problem, so we can focus on games in strategic form where N is the set of players, C_i denotes the set of strategies of player i, and $U_i(c)$ denotes the utility payoff to player i from strategy profile c in $C = \times_{j\in N} C_j$.

To a game theorist, an institutional reform means changing the structure of the game that people play in the institution. So it is common for game theorists to think that institutions are games. But Hurwicz (1998) observes that what we normally mean by institutions (or institutional arrangements) typically does not include the specification of individuals' preferences (nor does it typically include the beliefs that we specify in Bayesian games). So an institution for Hurwicz is more properly to be identified with a *game-form* of Gibbard (1973), specifying only the set of players N, the sets of strategies C_i for each player i, and an outcome function $\Theta : C \to Y$ that defines how outcomes in some set Y would depend on the players' strategies. Such game-forms are mechanisms in Hurwicz's sense. To analyze such a game-form or mechanism, however, we must specify each player i's preferences for outcomes by a utility function $u_i : Y \to \mathbb{R}$ on the outcome set Y. With these outcome-based utility functions, we can then define the strategy-based utility functions $U_i(c) = u_i(\Theta(c))$ that complete the structure of the strategic-form game which corresponds to the institution once preferences are given.

When we ask how an institution is established, we must embed it somehow in a larger game. For example, when two people play a game of chess, typically each of them is physically able to grab the other's king at any time, but is deterred from chess-illegal moves by the damage such behavior could do to one's reputation in the larger game of life. So the chess game seems supported by some kind of reputational equilibrium in a larger more fundamental game. But saying "games are equilibria of larger games" cannot be right, because if chess were embedded as an equilibrium in the game of life, then that equilibrium would specify each player's strategy in the chess game itself.

Hurwicz (1998) explains that, if our *legal game* $G = (N, (C_i)_{i\in N}, (U_i)_{i\in N})$ is embedded in some true game H, the structural relationship must be that $H = (N, (D_i)_{i\in N}, (U_i)_{i\in N})$ has a larger strategy spaces

$$D_i \supset C_i \quad \forall i \in N$$

and has utility functions that extend those of the legal game G to the larger domain $D = \times_{j \in N} D_j$. Hurwicz (1998) then suggests that a strong formulation of successful enforcement could require that, for each player i, each illegal strategy outside C_i should be dominated by some legal strategy in C_i, so that a player's best responses always take him into the legal game, even if others deviate.

Hurwicz (2008) remarks, however, that a normally law-abiding player might not want to remain law-abiding when others are acting illegally, and so a weaker concept of enforcement may be appropriate. Thus, I would suggest that the definition of institutional enforcement should be weakened, to say that G is enforceable in H when

$$\forall i \in N, \quad \forall c_{-i} \in \times_{j \in N-i} C_j, \quad \forall d_i \in D_i \backslash C_i, \quad \exists c_i \in C_i$$
$$\text{such that } U_i(c_{-i}, c_i) > U_i(c_{-i}, d_i),$$

so that each player's optimal actions are in his legal strategy set when all others' actions are expected to be in their legal sets. That is, G is enforceable when its strategy sets form a *curb set* in H, as defined by Basu and Weibull (1991) (Curb sets are closed under rational behavior).

This weaker definition of enforceability can admit a great multiplicity of enforceable institutions for a given environment, because a big true game H can contain many different minimal curb sets. This multiplicity may seem an annoying indeterminacy, to theorists who believe in economic determinism. But I would argue that the right mathematical model of institutions should admit such a multiplicity of solutions, because real institutions are manifestly determined by cultural norms and traditional concepts of legitimacy, which would have no scope for effect if the economic structure of the true game H admitted only one dominant solution.

For example, legal rules of a political constitution that are written on a piece of parchment in a museum may be enforced in a true game that involves millions of people on a large land-mass. What would prevent anyone from writing another set of rules (on a bigger piece of parchment) and acting according to them instead? Under any political constitution, such an act should be punished as sedition or treason by others who accept the given constitutional rules. But although treason never prospers, the definition of what is treason depends on an arbitrary social consensus. We all understand that a broad failure to agree about constitutional rules and authority can create an anarchy in which everyone suffers. So the social process of identifying what are the constitutional rules of politics and who are the legitimate leaders of our society has the basic structure of a coordination game with multiple equilibria, where the outcome must depend on culture and tradition through Schelling's (1960) focal-point effect.

The essential role of the focal-point effect in the foundations of our basic political institutions has been emphasized by Hardin (1989) and Myerson (2004, 2008). The new theoretical point here is that Schelling's focal-point effect can be extended to questions of selecting among multiple curb sets, just as among multiple equilibria. Once everyone understands that everybody else will be restricting themselves to strategies in one particular constitutional curb set, it becomes rational for each individual to stay in his or her respective portion of this curb set.

Although people may be symmetric in the true game *H*, this symmetry can be broken in the curb set *G*. Indeed, the enforcement of a constitutional curb set may depend crucially on a small group of specially designated individuals (called law-enforcement officials) whose curb-set strategies stipulate that they would punish any deviator who violated constitutional restrictions.

7 Moral hazard and privilege in sovereign political institutions

The preceding model of how institutions are enforced in larger games is very abstract. To move from broad abstractions to practical specifics, we need to think more carefully about the officials who are the guardians of our institutions, as Hurwicz (1998) has emphasized. Let me follow him now in examining the basic question of who guards these guardians, that is, who forces the enforcers to enforce our laws.

Consider again the problem of enforcing the fundamental political institution of a nation, such as the Constitution of the United States. A constitution can be effective only when there are agents who expect to be rewarded for implementing its rules. In particular, it must designate officials who are expected to prosecute sedition and other violations of the constitution, so as to deter the rest of the population from such illegal moves. But what makes these officials do their official function? Of course, a problem of getting people to do what they are supposed to do is what we call a moral-hazard problem. So the basic problem of getting government officials to enforce constitutional rules is a moral-hazard agency problem in the upper levels of government.

Such an agency model has been analyzed by Becker and Stigler (1974). They recognized that powerful officials have regular opportunities to profit from abuse of power, and that such abuse of power can be difficult for others to detect. For abuse of power to be deterred, the official must expect to do better by acting to enforce the rules correctly, and so must expect substantial rewards that would forfeited if evidence of abuse of power were discovered. Assuming risk-neutrality, the magnitude of these rewards must be at least the potential profit that the official could earn from abuse of power divided by the probability that such abuse of power would be discovered. So when temptations are large and probabilities of detection is small, powerful officials may need to be very well rewarded. Thus, we should expect the leaders of fundamental political institutions to be a very well-rewarded elite, highly motivated by the need to preserve their privileges, as Michels (1915) observed even of socialist political parties.

So our concept of a constitution is incomplete if we ignore the essential role of those who expect enjoy the privileges of high office under the constitution and are therefore well motivated to act to sustain it. From a purely structuralist perspective, it might seem that a political constitution could be fully defined by specifying (1) a set of political offices, (2) the powers, privileges, and responsibilities of these offices, and (3) the procedures for selecting future holders of these offices. But to fully characterize a political constitution as a self-enforcing dynamic system, embedded in a true game where people are symmetric, one must also specify (4) the privileged individuals who actually hold these offices at some initial time (or who expect to be on the short list of serious candidates for these offices in the first elections). In this sense, the specific identity of the small privileged group who are called "Founding Fathers" of the American Republic may be considered an essential component of the American Constitution, as essential as the words written on old parchment in Philadelphia.

If moral-hazard opportunities imply that responsible officials must be well rewarded, then people should be willing to pay for promotion to such offices. In Becker and Stigler (1974) theory, an efficient organization would pass the cost of an official's incentive rewards back to the official ex ante, by charging a fee for promotion to the office. In effect, a candidate for office would be asked to post a bond, which would be returned to the official on retirement if there is no evidence of malfeasance. Such a plan appears to be a simple efficient solution to the fundamental agency problem of government. But it creates a new moral-hazard problem at the highest level, because it implies that the leader who controls appointments to high offices will have an incentive to convict officials of malfeasance and resell their offices. The whole scheme depends on the promise that high officials will be appropriately judged, so that they can expect to be rewarded for correct service and punished for abuse of power, but there may be no neutral party to make such judgments. An official must always be worried that others in the power structure would be tempted to convict him of malfeasance and sell his position to someone else.

Thus, the organizational problem of metering rewards, which Alchian and Demsetz (1972) considered for economic producers, arises even more forcefully for political organizations. Indeed the terms of the problem may be sharpened in the political context, where there can be no question of looking to some higher court for adjudication of contractual relationships.

Hurwicz (1998) recognized that the guardian officials of a sovereign political institution must in some sense be organized into a circle of mutual monitoring and judgment, where the actions of each individual are monitored and judged by others in the circle. But when an individual i is called to monitor the actions of some individual j in such a circle, the monitored actions may include *j*'s monitoring of yet other individuals, which further broadens the scope of activity that individual i must be prepared to observe. So some collective aspect of the fundamental adjudication process seems unavoidable. Within a ruling political faction that admits no higher court of appeal, membership in the faction may require an individual to keep manifestly informed about the general status of other members, perhaps formally by attending regular factional meetings, or informally by staying current in a factional network of gossip.

So the survival of a political institution must depend on its being led by some faction or core group of powerful officials who share a basic trust in each others' judgments. In effect, the members of this group may form a court where they each have a right to be tried before any punishment or loss of privilege. In such a court, evidence of malfeasance against any of them would be commonly heard, so that all members of the group should be able to evaluate whether resulting judgment was reached appropriately. The collective sanction against wrongful judgments in this court could be that the members of this ruling faction would lose trust in each other, so that they would all switch to an equilibrium where each opportunistically abuses his individual power. We may assume that, in a competitive world, a faction would not long hold political power over a large society if members of the faction could not solve free-rider problems in collective actions to defend their power against challenges from other potential factions (Myerson 2008). With this assumption, any one member of a ruling faction could feel protected by the expectation that her colleagues could not mistreat her without risking a general loss of mutual trust within the faction, which would jeopardize all of their privileged positions.

8 Leadership and moral hazard at the center

To be more specific about how such factions are formed, we must recognize the role of leaders as entrepreneurs of institutions. Throughout history, governments have been formed by political leaders whose path to power began by gathering a trusted group of active supporters. When a faction has been organized in this way, privileges of membership in the faction are allocated by the leader. Then the circle of monitoring can be closed by a simple factional rule that the leader should never remove a member's privileges without a process of judgment that is collectively witnessed by other members of the faction. Indeed, rulers throughout history have generally maintained courts or councils, where high officials and others close to the ruler were regularly gathered, and where the ruler's treatment of any courtier could be witnessed and scrutinized by other courtiers. Thus, each individual courtier could feel confident of getting appropriate rewards from the leader, because of the leader's need to maintain a general reputation for appropriately rewarding all courtiers, who are the primary agents of his power.

Popular books on leadership have filled shelves in bookstores, but their descriptions of leadership are often focused on leadership as visionary strategic decision-making (Maxwell 2002). Of course, when people need to coordinate, they may look to a leader for strategic decisions about whether to attack at dawn, or at noon, or not at all. But when we ask what is really the essential function of a leader, I would suggest that the role of strategic planner may be generally less important than the role of honest monitor and reliable paymaster that Alchian and Demsetz (1972) identified. A leader makes a group into an effective team by his reputation for actively monitoring the contributions of individuals in the group and appropriately rewarding their efforts. Such a reputation with a group of supporters, small enough to be individually monitored but large enough to achieve competitive success by their collective actions, is the essential asset that defines a leader. If a leader loses this reputation for appropriately rewarding the members of his group, then the leader must be replaced or the group will lose its ability to compete with other teams that have better leadership.

This idea dates back at least to Xenophon, whose "Education of Cyrus" (c. 360 BCE) depicts a great leader who establishes a great empire by cultivating a reputation for honestly and generously rewarding captains who serve well in battle. While other leaders think that their power depends on the assets in their treasury, Cyrus understands that his power really depends on his credit with his captains, so that it can be better to pay out generously than to keep anything for himself. In another paper (Myerson 2008), I have analyzed a similar model of the foundations of the state by leaders whose ability to hold power depends on their reputation for reliably rewarding the captains who support them against their rivals in contests for power.

An economic entrepreneur must be able to credibly promise future payments both to the investors who supplied his initial capitalization and to the managers whose moral-hazard opportunities require promises of large future rewards. Similarly, a political leader must be able to credibly promise future rewards both to the supporters or captains whose efforts put him in power and to the high officials or governors through whom his power is exercised.

To further probe the difficulties of maintaining a reputation for appropriately rewarding agents in political institutions, let me describe one more model of moral hazard by high government officials, an extension of the Becker–Stigler model that I have analyzed (Myerson 2015). In this model we consider a high official, whom we may call a governor, in a state that is ruled by a single leader or monarch. At any time, the governor can behave well (govern appropriately), or misbehave (govern corruptly), or openly rebel. The leader cannot directly observe whether a governor is behaving or misbehaving, but he can observe any costly crises that occur in the governor's province. Crises occur as a Poisson process with a low expected rate α when governor behaves, but a high expected rate β when the governor misbehaves, where $\beta > \alpha$. Misbehavior also gives the governor a flow of additional hidden benefits that are worth γ per unit time. The governor observes any crisis in her province shortly before the leader does, but she can be called to court for a brief visit during which rebellion is impossible. Let D denote the expected payoff to the governor when she rebels (which is observable to the leader). Crises and rebellions are very costly for the leader, so he wants his governors to always behave well, that is, to never misbehave or rebel. Each individual is risk neutral and has discount rate δ.

To be deterred from rebellion, a governor must always expect rewards that are worth at least D. Candidates for governor can be asked to pay something for promotion to the office, but any candidate's ability to pay is limited by her wealth, which we denote by A. We assume that a governor's potential gains from rebellion are greater than the private wealth of any candidate for office, so $A < D$. On the other hand, the leader may feel tempted to free himself of his debts to a governor, by sacking the governor, and such moral hazard at the top is essential to the problem of political leadership. To admit it into our model as simply as possible, we assume an upper bound H on the debt that the leader can be trusted to owe a governor. These parameters $(\alpha, \beta, \gamma, D, \delta, A, H)$ characterize our model.

To minimize the leader's expected cost of paying governors, the optimal incentive plan (derived in Myerson 2015) can be characterized at any time by the expected present discounted value of all future rewards to the incumbent governor, which we may call the governor's credit. To deter hidden misbehavior, any crisis in the province must cause the governor's credit to decrease by a penalty that has expected value

$$\tau = \gamma/(\beta - \alpha).$$

Normally, the sanction for a crisis should be to reduce the governor's credit by this amount τ. But the governor would rebel if her credit ever went below D after a crisis, and so the governor's credit beforehand must never be less than $D + \tau$. So if a crisis occurs when the governor's credit U is less than $D + 2\tau$, then the governor should be called to the leader's court for a trial, where the outcome is either to reinstate the governor at the credit $D + \tau$ with probability $(U - \tau)/(D + \tau)$, or else to dismiss the governor (who thereafter gets 0) and instead appoint a new governor at the minimum feasible credit level $D + \tau$.

Thus, the need to deter both hidden misbehavior and open rebellion requires the leader to make randomized decisions about whether to dismiss or forgive a governor after a crisis. But the leader is not indifferent in such situations, because dismissing the incumbent governor would create an opportunity to resell the office to a new governor for the payment $A > 0$. So the process of deciding a governor's fate in such a situation must be actively monitored by others, because otherwise the leader's ex-post incentive would always be to dismiss the governor. That is, the leader needs to institutionalize a formal trial procedure where others (whose trust he needs to maintain) can observe that he has given the governor an appropriate chance of reinstatement before any dismissal.

The expected discounted value of the leader's cost, at any point in time, is equal to the credit U that he owes to the current governor, plus the expected discounted value of the leader's net cost of promises to other governors who will be promoted into the position after dismissals in the future ($D + \tau - A$ at each promotion). So the optimal plan for the leader should minimize the expected frequency of future dismissals, which can be achieved by keeping governors as far as possible from the low credit range (below $D + 2\tau$) where dismissals occur. Thus, in the optimal incentive plan, a governor should be paid only in credit, not in cash, until the credit bound H is reached. To keep promises to a governor, her credit should increase between crises at the rate $U' = \delta U + \alpha\tau$ until it reaches the bound H on what the leader can be trusted to owe. When the credit owed equals H, the governor should be paid $\delta H + \alpha\tau$ until the next crisis causes her credit to drop to $H - \tau$. In this solution, increasing the trust bound H would strictly decrease the leader's expected discounted cost, as assessed ex ante when a new governor is first appointed. But with very high H, the leader will ultimately incur large expensive debts to governors who become entrenched in their offices.

That is, even when the leader has the same discount rate as the high officials of his government, the need to deter them from abuse of power creates a motivation for the leader to become a debtor to these officials. Of course this conclusion is just an extension of the results of Becker and Stigler (1974) analysis. Our extended model has been designed only to show how problematic this debt relationship can be, because (to deter corruption) the leader must sometimes actually dismiss officials without paying them their promised rewards, but the circumstances of these dismissals cannot be simply predictable (to avoid rebellions) and so can be verified only by actively monitoring the judgment process, during which the leader's natural incentive is actually to dismiss rather than reinstate (because he can resell the office).

Thus, someone needs to actively monitor the leader's judgments of his high officials and constrain him to act according to an optimal random rule. But who can have such power over the leader of a sovereign political institution? The other high officials on whom his regime depends have such power, because they would rationally misbehave or rebel if they lost trust in the leader's promises of future rewards. (In particular, the leader's problem of deterring misbehavior and rebellion would become infeasible if the amount H that they trust him to owe ever became less than $D + \tau$.) So a sovereign political leader needs a court or council where high officials witness his appropriate treatment of other high officials. Such high councils of government seem universal in political systems. In them, the chief guardian's reputation for rewarding his supporters is collectively guarded by his chief supporters.

Thus, in our fundamental theory of institutions, we should recognize that political institutions are established by political leaders, and political leaders need active supporters. Like a banker, a leader's promises of future credit must be trusted and valued as rewards for current service. The leader's relationship of trust with his inner circle of high officials and supporters requires that they must act collectively to monitor and verify his judgments against any of them. Such a relationship of trust with a group of supporters, small enough for the leader to personally monitor but large enough to effectively control the larger institutions of government, is a political leader's most valuable asset. Furthermore, the members of this group must share a sense of identity, in that each must be confident that the leader's wrongly punishing any one of them could cause all others to lose trust in the leader.

So the establishment of fundamental institutions by political leaders may ultimately rely on a sense of identity among members of a group that is small enough to gather in a court of common judgment to hear a case against any one of them. From this perspective, we can make sense of cases throughout history where powerful political forces have been led by small groups of people who are connected by narrower forms of identity, such as family relationships, or old school ties, or bonds of personal loyalty to their leader, even though these personal connections may seem to have no intrinsic relationship with anyone's position on great questions of national policy. Like the nineteenth century socialists, we may dream of great utopian social reforms, but we should understand that the institutions of any such brave new world would be built on narrower factional foundations, organized by political leaders whose first imperative is to maintain their reputation for rewarding loyal supporters.

References

Alchian AA, Demsetz H (1972) Production, information costs, and economic organization. Am Econ Rev 62:777–795

Barone E (1908) The ministry of production in the collectivist state. In: Hayek FA (ed) Collectivist economic planning (Routledge, London, 1935); translation from Giornale degli Economisti

Basu K, Weibull JW (1991) Strategy subsets closed under rational behavior. Econ Lett 36:141–146

Becker G, Stigler G (1974) Law enforcement, malfeasance, and compensation of enforcers. J Legal Stud 3:1–18

Dewatripont M, Maskin E (1993) Centralization of credit and long-term investment. In: Bardhan PK, Roemer JE (eds) Market Socialism. Oxford University Press, Oxford, pp 169–174

Gibbard A (1973) Manipulation of voting schemes: a general result. Econometrica 41:587–601

Hardin R (1989) Why a constitution. In: Grofman B, Wittman D (eds) The federalist papers and the new institutionalism. Agathon Press, NY, pp 100–120

Hayek FA (1935) The present state of the debate. In: Hayek FA (ed) Collectivist economic planning. Routledge, London

Hayek FA (1945) The use of knowledge in society. Am Econ Rev 35:519–530

Hurwicz L (1972) On informationally decentralized systems. In: McGuire CB, Radner R (eds) Decision and organization: a volume in honor of Jacob Marshak. North-Holland, Amsterdam, pp 297–336

Hurwicz L (1973) The design of mechanisms for resource allocations. Am Econ Rev 63(2):1–30

Hurwicz L (1998) But who will guard the guardians. University of Minnesota paper, http://www.econ.umn.edu/workingpapers/hurwicz_guardians.pdf, revised for Nobel Lecture in American Economic Review 98(3):577–585 (2008)

Lange O (1938) On the economic theory of socialism. In: Lippincott BE (ed) On the economic theory of socialism. University of Minnesota Press, MN, USA

Maxwell JC (2002) Leadership 101. Thomas Nelson, Inc., Nashville

Makowski L, Ostroy J (1993) General equilibrium and market socialism: clarifying the logic of competitive markets. In: Bardhan K, Roemer JE (eds) Market socialism. Oxford University Press, Oxford, pp 69–88

Michels R (1915) Political parties: a sociological study of oligarchic tendencies in modern democracy. Hearst, NY

Myerson RB (1982) Optimal coordination mechanisms in generalized principal-agent problems. J Math Econ 10:67–81

Myerson RB (2004) Justice, institutions, and multiple equilibria. Chicago J Int Law 5:91–107

Myerson RB (2008) The autocrat's credibility problem and foundations of the constitutional state. Am Polit Sci Rev 102(1):125–139

Myerson RB (2015) Moral hazard in high office and the dynamics of aristocracy. Econometrica 83:2083–2126

Samuelson PA (1954) The pure theory of public expenditure. Rev Econ Stat 36:387–389

Schelling TC (1960) Strategy of conflict. Harvard University Press, Cambridge

Schotter A (1981) Economic theory of social institutions. Cambridge University Press, London

Tirole J (2006) Theory of corporate finance. Princeton Unviersity Press, Princeton

von Mises L (1920) Economic calculation in the socialist commonwealth. In: Hayek FA (ed) Collectivist Economic Planning (Routledge, London, 1935); translation of Die Wirtschaftsrechnung im sozialistischen Gemeinwesen. Archiv fuer Sozialwissenschaften 47

Xenophon (2001) The education of cyrus, translated by Wayne Ambler. Cornell University, Ithaca

The Hurwicz Program, Past and Suggestions for the Future

Andrew Postlewaite and David Schmeidler

1 Mechanism Design

The modern neoclassical consumer model was formulated in Scandinavia between the world wars, but modern theory started essentially with the publication by Arrow and Debreu (1954) of the proof of existence of competitive (Walrasian or price) equilibrium. The first conceptual contribution by Leo (Hurwicz 1972, and to some extent, 1960a) was to separate the economic variables into two groups: The *environment*, as he termed it, includes the characteristics of economic agents, initial endowments, preferences, and production sets, and the allocation mechanism, i.e., the methods or institutions by which the society organizes the exchange of commodities and makes production and consumption decisions. Next (in the same papers), he introduced and formally defined concepts like: *performance correspondence, implementation, incentive compatibility, informational decentralization, equilibrium of a mechanism*, etc. (and proved theorems relating these terms).[1] Until about the end of the sixties, the theory, still named mathematical economics, dealt mainly with the properties of competitive equilibria including stability and convergence. See, for example, Arrow and Hahn (1971).

In Leo's framework, the competitive mechanism is only one of many possible. In the theoretical literature, alternatives to price equilibrium were mostly considered

[1] These ideas evolved from simplified models, virtually special cases of his general model. This is evident from his papers before 1972 (Hurwicz 1951, 1955, 1960a,b, 1966, 1969, 1970, and 1971).

A. Postlewaite (✉)
University of Pennsylvania, Philadelphia, PA, USA
e-mail: apostlew@econ.upenn.edu

D. Schmeidler
Tel-Aviv University, Tel Aviv, Israel

W. Trockel (ed.), *Social Design*, Studies in Economic Design,
https://doi.org/10.1007/978-3-319-93809-7_4

in cases of market failure such as public goods, nonconvexities, etc.[2] Mechanism design has been also applied to allocations within firms and organization. However, the main goal of the mechanism design research was not just to design mechanisms but to investigate which combinations of desired properties of the performance correspondence can be implemented by a mechanism with desired properties. We should recall here that a mechanism is defined for a whole class of environments, such as neoclassical environments with some fixed number of commodities. The mapping that assigns equilibria outcomes (allocations) to environments is termed the performance correspondence (of this mechanism). Any correspondence from environments to outcomes is termed social choice correspondence (SCC). Thus, a performance correspondence is an SCC implemented by the above mechanism.

The noncooperative game theory developed during the fifties and the sixties, and in the seventies it entered the economic theory and partially or mostly replaced competitive price equilibria as its principal tool of research. In the seventies, Leo, and others who joined the mechanism design research program, replaced the abstract concept of mechanism with game forms, and the abstract equilibria with equilibria of strategic games like Nash equilibria, strong equilibria, dominant strategies equilibria Bayesian equilibria, etc.[3] Already by the late sixties, Shapley and Shubik constructed a descriptive model of the market: a game form whose strategic equilibria coincided asymptotically with competitive equilibria, but suggested more realistic equilibria for oligopolistic markets. Many variants of this game form were suggested and investigated.

A central feature of Leo's contribution was to reverse the search for desirable mechanisms. Instead of inquiring the properties of equilibria of a specific game form, he started from the desired properties of an SCC: Does there exist a mechanism whose performance correspondence satisfies these desiderata? In addition, some desired properties were sometimes prescribed, for example, the domain of the environments, the informational requirements of the mechanisms including the type of the equilibrium, etc.

An important related area where mechanism design extended and redefined the scope of research is the voting\social choice theory. It started with Arrow's cardinal impossibility result (1952) and continued with the by now classical Gibbard (1973) and Satterthwaite (1975) results showing the impossibility of a straightforward (Farquharson 1969) voting rule. Concurrently, majority voting rules modelled along the lines of those used in parliament were investigated. The mechanism design approach asks whether there are mechanisms (voting rules) whose strategic equilibria (i.e., the performance correspondences) satisfy certain desiderata. Maskin (1999) showed that "monotonicity" of an SCC is a necessary and

[2]Historically, alternatives to the market mostly originated from socialist utopias starting with utopian socialism, continuing with Marx and Engels' socialism and communism, and the attempts to implement them, from the USSR in 1917 to North Korea today (Jan 2018). In mainstream economics, these utopias were relegated to economic history and the history of economic thought.

[3]Leo was not very keen of mixed strategies, vNM utility or Bayesian equilibrium.

almost sufficient condition for it to be possible to implement the SCC via Nash equilibria.

While these results about social choice mechanisms are of first-order importance, the impetus for Leo's first work on mechanism design was motivated by the Lange–Lerner debates about the viability of socialism (Lange 1942, Lerner 1944, Hayek 1945). The debate was, to a large extent, whether a centralized system could uncover information dispersed among many agents, and use that information to achieve Pareto efficient outcomes. Formalizing this, Leo considered a set of pure exchange economies consisting of a finite number of agents, each of whom was characterized by a nonnegative initial endowment of the goods in an economy and a utility function over possible consumption bundles. A performance function on this set of pure exchange economies is then a function that maps each economy (a finite collection of agents) into an equilibrium redistribution of the agents' endowments. In Hurwicz (1972), Leo showed the impossibility of a performance function that lead to individually rational Pareto efficient allocations when the equilibrium notion was dominant strategy. The literature turned the question of whether there were performance functions with desirable properties when Nash equilibria was the solution concept, where quite general characterizations were obtained.

Much of the literature motivated by Leo's early work focused on the possibility of implementing Pareto efficient performance functions, motivated by the debates about planned economies. *What* performance functions can be implemented is obviously important, but *how* a performance function is implemented is no less important. Leo's conceptual framework separates the performance correspondence one might want to arise from the game form that governs individuals' behavior, allowing one to investigate the properties of the game form (the institutions that provide incentives for behavior) separately from the properties of the performance (the equilibria arising from the institutions).

There are two broad reasons to care about the properties of a game form that implements a given performance function: The analyst may care about the game form, and the agents participating in the game form may care.

The Analyst's Concern As noted above, much of the mechanism design literature asks whether a performance function can be implemented in Nash equilibrium, that is, is there a game form for which the Nash equilibrium outcomes coincide with the outcomes specified by the performance function. The reliance on Nash equilibrium as the solution concept did not arise because it was particularly compelling, but rather, because it seemed the "least flawed" solution concept that gave interesting insights.[4]

Maskin's (1999) seminal paper mentioned above illustrates potential conceptual problems with Nash equilibrium as a solution concept. Roughly, the paper gives sufficient and (nearly) necessary conditions on a performance function for it to be implementable in Nash equilibria. Sufficiency is shown by constructing a game form

[4]See Jackson (1992) for an early paper along this line.

whose equilibrium outcome is the outcome proposed by the performance function satisfying the sufficient conditions. An agent's strategy in the constructed game form includes a precise description of all agents' preferences, both other agents' preferences and her own. If all agents' agree on the profile of preferences, the outcome is that prescribed by the performance function. The game form is cleverly designed so that the only possible equilibrium is that the agents agree.

This works fine when (as implicitly assumed) the profile of preferences is common knowledge among the agents. But, the central idea is applied to pure exchange economies, the game form is highly discontinuous, and the slightest deviation by a single agent can lead to very bad outcomes.[5] So, while the Nash equilibrium solution captures the incentives among agents, it is unreasonable to think of it as being plausible for this game form implementing, say, the (constrained)[6] Walrasian performance function.[7,8]

An analyst might well think that while Nash equilibrium is appropriate for some game forms, but prefer a game form that was continuous as this would avoid disastrous outcomes when small deviations from equilibrium play. Postlewaite and Schmeidler (1978) analyze a *continuous* game form for pure exchange economies in which there are Nash equilibria arbitrarily close to Walrasian equilibrium allocations when the number of agents is sufficiently large.

One might rate this game form as preferable to a Maskin-type game form on this basis, but less desirable on two counts. First, Nash outcomes are not fully Pareto efficient.[9] Second, while when there are many agents there is a Nash outcome that is close to the Walrasian outcome, there are other equilibria that lead to no trade. This is in contrast to the performance of a Maskin-type game form which does not have such less desirable equilibrium outcomes.

There are other important characteristics of game forms besides continuity (or lack thereof) and multiplicity of equilibria. As mentioned above, in Maskin-type game forms an agent's strategy includes announcing the vector of preferences for the participating agents. As the number of agents gets large, the size of her messages grows proportionately. In addition to the implausibility that she would have this information, there is the sheer difficulty of acting upon it. In contrast, in

[5]See Hurwicz et al. (1995) for details.

[6]Constrained Walrasian equilibria are essentially price and allocations for which all agents are maximizing subject to their budget sets and the feasibility of their demands. For simplicity, we will drop the "constrained" and refer simply to the Walrasian correspondence.

[7]One might argue that if the problem is that agents may not, in fact, know precisely the preferences of all agents in the economy, one should then include in the basic model agents' *beliefs* about the preferences. This, however, does not really solve the underlying problem. Postlewaite and Schmeidler (1986) show that when following this path, one can accomplish the analogous Bayes Nash implementation for exchange economies with asymmetric information using a similarly discontinuous game form.

[8]See Eliaz (2002) for a discussion of this and related issues.

[9]Postlewaite and Wettstein (1989) demonstrate a continuous game form somewhat resembling Maskin-type game forms whose outcomes are constrained Walrasian.

the Postlewaite and Schmeidler (1978) paper mentioned above, an agent's strategy is a vector of amounts of goods she wishes to put up for sale and the amount she is willing to spend for each good she wishes to buy. This is a vector of dimension twice the number of goods independent of the number of agents.[10]

The discussion so far has outlined features of a game form under consideration that an analyst might look at in evaluating the plausibility of Nash equilibrium as a solution concept for the game form. Another feature is the information an agent needs to determine her best response to other agents' strategies. In Maskin-type game forms an agent needs to know precisely the strategies of each and every agent to determine her best response, while in the Postlewaite and Schmeidler (1978) game form, agents need only predict the *sum* of other agents strategies. That, along with the continuity, might give greater plausibility to the Nash outcome of the game form.

Before turning to participants' possible concerns about properties of the game forms employed in the design of mechanisms, it is useful to mention how analysts often derive optimal mechanisms. A common technique is to invoke the revelation principle.[11]

A Participant's Concern The discussion above dealt with the analyst's concerns, driven primarily on the suitability of Nash equilibrium as the solution concept. In addition to those concerns, the participants of the game form might have concerns that are separate from questions on Nash equilibrium. If I were an agent in a pure exchange economy who had a choice of what game form I would like to govern reallocation, I would care about many of the properties that the analyst cares about. I would like the game form to be continuous so I did not need to worry about small trembles on my part or by other agents; I would like a game form that would not entail my needing to predict all other agents' individual strategies in detail; I would prefer a game form, where I would also care about how complex my strategies in the game form, for example, am I to choose a finite-dimensional vector? Do I need to choose from more complicated sets when there are more agents involved?

In addition to the properties of interest to the analyst, I would like to know how "risky" the game form is that is, how badly off could I be in a worst-case event? Suppose there is a given game form that implements the Walrasian outcome for pure exchange economies. Suppose that I am an agent in a pure exchange economy and I play my part of a Nash equilibrium for this economy. For the game form in Hurwicz, Maskin, and Postlewaite, the outcome may be the worst possible—my endowment is confiscated and I consume nothing. I would prefer an alternative game form that implemented Walrasian outcomes, but in which I could guarantee an outcome that was at least as good as my initial endowment even if other agents did not play their Nash strategies (if such a mechanism existed). (Leo coined the term *non-confiscatory* for game forms which guaranteed that agents were guaranteed not

[10]See Hurwicz and Reiter (2006) and Mount and Reiter (1974) for a discussion of related issues.

[11]See Wikipedia https://en.wikipedia.org/wiki/Revelation_principle.

to be worse off than their initial endowment, but this constraint was on equilibrium outcomes not, as suggested here, that this constraint hold should other agents choose nonequilibrium strategies.)

There are (at least) three notions of how I might guarantee that I will not be worse off than at my initial endowment: (1) for any strategy vector of other agents, I have a strategy that guarantees an outcome at least as good as my initial endowment; (2) a stronger notion, that I have a strategy that for any strategy choice of other agents I will not be worse off than at my initial endowment; and (3) an even stronger notion, that my equilibrium strategy guarantees I will not be worse off than at my initial endowment.

The game form in Postlewaite and Schmeidler (1978) satisfies the stronger notion that an agent has a strategy that uniformly across all possible strategies other agents may choose leaves me as well off as with my initial endowment. Unfortunately though, that strategy leaves me with my initial endowment regardless of others' strategies. One would like a game form that has the desirable property (a strategy that leads to an outcome at least as good as my initial endowment) but also (at least sometimes) leads to gains relative to my endowment. One can imagine a game form analogous to that in Postlewaite and Schmeidler (1978), but one in which an agent chooses a demand function, and the outcome of the game form is a Walrasian equilibrium for the vector of demand functions chosen.[12] This has many of the desirable properties of the game form in Postlewaite and Schmeidler (1978)—the outcome function is (upper hemi) continuous, agents need only predict the sum of other agents' strategies to compute a best response, and the agent has a (natural) strategy that assures an outcome at least as good as her initial endowment (choose her honest demand function). But, unlike the strategy that guarantees an outcome as good as the initial endowment in Postlewaite and Schmeidler (1978), announcing my true demand function typically gives a gain relative to my initial endowment. In fact, if all agents announce their true demand functions, the outcome is the Walrasian outcome for the given exchange economy, and consequently Pareto efficient. The game form in which agents choose demand functions has a serious defect relative to the game form in Postlewaite and Schmeidler (1978), however. While for large economies agents (usually) gain little by deviating, whatever the other agents do, the Nash equilibrium outcome can be far from the Walrasian outcome. To our knowledge, it is not known whether the Walrasian correspondence can be implemented with a game form for which an agent who plays her equilibrium strategy can be guaranteed an individually rational outcome.

While both the analyst and the participant might care about the aspects above of a game form that implements a particular SCC, there are other aspects that the analyst might less interested in than a participant. I would like institutions that lead to efficient outcomes, but in addition I care about the process by which outcomes arise. For example, I prefer to share as little information about myself as possible, given the goal of implementing the Walrasian correspondence. An agent's

[12]Roberts and Postlewaite (1976) analyze such a game.

equilibrium strategy in the game form in Hurwicz, Maskin, and Postlewaite includes the agent's true preferences, while an agent's equilibrium strategy in the game form that also implements the Walrasian correspondence in Postlewaite and Wettstein has the agent revealing only his net trade at that Walrasian price. There might be instrumental reasons for wanting to reveal as little as possible in an implementing game form, such as a fear that information I reveal might be used to my detriment in the future. But separately from instrumental concerns, an agent might have a *direct* preference to maintain as much privacy as possible.

A participant may also care about the range of outcomes that he can effectuate when other agents play their part of a Nash equilibrium. By definition, at a Nash equilibrium, the outcome I get is as good or better than any of the others available to me given the play of others. But for two different game forms that give rise to the same equilibria, when others play their part of a Nash equilibrium the range of outcomes achievable as I choose different strategies in one may be larger than the range in another. For example, in the game form implementing the Walrasian correspondence in Hurwicz, Maskin, and Postlewaite, at a Nash equilibrium I can achieve all feasible allocations that are no better than my Nash equilibrium outcome. In the game form in Postlewaite and Wettstein that implements the Walrasian correspondence, at a Nash equilibrium I can achieve only the feasible outcomes that give me a bundle that cost less than my Walrasian equilibrium bundle (at the Walrasian equilibrium price),a (typically) strictly smaller set of outcomes. Hence, I would prefer the latter if I would like a smaller choice set available at equilibrium and the former if I like a larger set.

Dealing with the Walrasian correspondence as we have above is relatively easy for (at least) two reasons. First, an agent cares only about the bundle of goods that he consumes, and is indifferent about other agents' consumption. Second, agents' choices and the outcomes that result from those choices are precisely defined. When we move to interesting real-world mechanism design problems, we see limitations of this framework.

The creation of the constitution of the United States is a leading example of a real-world problem in which a set of agents met to design institutions for a new country. The actors in the venture were very intelligent and knowledgeable, and engaged in prolonged heated discussion about the institutions they were creating. The power of the to-be-formed central government to levy taxes was one of the most contested issues. It was imperative that taxes to support an army be included if the system was to survive. Previous central authorities relied on voluntary contributions of the independent states. Predictably, free riding crippled the central authority. While this was universally recognized, many of the delegates charged with designing the constitution were very apprehensive of granting the central government too much power given the recent experience under British rule.

The conflicts among the delegates writing the new constitution illustrate two problems in mechanism design that typically do not show up in the standard academic mechanism design literature. First, while the constitutional delegates had different preferences over the outcomes that would result from the new constitution, much of the debate centered not so much on which outcomes were preferable, but instead on which outcomes *were likely* to arise from different sets of rules. It was not possible to completely describe the actions available to various players, or even what outcome would result from a given set of actions agents might take. In our language, there was no general agreement about what would be the Nash equilibrium outcomes from any proposed constitution.

This is not an issue in the academic mechanism design literature, as the "rules of the game" for writing academic papers in this area more or less require that the game form be specified precisely.[13] A necessary step in transferring the mechanism design methodology to many real-world problems is to formalize participating agents' difficulty in predicting equilibrium outcomes for the proposed mechanism.[14]

References

Arrow, K., & Hahn, F. (1971). *General competitive analysis*. San Francisco: Holden-Day.

Arrow, K. J., & Debreu, G. (1954). Existence of an equilibrium for a competitive economy. *Econometrica, 22*(3), 265–290.

Eliaz, K. (2002). Fault tolerant implementation. *Review of Economic Studies, 69*, 589–610.

Farquharson, R. (1969). *Theory of voting*. New Haven: Yale University Press (Accepted for publication 1963).

Gibbard, A. (1973). Manipulation of voting schemes: A general result. *Econometrica, 41*, 587–602.

Hayek, F. A. (1945). The use of knowledge in society. *American Economic Review, 35*, 519–530.

Hurwicz, L. (1951). Theory of economic organization. *Econometrica,19*, 54.

Hurwicz, L. (1955). Decentralized resource allocation, Cowles Commission Discussion Paper: Economics No. 2112.

Hurwicz, L. (1960a). Optimality and informational efficiency in resource allocation processes. In K. J. Arrow, S. Karlin, & P. Suppes (Eds.), *Mathematical methods in the social sciences* (pp. 27–46). Stanford: Stanford University Press.

Hurwicz, L. (1960b). Conditions for economic efficiency of centralized and decentralized structures. In G. Grossman (Ed.), *Value and plan* (pp. 162–183). Berkeley: University of California Press.

Hurwicz, L. (1966). *On decentralizability in the presence of externalities*. Paper presented at the San Francisco meeting of the Econometric Society (unpublished).

Hurwicz, L. (1969). On the concept and possibility of informational decentralization. *American Economic Review, 59*, 513–534.

Hurwicz, L. (1970). Organizational structures for joint decision making: A designer's point of view. In M. Tuite, R. Chisholm, & M. Radnor (Eds.), *Interorganizational decision-making*. Chicago: Aldine Press.

[13]We are not arguing that there are *no* problems for which the game form can be precisely specified; there are, for example, problems in designing computer protocols where this issue may not arise. Rather, we suggest that there are important real-world problems where this issue is of first-order importance.

[14]One's first reaction might be to model agents' uncertainty by putting a prior over agents' uncertainty and utilizing Bayes equilibrium as the solution concept. This does not seem to be a realistic solution to problems like writing constitutions.

Hurwicz, L. (1971). Centralization and decentralization in economic processes. In A. Eckstein (Ed.), *Comparison of economic systems: Theoretical and methodological approaches*. Berkeley: University of California Press.
Hurwicz, L. (1972). On informationally decentralized systems. In R. Radner, & C. B. McGuire, *Decision and Organization*. Amsterdam: North-Holland.
Hurwicz, L., Maskin, E., & Postlewaite, A. (1995). Feasible implementation of social choice correspondences by Nash equilibria. In J. Ledyard (Ed.), *Essays in honor of Stanley Reiter* (pp. 367–433). Dordrecht: Kluwer Academic Publishers.
Hurwicz, L., & Reiter, S. (2006). *Designing economic mechanisms*. New York: Cambridge University Press.
Jackson, M. (1992). Implementation in undominated strategies: A look at bounded mechanisms. *Review of Economic Studies, 59*(4), 757–775.
Lange, O. (1942). The foundations of welfare economics. *Econometrica, 10*, 215–228.
Lerner, A. P. (1944). *The economics of control*, New York: Abba P. Lerner.
Maskin, E. (1999). Nash equilibrium and welfare optimality. *Review of Economic Studies, 66*, 23–38.
Mount, K., & Reiter, S. (1974). The informational size of message spaces. *Journal of Economic Theory, 8*, 161–192.
Postlewaite, A., & Schmeidler, D. (1978). Approximate efficiency of non-Walrasian Nash equilibria. *Econometrica, 46*(1), 127–137.
Postlewaite, A., & Schmeidler, D. (1986). Implementation in differential information economies. *Journal of Economic Theory, 39*, 14–33.
Postlewaite, A., & Wettstein, D. (1989). Implementing constrained Walrasian equilibria continuously. *Review of Economic Studies, 56*, 603–612.
Roberts, J., & Postlewaite, A. (1976). The incentives for price-taking behavior in large exchange economies. *Econometrica, 44*, 115–129.
Satterthwaite, M. (1975). Strategy-proofness and Arrow's conditions: Existence and correspondence theorems for voting procedures and welfare functions. *Journal of Economic Theory, 10*, 187–217.

Social Networks from a Designer's Viewpoint

Fernando Vega-Redondo

1 Introduction

By way of motivation, let me start with three examples.

As a **first example**, consider a financial system consisting of a large number of firms (call them "banks"), each of them holding some individual asset of identical market value (e.g., an equally priced portfolio of mortgages). For simplicity, let us assume that the induced returns for each bank are determined by random variables that are both stochastically uncorrelated and identical. Their assets, therefore, can be conceived as perfect substitutes from an ex ante perspective but typically exhibit different realizations ex post. In particular, some of those banks can be hit by individual-specific shocks of varying magnitude—which, if large, can lead to the bankruptcy of the affected banks.

Let us suppose, however, that the most likely shocks—the "ordinary" ones—are not very large in the following sense. If a bank diversifies its portfolio (by, say, securitizing its mortgage portfolio and exchanging a significant part of it with an equal portfolio share of some other bank), then no ordinary shock brings the bank under the bankruptcy threshold. This is, in essence, what provides banks with the incentive to conduct some risk-sharing asset exchange: it protects them from bankruptcy in the face of the most frequent shocks.

But, the problem then is that, through such an asset exchange, banks become also exposed to the risk of being *indirectly* affected by more large shocks. These shocks are much less likely than ordinary ones but not impossible. Furthermore, if one of them arrives and is indeed large, no asset exchange will work as an effective risk-sharing mechanism for the bank directly hit. And, for the others, it could in

F. Vega-Redondo (✉)
Bocconi University & IGIER, Milan, Italy
e-mail: fernando.vega@unibocconi.it

W. Trockel (ed.), *Social Design*, Studies in Economic Design,
https://doi.org/10.1007/978-3-319-93809-7_5

fact be counterproductive, leading through "contagion" to bankruptcy in a much wider scale than would have happened otherwise. Here is, therefore, an important trade-off. On the one hand, some small degree of risk-sharing is always optimal for individual banks because it helps them confront successfully their ordinary shocks. On the other hand, it is precisely because of the connections underlying such risk sharing that, if a truly big shock arrives, many banks can be severely affected, even to the extent of forcing them to go bankrupt.

In general, the optimal handling of the trade-off outlined calls for an exercise in network/mechanism design. Depending on the nature and magnitude of the shocks, the characteristics of the banks, or their information asymmetries, the issue is how to design the financial network so that it is as robust as possible. Often, this problem is conceived as involving the minimization of *systemic risk*, a term that has been commonly used in dealing with the latest world economic crisis... and in trying to avert the next one!

The problem, however, is not as "simple" as designing the optimal network. For, in a free-market economy, trade (naturally, also financial trade) is not imposed but is chosen by agents in line with their individual interests. Thus, in the end, the mechanism-design problem becomes one of setting the rules (the game-form, in Hurwicz's terminology) for the network-formation game played by the agents/banks involved. These rules should be chosen so that, assuming that the strategic behavior of the banks is correctly anticipated, the resulting network displays the desired characteristics.

As a **second example**, consider a context where there is a population of entrepreneurs whose main objective is to develop new products for a booming market. All of them share the same objective—come up with the "next big thing"—but they are otherwise very diverse. They operate in different markets (either geographically or sectorally), have different areas of expertise, enjoy different access to alternative resources (e.g., human, physical, or financial capital), or are of different age and experience. Suppose that one wants to harness such a diversity within the entrepreneur population to stimulate innovation. Thus, to this end, they are divided into groups and given the opportunity to interact among themselves. The following question then arises: What combination of individual (observable) characteristics is most productive to make the entrepreneurs most innovative? In other words, what are the dimensions in which diversity is beneficial for innovation (possibly, being even detrimental in some others)?

This question again raises a mechanism-design issue—but one that, unlike in the previous example, comes in two steps. First, there is the need to select the "right" profile of characteristics for each group. Second, one must decide on the rules of interaction and the incentives that shape agents' networking, communication, and cooperation decisions. The objective here is to arrive at a harmonized combination of group characteristics, rules, and incentives—all "externally" set by the designer—that in the end lead to an *endogenous* pattern of peer networking that promotes innovation.

Of course, the hypothesis underlying this approach is that since innovation is largely a *social* activity—be it in the arts or sciences, as much as in business or technology—it should profit substantially from diverse interaction, in particular among peers. The task at hand is undoubtedly complex, its complexity deriving from the fact that such peer interaction typically involves many different components: diffusion (in spreading information), complementarities (due to specialization), substitutabilities (as a result of competition), trust (underlying cooperation), or bridging (that closes interaction gaps).

At present, we are far from having a satisfactory model of social-based innovation that does justice to such richness and multidimensionality. The bulk of research on this topic has been mostly empirical, with little theoretical guidance. This has imposed significant limitations on our ability to better understand how innovation is, and must be, embedded in the social structure. Of course, innovation has always had an important social/network component to it. What is new in the current hyper-connected age is that the potential benefit of leveraging the social network, thus thinking of innovation as a genuinely distributed phenomenon, has become so much greater.

My **third example** is also motivated by the dense and wide connectivity enjoyed by modern societies. In this case, however, the focus is on how opinions and beliefs form in large populations. Consider a situation where individuals hold sociopolitical opinions that can be suitably represented along a left-right scale—say as a point in a given interval. Each individual starts with some initial position, which reflects some idiosyncratic information and previous experiences. Then, over time, all of them adapt their positions by combining their former opinions with those held by the peers (friends, relatives, and colleagues) with whom they usually interact. The relative importance given to every peer in such an updating process defines the prevailing (weighted) network of social (interagent) influence.

In this context, a natural question to ask is whether, through such a repeated combination of their opinions, individuals will end up approaching a common understanding of matters. This is important to the extent that some "common ground" may be required if the population is to act effectively in addressing the problems they face. Intuitively, such an opinion convergence could be expected if the society is globally and densely connected. But, in a large community (e.g., a country), a global and dense pattern of connection would be unfeasible. At best, one can hope that there is some path in the network that allows almost everyone to learn indirectly about the opinions of most others. But, this might be insufficient to bring about a sufficient integration of opinions if the population is large and those paths are long. Even worse, it is even possible that, in fact, the society is de facto split into several groups (say left- and right-minded groups, or democrats and republicans) with very little information and influence flowing across them. This would happen if, for example, individuals tend to pay attention (and hence be influenced by) only those who are quite similar to themselves—that is, if they display some so-called homophily.

Such a problematic situation is largely a result of the prevailing social network, and hence it is on this network that the solutions must be sought for. One natural route to address the problem might be to establish some new bridging links that could close gaps and thus favor integration. The establishment of these links, however, may not be easy nor cheap. For a large population, the number of new links required to have a significant effect may be prohibitively costly, unless one has a thorough understanding of the prevailing network, the social-learning process operating on it, and who are the key agents. To be effective, therefore, again requires the adoption of a sophisticated designer's viewpoint.

The challenge, however, does not pertain only to the creation of links, conceived as new channels of influence and information. For, even if additional links are established, it is up to the agents to rely effectively on them, attributing to those connections enough weight in the revision of their behavior and opinions. Influence weights, in other words, are endogenous. Then, the designer's task is to be viewed as that of shaping up the social network (opening some "channels," and perhaps closing others) so that the induced codetermination of opinions and influences leads to the desired outcome—say, a sufficiently cohesive society. In a world where social-learning processes evolve so fluidly, this task is indeed a tall order for even a powerful and sophisticated designer.

The remaining part of this chapter is motivated by, and structured along, the three examples discussed above. In Sect. 2 I, consider risk-sharing networks, in Sect. 3 social-innovation networks, and in Sect. 4 opinion-formation networks. The approach is different in each case. For the first case and the third (Sects. 2 and 4) the focus is theoretical, while in the second case (Sect. 3) it is empirical. I conclude in Sect. 5 with some final comments. In the end, the main message will be that, even though for some important contexts we have gained valuable insights, there is still much to do in understanding how to tackle the design problems entailed in a truly effective manner.

2 Risk-Sharing Networks

Risk is a key factor of socioeconomic environments. It impinges on trade, health, travel, farming, or investment. In some cases, there are institutions that can mitigate risk through explicit and formal insurance arrangements. But in many other situations, formal contracts are not viable due to frictions of different sorts—e.g., informational, legal, or/and strategic. Such frictions are particularly relevant in developing countries, where institutions are weaker and narrower in scope. Under these circumstances, informal ways to tackle the problem arise naturally, with social networks playing an important role. They underlie the transfers in money or kind by which the lucky help the unlucky when the latter are hit by individual shocks.

A rich literature, both theoretical and empirical, has been studying for long the problem of risk-sharing in social networks when individuals are hit by individual shocks and their aim is to smooth income and consumption—see, e.g., Townsend (1982), Fafchamps and Lund (2003), Genicot and Ray (2003), Bramoullé and Kranton (2007), Attanasio et al. (2012), or the recent Handbook chapter by Mobius and Rosenblat (2016). Quite recently, another strand of literature has developed whose focus is on financial contexts, where risk sharing has another important twin side to it: contagion.[1]

A natural way in which two financial institutions—I will call them banks—may share risk is by trading financial assets. So doing, they diversify their exposure away from individual shocks and hence can improve their situation from an ex ante viewpoint. Such asset exchanges, however, will also typically increase the *range* of shocks that can spread through the economy and in the end induce larger aggregate effects than would have occurred otherwise. It raises, in other words, some tension between

(a) *risk sharing*, which is generally beneficial at the local (individual) level, and
(b) *contagion*, which may induce large and very detrimental "systemic" effects.

Understandably, the study of financial networks from this network perspective has been much stimulated by the financial crisis sparked in 2007–2008 by the collapse of Bearn Stearns and the bankruptcy of Lehman Brothers—events that, in a relatively short time span, spread havoc through much of the financial system. The literature has advanced both on an empirical and a theoretical front. From the empirical angle, the main concern has been to first "map" the relevant financial networks, next turning to operationalize suitable measures of systemic risk, as well as identifying those crucial "nodes" whose systemic effect is largest.[2]

On the other hand, in the theoretical realm, much of the effort has been devoted to shedding light on the following three fundamental questions:

1. What is the optimal configuration of the network that best balances risk-sharing and contagion?
2. If banks are heterogenous in some relevant respect, how does this bear on the previous question?
3. Is social optimality consistent with agents' individual incentives to "connect" (or diversify)?

Naturally, the specific answers to these questions must depend on how optimality is defined, the sort of interbank heterogeneity considered, and what incentives are taken to drive individual behavior. In these three respects, the literature varies widely—see, for example, the Handbook chapter by Cabrales et al. (2016), which conducts a comparative study of the following recent papers on the topic: Elliott et al. (2014), Acemoglu et al. (2015), Glasserman and Young (2015), and Cabrales et al. (2017). In what follows, my discussion focuses on the latter paper, labelled CGV for conciseness.

[1]In fact, there are many other cases where shocks affecting individual agents can spill over to other agents with whom they are connected, directly or indirectly. This occurs, for example, in production networks, as studied, e.g., by Acemoglu et al. (2012) and Baqaee (2016) within a static general-equilibrium model. For a genuinely dynamic approach in a simpler context, see Brummitt et al. (2017).

[2]See, for example, Battiston et al. (2012), Denbee et al. (2017), or Elsinger et al. (2006).

Consider an environment with a set of n ex ante identical banks, $N = \{1, 2, \ldots, n\}$, every one of them initially associated to one individual-specific project. Usually (i.e., with high probability), each of those projects yields a return $R > 0$. Occasionally, however, it can be hit by a random shock that induces a loss given by the random variable $\widetilde{L}$. Suppose, for simplicity, that at most one bank is hit by a shock at any given point in time. If the shock is large enough, the bank in question would go bankrupt if it were to remain fully exposed to its own project. Thus, to try to avoid that drastic (and irreversible) outcome, all banks diversify their risk by exchanging shares on their asset/project for correspondingly equal shares of assets of other banks. In the end, this leads to a situation where each bank i faces a portfolio of exposures $(a_{ij})_{j \neq i}$ to the different assets of the other banks of the system. These portfolios can all be compactly described by a matrix $A = (a_{ij})_{i,j=1}^{n}$ where, along the main diagonal, each a_{ii} stands for the share that each bank $i \in N$ holds of its own original asset.

Given such a pattern of exposures, when a shock to asset k of magnitude L arrives, the return ρ_i to bank i is given by:

$$\rho_i = (R - L)\, a_{ik} + R \sum_{j \neq k} a_{ij} - M \tag{1}$$

where $M > 0$ represents some exogenous liabilities, assumed identical for every bank. Then, if it turns out that $\rho_i < 0$ for some i, this bank goes bankrupt. The issue is to understand how the nature/distribution of the shocks bear on optimality at the social and individual level (cf. points (1)–(3) above). Here, optimality, both at the individual and social level, is associated to survival: for a bank, it is identified with minimizing its probability of bankruptcy; for society at large, with the minimization of the expected number of defaults.

For simplicity, let us focus here on the case where shocks are distributed as follows. On the one hand, with a very high probability, the shock is relatively small. In this case, the bankruptcy of the affected bank can be averted by simply sharing risk (i.e., exchanging a suitable share of one's own asset) with at least one other bank. On the other hand, with some small but positive probability, the shock lies in the medium to large scale. In this second case, the magnitude of the shock is distributed on the interval $[\underline{L}, \infty)$ according to some *probability mixture* of power laws (i.e., Pareto distributions) with a continuous density of the form:

$$\phi(L) \propto L^{-\gamma} \tag{2}$$

for some $\gamma > 1$ and a common $\underline{L} > 0$.

Often, power laws of the sort specified in (2) are used to discuss in a stark manner the classical dichotomy of "thin versus fat tails." This distinction hinges upon whether γ is, respectively, higher or smaller than 2, thus leading to a first-order moment of the distribution that is finite or not.

To fix ideas, suppose that all distributions under consideration are mixtures of two given distributions—one with thin tails, and the other with fat tails—with respective weights p and $1 - p$. Then, the following statements bearing on questions (1)–(3) apply:

(a) Consider the homogeneous case where all banks are ex ante identical. Then, if p is large enough, the unique network configuration that is optimal at both the individual and social level is a complete network with uniform exposure. Instead, if p is sufficiently small, the optimal network (again, socially and individually) involves the segmentation into $n/2$ separate pairs.
(b) Still in a homogeneous context, there is some middle range for p such that, from the point of view of any individual bank, the optimal arrangement is to be part of a component of intermediate size $m \in [3, n - 1]$, uniformly connected. Generally, however, this is *not* socially optimal.
(c) If agents set their connections through a coalitional strategic game, the network configuration induced will usually *not* be socially optimal.
(d) When banks are heterogeneous in either their size or the risk conditions they face, individual and social optimality both require that components be homogeneous.

The intuition underlying (a)–(d) is not difficult to understand as follows:

(a′) First, one can make the general point that if shocks are likely to be small or large, then individual and social optimality must respectively require that either sharing risk or protection from contagion should be the more relevant consideration. This, when the shock distribution is given by a power law, makes everything drastically hinge upon what happens, asymptotically, in the tail of the distribution. If the tail is thin, risk-sharing becomes preeminent and the optimal network is a fully connected one; instead, if the tail is fat, the overpowering consideration is the protection against contagion and the optimal configuration is the minimally connected one that insures against small and frequent shocks, i.e., maximal segmentation in dyadic pairs.
(b′) Instead, if both medium and large shocks are relatively likely for some appropriate mixture of the two scenarios (fat- and thin-tail distributions), it is individually optimal for banks to be part of components of intermediate size. This, however, is not generally optimal at the social level. The reason is that (using a standard convexity argument) one can show that social optimality requires components of equal size. And, typically, this condition is incompatible with the former condition for individual optimality.
(c′) The indicated inefficiency is also reflected by the equilibria of a suitably defined coalitional network-formation game. In essence, the root of the problem is that configurations that are individually optimal for some individuals (and hence part of an equilibrium) impose a size externality on others, forcing them to share risk within suboptimally small groups.
(d′) Finally, the reason why optimality requires homogeneity for every component is simple: any heterogeneity within a given component would produce a

mismatch between the optimal size for this component prescribed for the different types included in it.

All of the above serves to underscore the general point that, in an uncertain environment, the key issue of how to share/diversify risk needs to be studied systemically. Inherent to this approach, of course, is the fact that externalities abound, as a consequence of the connections (e.g., risk-sharing arrangements) established by the agents (banks, in our example). This, in turn, can lead to large global effects with severe overall implications.

How to address then the problem in practice? Clearly, once should do it by influencing (restricting or encouraging) the connections that agents form. The problem is that, in many real-world contexts, such bilateral agreements are often kept private and unregulated—for example, by 2014, over-the-counter trading in the US stocks reached 40% of all such trades. Attempting to regulate them, therefore, is bound to be very challenging. What should a "Hurwicz designer" then do? He should proceed indirectly, given the informational constraints and limited resources at his disposal. That is, he should tackle a second-best design-optimization problem that takes into account the sharp informational asymmetries and large monitoring costs actually faced, not only by himself but also by the agents themselves. The modern theory of mechanism design can contribute the methodological perspective and some of the tools to address the problem. In doing so, however, it would seem that relaxing the standard paradigm of full rationality—that is, allowing for some extent of bounded rationality and behavioral assumptions—might well be in order.

3 Peer-Innovation Networks

Innovation is widely viewed as the main source of sustained economic growth. And, in modern economies, much of the innovation has indeed an important social—often peer—component. This is particularly true, for example, in some of the most dynamic industries such as IT, pharmaceuticals, or aerospace where it is common for firms to be involved in joint R&D collaboration—cf. Hagedoorn (2002). The importance of such collaboration is also apparent in the abundance of networking hubs and innovation parks, well epitomized by the "mother of them all," Silicon Valley, which attracts around one third of all venture capital investment in the USA. As explained by Saxenian (1994) and Castilla et al. (2000), much of its success is to attribute to the fluidity and pervasiveness of its peer networks as well as to their rich diversity.

To understand what really underlies the performance of peer networks is indeed a major challenge—only a few such networks succeed while many fail. The difficulty derives from the fact that, a priori, there are quite a number of distinct dimensions involved in how peer networks operate that could have a powerful effect on how they breed innovation. Among these, one can list:

- **Diffusion**: Peers represent a key channel through which information about new technologies, or simply different ways of doing things, spread in connected population.
- **Learning**: Relatedly, the performance of any novel technology or behavior has to be learned, and this often relies on the experience/observation of peers who have already taken the lead in adopting them.
- **Trust**: Transfer of information is often associated to the peer involvement in some joint project. This requires cooperation and hence trust on the other party.
- **Matching**: In a diverse population, finding peers that are stable and complementary often requires undergoing a costly and possibly lengthy search for a good match.
- **Competition**: Peers often collaborate but sometimes they can also be competitors (say, in the same or a related market), which induces a tension between cooperation and competition that may be hard to handle.

The above are inherent and essential components in many innovation networks. In fact, all five of them have been amply studied by the economic and network literature,[3] but largely in a separate fashion. To integrate them into a unified framework is a major challenge that must be faced by a suitable account and study of the problem.

In principle, one would like to proceed by first building some theory, possibly quite preliminary, then building upon it to design a well-founded empirical analysis of the phenomenon. Given, however, the little we know on the complex interplay among the different components listed above, proceeding in a reciprocal manner seems more promising. That is, it appears advisable to start with an empirical investigation that gathers systematically information on how real peer networks operate. Then, on the basis of this information, a better-informed theory may be developed that of course should feed back on subsequent empirical analysis. This is the viewpoint that motivates the field research whose preliminary results are reported in Vega-Redondo et al. (2018).

The research is based on a large random control trial including around 5000 entrepreneurs from all over Africa (45 different countries represented). To shed light on how peer networking impinges on innovation, the treatment involved the opportunity of interacting with other participants under different conditions (i.e., alternative treatment arms). More specifically, the treated participants were divided into randomly formed groups of 60 individuals, with the individuals of each of these groups then interacting among themselves in one of the following three different types of treatment.

[3]For diffusion, the reader can see Jackson and Yariv (2005), López-Pintado (2008), Young (2009), Lamberson (2016), Duernecker and Vega-Redondo (2018); for learning, Bala and Goyal (1998), Golub and Jackson (2010), Golub and Sadler (2016); for cooperation and trust, Jackson, Rodriguez-Barraquer and Tan (2012), Mobius and Rosenblat (2016), Fainmesser and Goldberg (2018); for competition in networks, Fournier and Scarsini (2014), Heijnen and Soetevent (2018); and for matching, Gale and Shapley (1962), Roth and Sotomayor (1990), Liu et al. (2014).

- A *face-to-face* treatment, based in Uganda, where entrepreneurs *met personally* at scheduled times to discuss their business ideas and plans.
- A virtual treatment, labelled *virtual-within*, where the interaction groups were nationally homogeneous, i.e., consisted of individuals of the same country. In this case, individuals communicated with others in their group through a chatting platform that allowed a discretionary structure (organized by the individuals themselves, in channels admitting selected peers or/and focused on different topics) and versatile (e.g., could be modulated, also discretionarily, at desired levels of privacy).
- Another virtual treatment, labelled *virtual-across*, where the interaction groups used the same chatting platform as explained before but their composition was nationally heterogeneous, i.e., included entrepreneurs originating in different countries.

As usual, the population was randomly assigned to being part of the control population or one of the three former treatment conditions. The treatment lasted two and a half months. In parallel to it and with the same duration, all entrepreneurs (treatment and control alike) followed an online business course specifically tailored to the experiment.

At the end of the experiment, all entrepreneurs (again, control and treatment) were asked to submit their business proposals for evaluation and possible funding. The evaluation was conducted in two stages. First, the proposals went through a first stage in which a 15-member panel of African professionals evaluated them in a 1–5 scale according to a wide range of different criteria (innovativeness, market potential, sales strategy, cost assessment, social impact, etc.). Based on this evaluation, the proposals were ranked and the 600 best passed to the next stage. In this second stage, the selected proposals were again evaluated and ranked by our 32 financial partners (VCs, angel investors and institutional ones), who also chose the subset on which they wanted to conduct further discussion for possible funding. The results from these nested rounds of evaluation constituted two of the key outcomes used in the analysis of the experiment.

The RCT outlined has generated valuable information on two interrelated fronts:

- First, the professional evaluation of the submitted business proposals permits an econometric estimation of the size and significance of the treatment (peer-networking) effect under different circumstances (more or less national diversity) and alternative interaction mechanisms (face-to-face and virtual).
- Second, the exhaustive recording of the whole activity displayed in the virtual treatment provides a rich collection of panel data on the following two complementary dimensions: (a) the networking unfolding over time; (b) the communication exchanged throughout.

The networking and communication undertaken by the treated entrepreneurs in the two setups with virtual peer interaction are of course *endogenous* to the experiment. They cannot be directly attributed, therefore, a causal impact on the entrepreneurial performance observed in this context. They can provide, however, important insights on how and why the treatment has, or does not have, an effect on that performance. I shall return to this point below, once the preliminary results obtained so far have been summarized in what follows.

The results available at the time of writing show a positive and highly significant treatment effect when peer interaction takes place virtually among groups of the same nationality (i.e., what we have called the virtual-within treatment). The effect, however, is not statistically significant when the treatment is face-to-face or among individuals of mixed nationalities (the treatment labelled virtual-across). This suggests that virtual interaction is effective when the groups are homogeneous on the nationality dimension, in which case it is also better than face-to-face interaction. Thus, to sum up, we may conclude that virtual interaction dominates interaction that is conducted face-to-face, but its positive effects get blurred when the virtual interaction occurs among individuals who are nationally diverse.

The results outlined raise at least two related questions. One is what sources of peer diversity in networks fruitfully breed innovation rather than being counterproductive. Clearly, an indispensable requirement for peer interaction to be a source of innovation is that the set of individuals involved span sufficient, and complementary, diversity. For it is the combination of different types of knowledge, skills, personalities, and backgrounds that can render interaction valuable for this purpose. Individuals who are very similar, quasi "clones" of each other, can hardly generate much novelty through interaction.

The second question concerns the conclusion that intercountry diversity is detrimental to innovation. Why is it? Does it have to do with how entrepreneurs establish connections in this case? Or does it have to do with the type or amount of communication that takes place? Our experiment gives the possibility of answering these questions by looking into the black box of how peer interaction develops over time in the different virtual contexts. A joint analysis of the networking dynamics and the messages exchanged in each case has the potential of shedding much light on this issue. Concerning the messages, for example, ongoing investigation that relies on the powerful techniques customarily applied by the so-called NLP (natural language processing) booming literature should be able to unveil whether, and why, in the different treatments peer interaction leads to more or less cooperative, focused, or substantive communication. In this respect, an interesting observation is that, unlike what might have been anticipated, it is not true in our case that entrepreneurs involved in the virtual-across treatment communicate less (in the sense of number of messages exchanged) than those in the virtual-within scenario—in fact, it is quite the opposite!

The previous discussion points into a direction that is at the heart of Hurwicz's research program, as applied here to the problem of entrepreneurship. The field experiment described should contribute to it by identifying features of the *mechanism to be designed* that are most relevant for improved performance—e.g., the extent and dimensions of individual diversity spanned, or the protocol and incentives governing communication. A systematic study of the rich panel evidence gathered on networking and communication should go a long way in meeting this objective.

A final consideration—also very much in line with Hurwicz's vision of economic research—is the following. Whatever insights are gathered from the experiment would not be useful only for the construction of a better-informed theory of the problem, as suggested before. The experiment itself provides a "proof of concept" that those insights and the improvements in design that may follow from it are implementable in a feasible and cost-effective way at a potentially large scale. Thus, in this sense, it would reflect as well the key motivation underlying Hurwicz's long academic career, namely, that economics be anchored in the objective of delivering answers to practical (even if theoretically formulated) concerns.

4 Social-Learning Networks

In a social environment, opinions (or beliefs) and behavior are formed through a variety of different channels, with social networks having always played an important role. Nowadays, this role has become more prominent than ever. For, by virtue of Internet and the social media, individuals can increase substantially the range and density of their connections to others. Furthermore, they are also able to change those connections very flexibly, by creating and destroying their links as they search and learn, modifying as well their opinions and behavior over time.

The fact that our opinions and behavior are affected by those displayed by our peers in the social network is hardly controversial. This is particularly true in the political arena, as illustrated, for example, by Bond et al. (2012) and Algan et al. (2015). The first paper involves a massive experiment conducted in Facebook (61 million users) that shows, for the 2010 US congressional elections, that the probability that any given individual decided to vote was strongly influenced by her friends' behavior. The second paper, on the other hand, focuses on how social influence shapes the political opinions of students at Sciences Po, the elite French school attended by many French politicians. It shows, specifically, that the friendships established by students early on lead to a substantial convergence among friends' opinions in a range of social and political issues—in fact, it also affects the probability that they join the same political party.

One important consideration, however, that bears on the issue is that, naturally, the social network is itself endogenous. This raises delicate econometric problems in the proper identification of the "true" peer-influence effect at work. For example, in the French school studied by Algan et al. (2015), it is quite conceivable that having similar political opinions should affect positively the probability that two students become friends. This is what is known as *homophily*, which is well known to be an inherent feature of human behavior.[4] Algan et al. (2015) tackle this problem through

[4]See, for example, the seminal paper on this topic by Lazarsfeld and Merton (1954) and the modern survey of recent literature provided by McPherson et al. (2001). On the other hand, for the question of how homophily affects learning, a good example is provided by the model studied by Golub and

an instrumental-variable approach that takes advantage of the fact that the initial assignment of students in different classes is done exogenously at the beginning of their studies.

Homophily, however, also raises a conceptual issue with important practical implications. If the social network is endogenous, the interplay of opinion and link adjustment can lead to an outcome (in particular, concerning the final opinions) that could be very different from the one that would have taken place in a fixed network. For example, it could produce acute segmentation on the population, in which only individuals who end up sharing a similar opinion are connected. This, of course, would perpetuate a partition of the population into essentially independent "opinion groups," thus leading to what has been called the "echo chamber effect." That is, a situation where every group only receives the echo of its own opinion. The implications of this segmented state of affairs could be quite detrimental if it is important that the population integrates to some extent the opinion of all its members.

The empirical relevance of such an echo-chamber phenomenon has been discussed, among others, by Adamic and Glance (2005) and Boutyline and Willer (2017). Both document the split of the US population along ideological lines. For example, Adamic and Glance (2005) focus on the blogs active around the 2004 US Presidential Election and classifies them as belonging to either the liberal or the conservative camp. The main point they make is that, in fact, they operate in a largely disconnected manner. That is, they hardly refer to any content generated in the other side, hence de facto preventing that cross-feedback may lead to any exchange of views and some genuine aggregation of information.

In a recent paper, Polanski and Vega-Redondo (2018) study the problem in a theoretical framework that generalizes one of the classical frameworks used to study processes of learning in social networks: the model originally formulated by DeGroot (1974), which has been revisited, among others, by DeMarzo et al. (2003) and Golub and Jackson (2010). The setup involves a finite population of agents, $N = \{1, 2, \ldots, n\}$, who are connected by a given network of influence formalized by a weighted adjacency matrix $T = (t_{ij})_{i,j=1}^{n}$. Off the main diagonal, the entries t_{ij} $(i \neq j)$ specify the extent to which each individual i is influenced by her peers j. Along the main diagonal, the entries t_{ii} reflect the persistence of i's opinions, i.e., how much weight i attributes to her own previous opinions. For simplicity, it is assumed that the matrix T is row-stochastic, so that all influence weights impinging on any agent i are normalized to add up to unity.

The learning dynamics operates in discrete time, $s = 0, 1, 2, \ldots$, in a very simple manner. At $s = 0$, each agent i starts with some initial opinions $x_i(0) = x_i^0$. Then, over time, the opinion updating proceeds by every agent $i \in N$ simply combining own and others' previous opinions according to the pattern influence embodied by

Jackson (2012), who focus on how homophily affects the speed of learning in the context of the DeGroot's model (see below for a description of this model).

T. In matrix form, this can be concisely written as follows:

$$\mathbf{x}(s) = T\,\mathbf{x}(s-1) \qquad (s = 1, 2, \ldots, \kappa), \tag{3}$$

where $\mathbf{x}(s)$ represents the column vector of opinions at s and κ is a parameter determining the number of updating periods involved in each learning spell.

The model proposed by Polanski and Vega-Redondo (2018) differs from the classical model of DeGroot (1974) in two respects: (a) the learning spell is finite; (b) opinions are multidimensional. On the one hand, the finiteness of the learning spell is motivated by the assumption that, as it indeed happens in the real world, the underlying environment changes relatively fast and therefore asymptotic results may not be very relevant. On the other hand, multidimensionality of opinions is taken to reflect the fact that as, say in the economic sphere, agents have separate opinions on a number of different dimensions, e.g., employment, inflation, public deficit, subsidies to education, or clean energies. Thus, the opinions of an agent i must be taken to be vectors of the form $x_i = [x_i(\omega)]_{\omega \in \Omega}$ where Ω stands for the set of different issues on which opinions are defined and each $x_i(\omega)$ is the opinion of agent i on issue $\omega \in \Omega$. Formally, it becomes convenient to think of such multidimensional opinions as real random variables, $x_i : \Omega \to \mathbb{R}$, with the set of issues playing a role *analogous* to a state space. The usefulness, however, of such a random-variable specification is essentially *formal*: even though opinions are conceived as *deterministic* (multidimensional) objects, the correlation between the opinions of different agents can then be readily defined.

Besides the added realism reflected by features (a)–(b) above, their main motivation is in terms of modelling strategy. To understand this point, consider the particular case where opinions are one-dimensional (i.e., Ω is a singleton) and the learning spell is unbounded ($\kappa = \infty$). Then, under mild regularity conditions on the matrix T, it is well known that the learning process converges to a situation of consensus where everyone holds the same (one-dimensional) opinion. Admittedly, this conclusion may be largely conceived as a theoretical benchmark rather than a prediction. However, the unfortunate consequence is that one can hardly build on it to define some nontrivial notion of similarity across agents' opinions. This, in turn, stands in the way of the key objective of Polanski and Vega-Redondo (2018), which is to rely on homophily to endogenize the influence network. Next, I turn to explaining how this can be done.

Suppose that, across (finite) learning spells of duration κ, every agent i updates from t_{ij} to t'_{ij} the weight/influence that she attributes to each of her peers j. Based on a notion of homophily that is opinion-based, the postulated updating criterion may be informally described as follows.

(H) The revised t'_{ij} is proportional to the *correlation* between the opinions of i and j induced, after κ periods of learning, by the previous influence matrix $T = (t_{ij})_{i,j=1}^{n}$.

The motivation for the previous updating rule is based on the idea that the opinions displayed by any two individuals provide a suitable basis to assess their similarity.

This, in fact, is in line with the approach commonly used with considerable success on the Internet, e.g., by Amazon.com, Booking.com, or Netflix.com.[5] Based on such measure of interagent similarity, (H) simply formalizes the idea that homophily drives the revision of interagent influence. That is, agents are postulated to revise the influence weights assigned to others so as to match (be proportional to) their corresponding similarity.

In general, besides the weighted influence network T, it is interesting to consider some separate *observation network* that restricts influence. For, naturally, an individual can only be influenced by those whom she observes. Let this observation network, exogenously given,[6] be formalized by a binary adjacency matrix $L = (l_{ij})_{i,j=1}^n$ where each $l_{ij} \in \{0, 1\}$. The interpretation here is that, for every pair of individuals i and j, $l_{ij} = 1$ if, and only if, i "observes" j. Then, formally, the aforementioned observational restriction simply amounts to the following implication:

$$\forall i, j \in N, \quad l_{ij} = 0 \Rightarrow t_{ij} = 0. \tag{4}$$

Given (H) and (4), the equilibrium condition that endogenizes the influence network can then be simply defined as embodying a fixed point in a process of influence adjustment given by:

$$t_{ij}(r+1) = \frac{l_{ij}\, \rho_{ij}(T(r))}{\sum_{k \in N} l_{ik}\, \rho_{ik}(T(r))} \qquad (i, j = 1, 2, \ldots, n;\ r = 0, 1, 2, \ldots), \tag{5}$$

where $r = 0, 1, 2, \ldots$ indexes influence-adjustment time, and $\rho_{ij}(T)$ stands for the Pearson correlation coefficient between the opinions of i and j under influence matrix T at the end of a learning spell of given duration κ for fixed initial opinions $\mathbf{x}^0$. Thus, if we denote by $F(\cdot)$ the vector field defined by the right-hand side of (5), we may simply write in matrix form the equilibrium condition stated in (H) as follows: $T^* = F(T^*)$.

It is relatively easy to show that an EIM always exists and to characterize it for some extreme benchmark cases. In general, however, one has to face a vast range of equilibrium multiplicity. For example, if the observation network is complete and agents' initial opinions are not correlated, then every arrangement where the population is partitioned into any number of independent cliques[7] defines an EIM. This suggests directing the analysis into two different, and complementary directions: on the one hand, moving beyond the unrealistic setups where the

[5] See, in particular, the so-called collaborative filtering ones as discussed, e.g., by Jannach et al. (2010) and Ricci et al. (2011).

[6] For example, it could reflect considerations related to geography, language, or age that are fixed and affect the relevant set of peers.

[7] A clique is defined as a completely connected subset of nodes that has no links to nodes outside this subset.

observation network is taken to be complete; on the other hand, discriminating across equilibria on the basis of their dynamic robustness.

First, still remaining within a static (equilibrium) approach, one important result relates the weight t^*_{ij} of any observational link from i to j to the extent to which both agents *share* many influential peers. This statement is quite intuitive. If the two agents in question have a strong neighborhood overlap,[8] they must receive common (and hence perfectly correlated) influence from influential third parties. This indirect route leads their opinion to be substantially correlated and hence, at equilibrium, their link must carry a substantial weight as well. One concludes, therefore, that in order for significant influence to flow through any given link it is essential that this link be suitably *supported* by (or "well-embedded" into) the overall network.

The previous discussion has an important bearing on one of the central questions that motivate the research, namely, when is it that the creation of some observation links among previously disconnected groups may lead to an integrated overall population—i.e., to a situation where all individual opinions have some influence in the final outcome. Or, formulating matters reciprocally, we may ask when will it be the case that the originally segmented configuration is robust enough to prevail despite the creation of (possibly many) "observation bridges" among the different groups.

First, let us stress that, as suggested before, in general one cannot discard the possibility that a segmented influence matrix may prevail at equilibrium (i.e., define an EIM), even after the creation of new observation links. So, the really interesting questions here are the following:

(a) whether, still from an *equilibrium/static* viewpoint, there are *also* other different EIMs that use effectively the new open channels to bring about social integration;
(b) whether, from a *dynamic* perspective, a segmented EIM is unstable for the influence-adjustment dynamics outlined before, i.e., any small perturbation leads to the breakdown of segmentation.

As it turns out, both questions largely hinge upon the notion of support introduced before. That is, for a bridge to be effective in channeling influence, it must be strongly embedded in the social (observation) network. Thus, it must have the possibility of being suitably supported by other links. The problem, in fact, could be posed in Hurwicz's classical terms as follows. Assuming social integration is the desired objective, what bridging links must be established (the analogue of a revised game form) so that, under the correct anticipation of how agents will react (either at equilibrium or as part of a dynamic adjustment process), social integration

[8]The measure of neighborhood overlap used here is akin to the notions of support that have been used in the network literature—see, for example, Jackson et al. (2012), where it is used to characterize conditions where cooperation can be sustained at equilibrium in binary networks. In contrast, however, the present notion is defined for weighted networks and takes into account the influence *weight* of the links that connect to common third parties.

is attained. This means, therefore, that the "planner" creating the links must be sophisticated enough to adopt the designer's viewpoint.

Again, we close with a word of warning. In a world where the scale at which social learning unfolds is truly global, one cannot ignore that fine-tuning possibilities are typically not available. Thus, to be effective, the designer's problem must suitably take into account the actual constraints faced, stringent as they may be in view of the underlying complexity. The planner must, in other words, solve a second-best design problem where computational, informational, and monitorial restrictions will typically abound.

5 Conclusion

The work of Leonid Hurwicz has had a wide and lasting impact on economics. As I have stressed here, for him the mechanism was not a datum of the problem but a variable of design. This methodological viewpoint, which is at the core of how we think today about most economic problems, has become so pervasive in our discipline that there is the risk that its revolutionary novelty at the time might be downplayed or passed unnoticed. Somewhat paradoxically, this is indeed a telling testimony to Hurwicz's influence on the way in which we, modern economists, conceive and advance our discipline.

In this brief piece, I have summarized how all this bears on three instances of my recent research on socioeconomic networks. For each of the three kinds of networks considered—risk-sharing networks, peer-innovation networks, and social-influence networks, I have made the following point. A truly useful formulation of the designer's problem requires taking into account the limitations brought about by the huge complexity of modern economies, which are relevant not only to the agents but to the designer herself. What information the mechanism can really use, what computational abilities it is reasonable to assume, or the extent of monitoring that is feasible to implement are important "practical" concerns that the designer cannot ignore. But, our economies and societies are not just very complex—they also change at a fast pace. They often evolve too rapidly for a reactive strategy to be successful. That is, the attempt to implement a fine-tuning strategy can prove counterproductive, if it arrives too late and addresses the key problem of yesterday. All these considerations delineate what in my view is one of the next important steps that must be undertaken by Hurwicz's research program: to meet the challenge of fast-evolving complexity. Conceivably, to tackle this challenge effectively, a blend of behavioral theory and empirical analysis might be one of the right ways to go. Modestly, this is largely the approach explored, more or less directly, in the research reviewed here.

References

Acemoglu, D., Carvalho, V. M., Ozdaglar, A., & Tahbaz-Salehi, A. (2012). The network origins of aggregate fluctuations. *Econometrica, 80*, 1977–2016.

Acemoglu, D., Ozdaglar, A., & Tahbaz-Salehi, A. (2015). Systemic risk and stability in financial networks. *American Economic Review, 105*, 564–608.

Adamic, L., & Glance, N. (2005). The political blogosphere and the 2004 U.S. election: Divided they blog. In *LinkKDD05, Proceedings of the 3rd International Workshop on Link Discovery* (pp.36–43)

Algan, A., Do, Q.-A., Dalvit, N., Le Chapelain, A., & Zenou, Y. (2015). How social networks shape our beliefs: A natural experiment among future French politicians, working paper, Sciences Po.

Attanasio, O., Barr, A., Cardenas, J. C., Genicot, G., & Meghir, C. (2012). Risk pooling, risk preferences, and social networks. *American Economic Journal: Applied Economics, 4*, 134–167.

Bala, V., & Goyal, S. (1998). Learning from neighbours. *Review of Economic Studies, 65*, 595–621.

Baqaee, D. R. (2016). Cascading failures in production networks. *Econometrica, 86*, 1819–1838.

Battiston, S., Puliga, M., Kaushik, R., Tasca, P., & Caldarelli, G. (2012). DebtRank: Too central to fail? Financial networks, the FED and systemic risk. *Scientific Reports, 2*, 541.

Bond, R. M., Fariss, C. J., Jones, J. J., Kramer, A. D. I., Marlow, C., Settle, J. E., et al. (2012). A 61-million-person experiment in social influence and political mobilization. *Nature, 489*, 295–298.

Boutyline, A., & Willer, R. (2017). The social structure of political echo chambers: Ideology and political homophily in online communication networks. *Political Psychology, 38*, 551–569.

Bramoullé, Y., & Kranton, R. (2007). Risk sharing across communities. *American Economic Review, 97*, 70–74.

Brummitt, C. D., Huremovic, K., Pin, P., Bonds, M., & Vega-Redondo, F. (2017). Contagious disruptions and complexity traps in economic development. *Nature Human Behavior, 1*, 665–672.

Cabrales, A., Gale, D., & Gottardi, P. (2016). Financial contagion in networks. In Y. Bramoulle, A. Galeotti, & B. W. Rogers (Eds.), *The Oxford Handbook of Network Economics*. Oxford: Oxford University Press.

Cabrales, A., Gottardi, P., & Vega-Redondo, F. (2017). Risk-sharing and contagion in networks. *The Review of Financial Studies, 30*, 3086–3127.

Castilla, E., Hokyu, H., Granovetter, E., & Granovetter, M. (2000). Social networks in Silicon Valley. In C.-M. Lee, W. F. Miller, M. G. Hancock, & H. S. Rowen (Eds.), *The silicon valley edge*. Stanford: Stanford University Press.

DeGroot, M. H. (1974). Reaching a consensus. *Journal of the American Statistical Association, 69*, 118–121.

DeMarzo, P. M., Vayanos, D., & Zwiebel, J. (2003). Persuasion bias, social influence, and unidimensional opinions. *The Quarterly Journal of Economics, 118*, 909–968.

Denbee, E., Julliard, C., Li, Y., & Yuan, K. (2017). Network risk and key players: A structural analysis of interbank liquidity, working paper, London School of Economics.

Duernecker, G., & Vega-Redondo, F. (2018). Social networks and the process of globalization. *The Review of Economic Studies* (forthcoming)

Elliott, M., Golub, B., & Jackson, M. O. (2014). Financial networks and contagion. *American Economic Review, 104*(10), 3115–3153.

Elsinger, H., Lehar, A., & Summer, M. (2006). Risk assessment for banking systems. *Management Science, 52*, 1301–1314.

Fafchamps, M., & Lund, S. (2003). Risk-sharing networks in rural Philippines. *Journal of Development Economics, 71*, 261–87.

Fainmesser, I. P., & Goldberg, D. A. (2018). Cooperation in partly observable networked markets. *Games and Economic Behavior, 107*, 220–237.

Fournier, G., & Scarsini, M. (2014). Hotelling games on networks: Efficiency of equilibria, SSRN 2423345.
Gale, D., & Shapley, L. S. (1962). College admissions and the stability of marriage. *American Mathematical Monthly, 69*, 9–15.
Genicot, G., & Ray, D. (2003). Group formation in risk-sharing arrangements. *Review of Economic Studies, 70*, 87–113.
Glasserman, P., & Young, H. P. (2015). How likely is contagion in financial networks? *Journal of Banking and Finance, 50*, 383–399.
Golub, B., & Jackson, M. O. (2010). Naïve learning in social networks and the wisdom of crowds. *American Economic Journal: Microeconomics, 2*, 112–149.
Golub, B., & Jackson, M. O. (2012). How homophily affects the speed of learning and best response dynamics. *Quarterly Journal of Economics, 127*, 1287–1338.
Golub, B., & Sadler, E. (2016). Learning in social networks. In Y. Bramoulle, A. Galeotti, & B. W. Rogers (Eds.), *The Oxford handbook of network economics*. Oxford: Oxford University Press.
Hagedoorn, J. (2002): Interfirm R&D partnerships: An overview of major trends and patterns since 1960. *Research Policy, 31*, 477–492.
Heijnen, P., & Soetevent, A. R. (2018). Price competition on graphs. *Journal of Economic Behavior and Organization, 146*, 161–179.
Jackson, M. O., Rodriguez-Barraquer, T., & Tan, X. (2012). Social capital and social quilts: Network patterns of favor exchange. *The American Economic Review, 102*, 1857–1897.
Jackson, M. O., & Yariv, L. (2005). Diffusion on social networks. *Économie Publique, 16*, 69–82.
Jannach, D., Zanker, M., Felfernig, A., & Friedrich, G. (2010). *Recommender systems: An introduction*. Cambridge: Cambridge University Press.
Lamberson, P. J. (2016). Diffusion in networks. In Y. Bramoulle, A. Galeotti, & B. W. Rogers (Eds.), *The Oxford Handbook of Network Economics*. Oxford: Oxford University Press.
Lazarsfeld, P. F., & Merton, R. K. (1954). Friendship as a social process: A substantive and methodological analysis. In M. Berger (Ed.), *Freedom and control in modern society*. New York: Van Nostrand.
Liu, Q., Mailath, G. J., Postlewaite, A., & Samuelson, L. (2014). Stable matching with incomplete information. *Econometrica, 82*, 541–587.
López-Pintado, D. (2008). Diffusion in complex social networks. *Games and Economic Behavior, 62* 573–590.
McPherson, M., Smith-Lovin, L., & Cook, J. M. (2001). Birds of a feather: Homophily in social networks. *Annual Review Sociology, 27*, 415–444.
Mobius, M., & Rosenblat, T. (2016). Informal transfers in social networks. In Y. Bramoulle, A. Galeotti, & B. W. Rogers (Eds.), *The Oxford Handbook of Network Economics*. Oxford: Oxford University Press.
Polanski, A., & Vega-Redondo, F. (2018). Homophily and influence, working paper, University of East Anglia and Bocconi University.
Ricci, F., Rokach, L., & Shapira, B. (2011). Introduction to recommender systems handbook. In *Recommender systems handbook* (pp. 1–35). Boston: Springer.
Roth, A. E., & Sotomayor, M. A. O. (1990). *Two-Sided Matching*. Cambridge: Cambridge University Press.
Saxenian, A. (1994). *Regional advantage: Culture and competition in silicon valley and route 128*. Cambridge, MA: Harvard University Press.
Townsend, R. M. (1982). Optimal multiperiod contracts and the gain from enduring relationships under private information. *The Journal of Political Economy, 90*, 1166–1186.
Vega-Redondo, F., Benedetti-Fasil, C., Brummitt, C., Pin, P., Rubera, G., Ubfal, D., Hovy, D., & Fornaciari, T. (2018). *Peer networks and entrepreneurship: A Pan-African RCT*. Working paper. Bocconi University.
Young, P. (2009). Innovation diffusion in heterogeneous populations: contagion, social influence, and social learning. *The American Economic Review, 99*, 1899–1924.

Part II
Design Under Uncertainties

Some Remarks on Bayesian Mechanism Design

Claude d'Aspremont and Jacques Crémer

1 Introduction

Although the problem of revelation of preferences for public goods had already been brought up in several instances, it was surely the merit of Leo Hurwicz to show that providing incentives was a fundamental problem in the design of any institution for collective decision based on decentralized information and control. In particular, the introduction by Hurwicz (1972) of the concept of incentive compatibility for correct revelation as the Nash equilibria of a family of noncooperative games with complete information, and the difficulty to enforce Pareto-optimality in such a model, suggested to view institutions as rules of a game with incomplete information, as formally defined by Harsanyi (1967, 1968a,b) and to apply his concept of Bayesian equilibrium.

Bayesian mechanism design highlights the importance of individual agents' beliefs in environments where information is incomplete. The prior contribution of Vickrey (1961) had already made this clear for auction design and, more generally for market design, and had, as well, identified the budget balance problem that a marketing agency would encounter in solving the demand revelation problem.

A resulting issue has been to look for the most general class of beliefs allowing to solve the revelation problem without weakening the collective efficiency

C. d'Aspremont
CORE, Université Catholique de Louvain, Louvain-la-Neuve, Belgium
e-mail: claude.daspremont@uclouvain.be

J. Crémer (✉)
Toulouse School of Economics, Toulouse, France
e-mail: jacques.cremer@tse-fr.eu; jacques@cremerline.com

W. Trockel (ed.), *Social Design*, Studies in Economic Design,
https://doi.org/10.1007/978-3-319-93809-7_6

requirement. The objective is to identify the most general conditions on agents' beliefs allowing for the implementation of collective decision rules. Many of these conditions have been presented in their "dual" form, following some variant of the theorem of the alternative. We shall concentrate on the two most general conditions (each in some respect) and show that they can usefully be used in their "primal" form, leading to more constructive arguments. We discuss the issue and analyze some interesting cases.

In Sect. 2, we present the model and formulate both the incentive compatibility problem à la Hurwicz and the Bayesian incentive compatibility problem as the problem of solving a system of linear inequalities and look at the dual consistency conditions. In Sect. 3, we adopt what may be called the "primal approach" showing the prominence of two generic conditions that can be fruitfully imposed on the beliefs. In Sect. 4, we indicate that these two conditions remain central for team adverse selection with participation constraints. This is also true for team moral hazard with stochastic outcome functions, as we mention in the concluding section, where we point out some areas for further research.

2 The Model and the Basic System of Linear Inequalities

The public (or collective) decision-making for which a mechanism has to be designed is very general. It may be a mechanism to provide public goods (or reduce a public bad, such as pollution), or to allocate goods, as in auctions. It is just defined by a set of agents $N = \{1, 2, \cdots, i, \cdots, n\}$, with $n \geq 2$ (for some results we will assume $n \geq 3$) and a compact set X of outcomes. We assume incomplete information about the characteristics of the agents. Adopting the Bayesian point of view (Harsanyi, 1967, 1968a,b), to each agent i is associated a finite set of possible types A_i as well as the beliefs of agent i concerning the types of the other agents. For every $\alpha_i \in A_i$, these beliefs $p_i(\alpha_{-i} \mid \alpha_i)$ are defined on $A_{-i} = \prod_{j \neq i} A_j$. For simplicity, we will assume that these individual beliefs are consistent, i.e., generated from a common prior $p(\alpha)$ defined on the set of states of nature, $A = \prod_j A_j$, but many results do not require this assumption. We call (N, A, p) the *information structure*.

We assume transferable utility: when the public decision is x and he receives a transfer t_i, the utility of agent i of type α_i is $u_i(x; \alpha_i) + t_i$. We will take some liberty with the terminology and call u_i the utility function of the agent. In general, the type α_i determines both the beliefs and the utility of agent i. We call $(N, A, p, X, \{u_i\})$ the *environment*. It is composed of an information structure, a set of outcomes, and the utility functions of the agents.

We study *direct (revelation) mechanisms*, where agents are given incentives to truthfully reveal their types. Formally, a *mechanism* is a pair (s, t), where s is an *outcome function* from $A \equiv \prod_{i=1}^{n} A_i$ to X and where t is a *transfer function* from A to $\Re^N$. The outcome function is *ex post efficient* if, for all $\alpha \in A$,

$$\sum_{i=1}^{n} u_i(s(\alpha); \alpha_i) = \max_{x \in X} \sum_{i=1}^{n} u_i(x; \alpha_i),$$

and *Pareto-optimal* if, in addition, the transfer function is *balanced*, that is if, for all $\alpha \in A$, the following budget balance condition holds:

$$\sum_{i=1}^{n} t_i(\alpha) = 0.$$

A mechanism defines a game of incomplete information, namely a collection of normal form games $\Gamma(\alpha)$, one for each vector of types, where the payoff function of agent i is $u_i(s(\cdot); \alpha_i) + t_i(\cdot)$ and his strategy space is A_i. For each agent i, we can define a set of "normalized" strategies, i.e., functions $a_i(\alpha_i)$ from A_i to itself, and focus on the "truth-telling strategy" : $a_i^*(\alpha_i) = \alpha_i$ for all $\alpha_i \in A_i$.

A first way to introduce incentives, based on the seminal contribution of Hurwicz (1972), is to require truth-telling to be what is now usually called an ex post equilibrium,[1] namely to require the vector $\left(a_1^*(\alpha_1), \ldots, a_n^*(\alpha_n)\right)$ to be a Nash equilibrium of $\Gamma(\alpha)$ for $\alpha \in A$. Since α can be any vector in A, a mechanism (s, t) is *incentive compatible* (abbreviated *IC*) if, for all $i \in N$, all $\alpha_i \in A_i$, all $\tilde{\alpha}_i \in A_i$, and all $\alpha_{-i} \in A_{-i}$)

$$u_i(s(\alpha_i, \alpha_{-i}); \alpha_i) + t_i(\alpha_i, \alpha_{-i}) \geq u_i(s(\tilde{\alpha}_i, \alpha_{-i}); \alpha_i) + t_i(\tilde{\alpha}_i, \alpha_{-i}).$$

In the present framework, where the utility function $u_i(x; \alpha_i)$ of agent i depends only on type α_i (the private value case), IC is equivalent to dominant strategy incentive compatibility (i.e., $\left(a_1^*(\alpha_1), \ldots, a_n^*(\alpha_n)\right)$ are dominant strategies in $\Gamma(\alpha)$ for every $\alpha \in A$) or strong incentive compatibility (SIC) to use the terminology of Green and Laffont (1977). The focal[2] class of efficient SIC mechanisms is the class of Vickrey–Clarke–Groves (VCG) mechanisms where s is an efficient outcome

[1] An ex post equilibrium is called a uniform equilibrium in d'Aspremont and Gérard-Varet (1979a) and incentive compatibility is called uniform incentive compatibility by Holmström and Myerson (1983). See Bergemann and Morris (2005) for more discussion and references.

[2] If the set of types is large enough (i.e., connected), Groves mechanisms are the only efficient SIC mechanisms. This result does not hold in our discrete types of framework. See Green and Laffont (1977), Walker (1978), and Holmström (1979).

function and the transfers are defined by:

$$t_i(\alpha_i, \alpha_{-i}) = \sum_{j \neq i} u_j(s(\alpha_i, \alpha_{-i}); \alpha_j) + h_i(\alpha_{-i}),$$

with h_i any real-valued function defined on A_{-i}. However, as known since Green and Laffont (1979) and Walker (1980), these transfers are in general not balanced, so that Pareto-optimality is not obtained. Finding Pareto-optimal SIC mechanisms is a difficult task even when utility functions are quasi-linear. In this framework, given an efficient outcome function, finding balanced SIC transfers amounts to solving a finite system of linear inequalities and we can apply theorems of the alternative to characterize the consistency of such systems. Using such a method, d'Aspremont et al. (1990) derive the following necessary and sufficient condition imposed on the environment (without involving the beliefs) and the outcome function for an SIC mechanism to exist: for all $\lambda : A_i^2 \times A_{-i} \to \Re_+$ and all $\mu : A \to \Re$ which satisfy

$$\sum_{\tilde{\alpha}_i \neq \alpha_i} \lambda_i(\tilde{\alpha}_i, \alpha_i, \alpha_{-i}) - \sum_{\tilde{\alpha}_i \neq \alpha_i} \lambda_i(\alpha_i, \tilde{\alpha}_i, \alpha_{-i}) = \mu(\alpha) \text{ for all } i \in N \text{ and } \alpha \in A,$$

then

$$\sum_{i=1}^{n} \sum_{(\tilde{\alpha}_i, \alpha_i) \in A_i^2} \sum_{\alpha_{-i}} \lambda_i(\tilde{\alpha}_i, \alpha_i, \alpha_{-i})[u_i(s(\tilde{\alpha}_i, \alpha_{-i}); \alpha_i) - u_i(s(\alpha_i, \alpha_{-i}); \alpha_i)] \leq 0.$$

The family of parameters λ and μ are the dual variables associated, respectively, with the incentive inequalities and the budget balance condition. To illustrate the usefulness of this duality result, that same article shows that a Pareto-optimal SIC mechanism exists whatever the environment and the efficient outcome function, provided that each agent has only two types ($|A_i| = 2$, for any i)[3]—this mechanism may have to be chosen outside the class of VCG mechanisms. The fact that we obtain such a strong result when all agents have only two types should at the minimum make us cautious about the use of the two-type assumption in applied theory.

A second way to introduce incentives for truth revelation in such environments is to reverse the point of view. Instead of looking at incentive compatibility by imposing conditions on the utilities valid whatever the beliefs, one can look at incentive compatibility by imposing conditions on the beliefs valid whatever the utilities. Given some information on structure, the payoff of player i of type α_i is evaluated, for every strategy vector $(a_1(\cdot), \cdots, a_n(\cdot))$, as the conditional

[3]This generalizes a result of Maskin (1986), proved in the case of two agents.

expected utility

$$\sum_{\alpha_{-i}}\Big[u_i(s(a_1(\alpha_1),\cdots,a_n(\alpha_n));\alpha_i)+t_i(a_1(\alpha_1),\cdots,a_n(\alpha_n))\Big]p_i(\alpha_{-i}\mid\alpha_i).$$

A mechanism (s,t) is *Bayesian Incentive Compatible* (BIC) if the truth-telling strategy $a^*(.)$ is a *Bayesian equilibrium* in the sense of Harsanyi; it is characterized by the following inequalities, for all $i\in N$ and all $(\tilde{\alpha}_i,\alpha_i)\in A_i^2$,

$$\sum_{\alpha_{-i}}[u_i(s(\alpha_i,\alpha_{-i});\alpha_i)+t_i(\alpha_i,\alpha_{-i})]\,p_i(\alpha_{-i}\mid\alpha_i)$$
$$\geq\sum_{\alpha_{-i}}\big[u_i(s(\tilde{\alpha}_i,\alpha_{-i});\alpha_i)+t_i(\tilde{\alpha}_i,\alpha_{-i})\big]\,p_i(\alpha_{-i}\mid\alpha_i).$$

Comparing these inequalities with the inequalities for IC, we observe that BIC holds for any information structure such that IC holds.

As above, given an environment and some efficient outcome function, finding balanced BIC transfers amounts to solving a finite system of linear inequalities, but in this case, they depend on the beliefs. Applying a theorem of Fan (1956), d'Aspremont and Gérard-Varet (1979a) obtained a necessary and sufficient condition to be imposed on the environment and the outcome function for a balanced BIC mechanism to exist: for all $\lambda: A_i^2\to\Re_+$ and all $\mu: A\to\Re$, such that for all $i\in N$ and all $\alpha\in A$,

$$p_i(\alpha_{-i}\mid\alpha_i)\sum_{\tilde{\alpha}_i\neq\alpha_i}\lambda_i(\tilde{\alpha}_i,\alpha_i)-\sum_{\tilde{\alpha}_i\neq\alpha_i}\lambda_i(\alpha_i,\tilde{\alpha}_i)p_i(\alpha_{-i}\mid\tilde{\alpha}_i)=\mu(\alpha)$$

we have

$$\sum_{i=1}^{n}\sum_{(\tilde{\alpha}_i,\alpha_i)\in A_i^2}\lambda_i(\tilde{\alpha}_i,\alpha_i)\times$$
$$\left[\sum_{\alpha_{-i}}\big[u_i(s(\tilde{\alpha}_i,\alpha_{-i});\alpha_i)-u_i\,(s(\alpha_i,\alpha_{-i});\alpha_i)\big]p_i(\alpha_{-i}\mid\alpha_i)\right]\leq 0.$$

Again, the λs and μs are the dual variables associated, respectively, with the incentive inequalities and the budget balance condition. The main application of this result is to allow for a first formulation of a condition (that we will call condition

C*)[4] imposed only on the information structure and sufficient to guarantee the existence of Pareto-optimal BIC mechanism whatever the utility functions and the *efficient* outcome function. As we will see below, this condition holds generically, is necessary for ensuring the budget balance condition, and, as such, is weaker (or equivalent) to the other conditions that have been proposed in the literature.

An information structure (N, A, p) satisfies Condition C* if whenever $\lambda : A_i^2 \to \Re_+$ and $\mu : A \to \Re$ satisfy

$$p_i(\alpha_{-i} \mid \alpha_i) \sum_{\tilde{\alpha}_i \neq \alpha_i} \lambda_i(\tilde{\alpha}_i, \alpha_i) - \sum_{\tilde{\alpha}_i \neq \alpha_i} \lambda_i(\alpha_i, \tilde{\alpha}_i) p_i(\alpha_{-i} \mid \tilde{\alpha}_i) = \mu(\alpha)$$

$$\text{for all } i \in N \text{ and all } \alpha \in A$$

then $\mu(\alpha) = 0$ for all $\alpha \in A$.

That BIC and budget balance are ensured thanks to condition C is immediate since by efficiency of the outcome function:

$$\begin{aligned}
&\sum_{i=1}^{n} \sum_{(\tilde{\alpha}_i, \alpha_i) \in A_i^2} \lambda_i(\tilde{\alpha}_i, \alpha_i) \sum_{\alpha_{-i}} \left[u_i(s(\tilde{\alpha}_i, \alpha_{-i}); \alpha_i) - u_i(s(\alpha_i, \alpha_{-i}); \alpha_i)\right] p_i(\alpha_{-i} \mid \alpha_i) \\
&\quad \leq \sum_{i=1}^{n} \sum_{\alpha} \sum_{j \neq i} u_j(s(\alpha); \alpha_j) \times \\
&\qquad \left[p_i(\alpha_{-i} \mid \alpha_i) \sum_{\tilde{\alpha}_i \neq \alpha_i} \lambda_i(\tilde{\alpha}_i, \alpha_i) - \sum_{\tilde{\alpha}_i \neq \alpha_i} \lambda_i(\alpha_i, \tilde{\alpha}_i) p_i(\alpha_{-i} \mid \tilde{\alpha}_i)\right] \\
&\quad = 0.
\end{aligned}$$

Note that, as can be easily verified, Condition C* is satisfied as soon as one agent i has "free beliefs," i.e., as soon as $p(\cdot \mid \tilde{\alpha}_i) \equiv p(\cdot \mid \alpha_i)$, for some i and all pairs $(\alpha_i, \tilde{\alpha}_i)$, and includes, among many others, the so-called independent case where all agents have free beliefs.

3 The "Primal" Approach

As shown by d'Aspremont et al. (2003), Condition C* has an equivalent "primal" version, Condition C, which turns out to be very useful. An information structure (N, A, p) satisfies Condition C if and only if for every function $R : A \to \Re$, there

[4]"C" for "Compatibility condition" the name given by d'Aspremont and Gérard-Varet (1979a) and the star in C* to indicate that it is the "dual" version of the condition. The "primal" version is studied in the next section.

exists a transfer rule t^C such that

$$\sum_{i \in \mathcal{N}} t_i^C(\alpha) = R(\alpha) \text{ for all } \alpha \in A,$$

and

$$\sum_{\alpha_{-i} \in \mathcal{A}_{-i}} t_i^C(\alpha_i, \alpha_{-i})\, p(\alpha_{-i} \mid \alpha_i) \geq \sum_{\alpha_{-i} \in \mathcal{A}_{-i}} t_i^C(\widetilde{\alpha}_i, \alpha_{-i})\, p(\alpha_{-i} \mid \alpha_i)$$

$$\text{for all } i \in N \text{ and all } (\alpha_i, \widetilde{\alpha}_i) \in A_i^2.$$

Using the same theorem of the alternative as above, one can show that this condition is satisfied if and only if, for every $R : A \to \Re$, whenever, for $\lambda : A_i^2 \to \Re_+$ and $\mu : A \to \Re$

$$p_i(\alpha_{-i} \mid \alpha_i) \sum_{\tilde{\alpha}_i \neq \alpha_i} \lambda_i(\tilde{\alpha}_i, \alpha_i) - \sum_{\tilde{\alpha}_i \neq \alpha_i} \lambda_i(\alpha_i, \tilde{\alpha}_i) p_i(\alpha_{-i} \mid \tilde{\alpha}_i) = \mu(\alpha)$$

$$\text{for all } i \in N \text{ and for all } \alpha \in A.$$

then

$$\sum_{\alpha \in \mathcal{A}} \mu(\alpha) R(\alpha) \leq 0.$$

This implies that μ must be identically equal to zero, and therefore C is equivalent to C*.

There is a nice story developed in d'Aspremont et al. (2003) to explain why Condition C guarantees the existence of a Pareto-optimal BIC mechanism whatever the utility functions and the *efficient* outcome function. Imagine a planner consulting two bureaus, the "preference bureau" and the "beliefs bureau." The preference bureau uses a VCG mechanism $t_i^{VCG}(\alpha_i, \alpha_{-i}) = \sum_{j \neq i} u_j(s(\alpha_i, \alpha_{-i}); \alpha_j)$ so that SIC (or IC) is satisfied thanks to efficiency of the outcome function s. But, this leads to a deficit $R(\alpha) = -\sum_{i \in \mathcal{N}} t_i^{VCG}(\alpha)$, for every α. Under condition C, the beliefs bureau can recommend a transfer t^C to cover this deficit while preserving incentives, so that the transfer function $t = t^{VCG} + t^C$ is balanced and ensures BIC and Pareto-optimality. This argument shows more. It shows that, under condition C, it is always possible to supplement transfers ensuring BIC by transfers to balance the budget, while keeping BIC. And, conversely, an information structure guarantees budget balance in that sense only if condition C holds.[5] Using this property of Condition C,

[5]For the proof, see d'Aspremont et al. (2004). Forges et al. (2002) call this property "automatic balance."

d'Aspremont et al. (2004) show that condition C is the weakest among all conditions that have been proposed[6] to implement efficient mechanisms.

Now, in some contexts, one might be interested in implementing mechanisms that are not Pareto-optimal (the outcome function is not efficient). From a normative viewpoint, this would be justified if the decision generates externalities beyond the set of agents who participate in the mechanism. From a positive viewpoint, this would be the case if the principal is at the same time budget constrained (so that he cannot transfer funds to the agents) but had his own, state of nature dependent, preferences over outcomes. In these cases, the mechanism (s, t) should satisfy BIC and the transfer function should be balanced. Assuming $n \geq 3$, Condition B, defined as follows, can be used for that purpose.[7] An information structure (N, A, p) satisfies Condition B if and only if there exists a transfer rule t^B such that for all $\alpha \in A$,

$$\sum_{i \in \mathcal{N}} t_i^B(\alpha) = 0,$$

and, for all $i \in A$ and all $\widetilde{\alpha}_i \in A_i$,

$$\sum_{\alpha_{-i} \in \mathcal{A}_{-i}} t_i^B(\alpha_i, \alpha_{-i})\, p(\alpha_{-i} \mid \alpha_i) > \sum_{\alpha_{-i} \in \mathcal{A}_{-i}} t_i^B(\widetilde{\alpha}_i, \alpha_{-i})\, p(\alpha_{-i} \mid \alpha_i).$$

By applying the same kind of duality argument as above d'Aspremont and Gérard-Varet (1982) (see lemma 3), B is equivalent to the dual condition B* which states that there exist no $\lambda : A_i^2 \to \Re_+$, $\lambda \neq 0$, and no $\mu : A \to \Re$ which satisfy

$$p_i(\alpha_{-i} \mid \alpha_i) \sum_{\tilde{\alpha}_i \neq \alpha_i} \lambda_i(\tilde{\alpha}_i, \alpha_i) - \sum_{\tilde{\alpha}_i \neq \alpha_i} \lambda_i(\alpha_i, \tilde{\alpha}_i) p_i(\alpha_{-i} \mid \tilde{\alpha}_i) = \mu(\alpha)$$

for all $i \in N$ and for all $\alpha \in A$.

This immediately shows that B* implies C*, hence that C holds whenever B holds. Moreover, for any environment with an information structure satisfying B, and any outcome function, the BIC constraints can be (strictly) satisfied. It suffices to premultiply the (balanced) transfer function t^B by some large positive number. But, the converse is also true: for any environment, the BIC constraints can be (strictly) satisfied for any outcome function only if the information structure satisfies B (d'Aspremont et al., 2003).

[6]It is strictly weaker than Chung (1999) weak regularity (hence than Matsushima (1991) regularity condition) and Fudenberg et al. (1994) pairwise identifiability. It is equivalent to Johnson et al. (1990) condition called LINK.

[7]Condition B was first defined by d'Aspremont and Gérard-Varet (1982). When $n = 2$, condition C is equivalent to independence of types, and condition B never holds.

The relationship between conditions C and B can be further clarified since condition B implies "No-Freeness": it cannot be the case that two types of an agent generate the same probability distribution over the types of the other agents. Formally, for all $i \in N$ and any two types α_i and $\widetilde{\alpha}_i$, we have $p(\cdot \mid \alpha_i) \neq p(\cdot \mid \widetilde{\alpha}_i)$. Furthermore, it can be shown that condition B is equivalent to C plus No-Freeness: an information structure (N, A, p) satisfies condition B if and only if it satisfies condition C and No-Freeness. Because for $n = 2$, condition C is equivalent to independence of types, that is freeness for any two types of any agent, condition B never holds with $n = 2$. Because, as we will see shortly, B holds generically with $n \geq 3$, this reinforces our words of caution about the generality of results obtained from models with only two types for each agent.

Last, but not least, in "nearly all" environments, we can construct explicitly the transfer function t^B used in the definition of condition B. Consider an information structure such that for all $i \neq j$, all $\alpha_i \in A_i$, and all $\alpha_j \in A_j$, $p(\alpha_j \mid \alpha_i) \equiv \sum_{\left\{\alpha'_{-i} \mid \alpha'_j = \alpha_j\right\}} p\left(\alpha'_{-i} \mid \alpha_i\right) > 0$. Define addition and subtraction on the indices of the agent modulo n, so that $n + 1 \equiv 1$ and $1 - 1 \equiv n$. We let

$$t_i^B(\alpha) = \log p\left(\alpha_{i+1} \mid \alpha_i\right) - \log p\left(\alpha_{i+2} \mid \alpha_{i+1}\right).$$

The negative term is constant in α_i and does not influence the incentives of agent i, but will ensure budget balance. The strict concavity of the function log implies that for all i and all $(\alpha_i, \widetilde{\alpha})$,

$$\sum_{\alpha_{-i}} \left[\log p\left(\alpha_{i+1} \mid \widetilde{\alpha}_i\right) - \log p\left(\alpha_{i+1} \mid \alpha_i\right)\right] p_i\left(\alpha_{-i} \mid \alpha_i\right)$$

$$= \sum_{\alpha_{i+1}} \left[\log \frac{p\left(\alpha_{i+1} \mid \widetilde{\alpha}_i\right)}{p\left(\alpha_{i+1} \mid \alpha_i\right)}\right] p\left(\alpha_{i+1} \mid \alpha_i\right)$$

$$< \log \sum_{\alpha_{i+1}} \frac{p\left(\alpha_{i+1} \mid \widetilde{\alpha}_i\right)}{p\left(\alpha_{i+1} \mid \alpha_i\right)} p\left(\alpha_{i+1} \mid \alpha_i\right) = 0,$$

whenever $p(\alpha_{i+1} \mid \widetilde{\alpha}_i) \neq p(\alpha_{i+1} \mid \alpha_i)$, for some α_{i+1}. Hence, in these cases, condition B holds, which proves that it holds generically. This also implies that condition C holds generically when $n \geq 3$.

4 Individual Rationality

This short review shows that, for Bayesian incentive compatibility and budget balance to hold whatever the utility functions, two conditions on the information structure seem to emerge. One, condition C (or equivalently C*) requires that the outcome function be efficient and is necessary and sufficient to ensure budget balance. The other, condition B (or equivalently condition B*) is stronger and is necessary and sufficient to implement any outcome function. Both the dual and primal approaches are useful in deriving the results.

We have not, in this note, explicitly introduced individual participation constraints, that is constraints which ensure that the agents are willing to participate in the mechanism. There are two types of individual rationality constraints that correspond to different extensive form games:

- ex ante participation constraints correspond to a game in which: (1) the mechanism is announced; (2) the agents decide whether they are willing to participate; (3) the agents learn their types; and (4) the agents play according to the rules of the mechanism. In this case, the participation constraints are written[8]

$$\sum_{\alpha \in A} \big[u_i(s(\alpha); \alpha_i) + t_i(\alpha)\big] p(\alpha) \geq 0 \text{ for all } i \in N.$$

- *interim* participation constraints correspond to the same game with stages 2 and 3 switched. The participation constraints are written

$$\sum_{\alpha_{-i} \in A_{-i}} \big[u_i(s(\alpha); \alpha_i) + t_i(\alpha)\big] p(\alpha_{-i} \mid \alpha_i) \geq 0 \text{ for all } i \in N \text{ and } \alpha_i \in A_i.$$

In the case of ex ante participation constraints, the agents learn their types only after having accepted to participate in the mechanism, whereas in the case of interim constraints, they learn their types before accepting to participate.

Both types of participation constraint require an ex ante nonnegative aggregate expected surplus condition:

$$\sum_{i \in N} \sum_{\alpha \in A} \big[u_i(s(\alpha); \alpha_i)\big]\, p(\alpha) \geq 0. \tag{1}$$

Actually, this condition is also sufficient, along with either condition C or condition B, for implementation if we impose *ex ante* individual rationality (d'Aspremont et al., 2003).

In the context of a bargaining problem (where the public decision is the final owner of the good) with two agents and independent types, Myerson and Satterthwaite (1983) showed their justly celebrated impossibility result: there exists no Bayesian incentive compatible, efficient, *interim* individually rational mechanism.

Despite the importance of this justly celebrated result, it is important to note that it only holds when there are two agents. In an important paper, Makowski and Mezzetti (1994) assume that there are at least three agents, that each agent has an infinite connected set of types and that types are independent, so that the following

[8]We are assuming that the reservation utilities are equal to 0, both in the case of ex ante and interim participation constraints. It is quite easy to prove that this does not entail any loss of generality.

result[9] holds: all efficient mechanisms are VCG in expectation, i.e., s is an efficient outcome function and the transfers are defined by:[10]

$$t_i(\alpha_i, \alpha_{-i}) = \sum_{\alpha_{-i} \in \mathcal{A}_{-i}} \left[\sum_{j \neq i} u_j(s(\alpha_i, \alpha_{-i}); \alpha_j) \right] p(\alpha_{-i}) + h_i(\alpha),$$

where $\sum_{\alpha_{-i}} h(\alpha_i, \alpha_{-i}) p(\alpha_{-i})$ does not depend on α_i; therefore, it does not affect the incentives of agent i. Then, they show that an efficient *interim* individually rational mechanism holds if, translated in our finite sets of types notation, the following condition holds (with s an efficient outcome function)[11,12]:

$$\begin{aligned}
&(n-1) \sum_{\alpha \in A} \left[\sum_{i \in N} u_i(s(\alpha); \alpha_i) \right] p(\alpha) \\
&\leq \sum_{i \in N} \sum_{\alpha_i \in A_i} p_i(\alpha_i) \times \\
&\qquad \min_{\alpha_i' \in A_i} \sum_{\alpha_{-i}} \left[u_i(s(\alpha_i', \alpha_{-i}); \alpha_i') + \sum_{j \neq i} u_j(s(\alpha_i', \alpha_{-i}); \alpha_j) \right] p_i\left(\alpha_{-i}\right).
\end{aligned}$$

We refer the reader to Makowski and Mezzetti (1994) and d'Aspremont and Crémer (2018) for more details.

Matsushima (2007) and Kosenok and Severinov (2008) have studied the existence of *interim* incentive compatibility for not necessarily efficient decision functions. Both papers produce conditions on the information structure which guarantee implementation when condition (1) is satisfied, with Kosenok and Severinov's being necessary and sufficient and therefore less restrictive than Matsushima's. Both conditions are strictly more restrictive than condition B.

More discussion of these conditions, showing how conditions B and C are still prominent, as well as exploration of the case where independence of types does not hold, but where there is still some freeness, can be found in d'Aspremont and Crémer (2018).

It may be useful to point out that the auction literature presents other examples of mechanisms where *interim* participation constraints are at the core of the problem. There, the types of the agents are their willingness to pay for the object. At the time at which they decide to participate in the auction, the potential buyers know their

[9]This result was first proved in d'Aspremont and Gérard-Varet (1979b) and generalized in Holmström (1977, 1979).

[10]Because of the independence of types, we can write $p(\alpha_{-i})$ instead of $p\left(\alpha_{-i} \mid \alpha_i\right)$.

[11]Note that with independence, we can write $p(\alpha_{-i})$ without ambiguity as $p(\alpha_{-i} \mid \alpha_i) = p(\alpha_{-i} \mid \widetilde{\alpha}_i)$ for all $\alpha_i, \widetilde{\alpha}_i, \alpha_{-i}$.

[12]To be totally clear, this condition is not necessary in the case of finite sets of types. We are writing it in this way to avoid introducing more notation.

types. One issue that the literature has tackled is the identification of information structures that guarantee that the seller can do as well as if he had full information. He can do so only if he sells the good to the agent with the highest valuation, hence the final allocation is efficient. However, the fact that he desires to extract the whole surplus rather than simply balance the budget is incompatible with free beliefs. See Crémer and McLean (1985, 1988). Interestingly, if Kosenok and Severinov's (2008) condition holds, the condition of Theorem 2 of Crémer and McLean (1988) must hold.

5 Concluding Remarks

The methods used to study "team adverse selection" can also be used to study "team moral hazard," say the sharing of an output (measured in money) depending on individual actions that are not perfectly observable (or controllable). This generates a noncooperative game and the collectively optimal level of output may be unenforceable whatever the proposed sharing rule.[13] A context, where this negative conclusion can be avoided, is when the outcome function is stochastic. We then get a system of linear inequalities in the transfers similar to the ones above, and conditions C (or C^*) and B (or B^*) are transposed as conditions on the stochastic outcome function to allow for a collective optimum to be noncooperatively enforceable (see d'Aspremont and Gérard-Varet, 1998).

Finally, we should point out that in many cases agents are not provided with information exogenously—they need to spend resources in order to acquire information and may choose either to do so or not to do so. There has been a substantial and still increasing body of work on this issue in the principal agent framework (see Crémer and Khalil, 1992; Crémer, Khalil, and Rochet, 1998a,b; Szalay, 2008, and the subsequent literature). It would be of great interest to understand better how the fact that the acquisition of information is endogenous affects the design of multi-agent Bayesian mechanisms.

References

Alchian, A., & Demsetz, H. (1972). Production, information costs and economic organization. *American Economic Review, 62*, 777–805.

Bergemann, D., & Morris, S. (2005). Robust mechanism design. *Econometrica, 73*, 1771–1813.

Chung, K.-S. (1999). A note on Matsushima's regularity condition. *Journal of Economic Theory, 87*, 429–433.

Crémer, J., & Khalil, F. (1992). Gathering information before signing a contract. *American Economic Review, 82*, 566–578.

[13]See Alchian and Demsetz (1972) and Holmström (1982).

Crémer, J., Khalil, F., & Rochet, J.-C. (1998a). Contracts and productive information gathering. *Games and Economic Behavior, 25*, 174–193.

Crémer, J., Khalil, F., & Rochet, J.-C. (1998b). Strategic information gathering before a contract is offered. *Journal of Economic Theory, 81*, 163–200.

Crémer, J., & McLean, R. P. (1985). Optimal selling strategies under uncertainty for a discriminating monopolist when demands are interdependent. *Econometrica, 53*, 345–361.

Crémer, J., & McLean, R. P. (1988). Full extraction of the surplus in Bayesian and dominant strategy auctions. *Econometrica, 56*, 1247–1258.

d'Aspremont, C., & Crémer, J. (2018). Bayesian incentive compatibility with and without free beliefs (in press).

d'Aspremont, C., Crémer, J., & Gérard-Varet, L.-A. (1990). Incentives and the existence of Pareto-optimal revelation mechanisms. *Journal of Economic Theory, 51*, 233–254.

d'Aspremont, C., Crémer, J., & Gérard-Varet, L.-A. (2003). Correlation, independence, and Bayesian incentives. *Social Choice and Welfare, 21*, 281–310.

d'Aspremont, C., Crémer, J., & Gérard-Varet, L.-A. (2004). Balanced Bayesian mechanisms. *Journal of Economic Theory, 115*, 385–396.

d'Aspremont, C., & Gérard-Varet, L.-A. (1979a). Incentives and incomplete information. *Journal of Public Economics, 11*, 25–45.

d'Aspremont, C., & Gérard-Varet, L.-A. (1979b), On Bayesian incentive compatible mechanisms. In J.-J. Laffont (Ed.), *Aggregation and Revelation of Preferences* (pp. 269–288). Amsterdam: North-Holland.

d'Aspremont, C., & Gérard-Varet, L.-A. (1982). Bayesian incentive compatible beliefs. *Journal of Mathematical Economics, 10*, 25–45.

d'Aspremont, C., & Gérard-Varet, L.-A. (1998). Linear inequality methods to enforce partnerships under uncertainty: An overview. *Games and Economic Behavior, 25*, 311–336.

Fan, K. (1956). On systems of linear inequalities. In H. W. Kuhn & A. W. Tucker (Eds.), *Linear inequalities and related systems*. Princeton, NJ: Princeton University Press.

Forges, F., Mertens, J.-F., & Vohra, R. V. (2002). The ex ante incentive compatible core in the absence of wealth effects. *Econometrica, 70*, 1865–1892.

Fudenberg, D., Levine, D. K., & Maskin, E. (1994). The folk theorem with imperfect public information. *Econometrica, 62*, 997–1039.

Green, J., & Laffont, J.-J. (1977). Characterization of satisfactory mechanisms for the revelation of preferences for public goods. *Econometrica, 45*, 427–438.

Green, J., & Laffont, J.-J. (1979). *Incentives in public decision making*. Amsterdam: North-Holland.

Harsanyi, J. C. (1967). Games with incomplete information played by "Bayesian" players, I-III. Part I. The basic model. *Management Science, 14*, 159–182.

Harsanyi, J. C. (1968a). Games with incomplete information played by "Bayesian" players, Part II. Bayesian equilibrium points. *Management Science, 14*, 320–334.

Harsanyi, J. C. (1968b). Games with incomplete information played by "Bayesian" players, Part III. The basic probability distribution of the game. *Management Science, 14*, 486–502.

Holmström, B. (1977). *On incentive problems in organizations*. Ph.D. thesis, Stanford University.

Holmström, B. (1979). Groves' scheme on restricted domains. *Econometrica, 47*, 1137–1144.

Holmström, B. (1982). Moral hazard in teams. *Bell Journal of Economics, 13*, 324–340.

Holmström, B., & Myerson, R. B. (1983). Efficient and durable decision rules with incomplete information. *Econometrica, 51*, 1799–1819.

Hurwicz, L. (1972). On informationally decentralized systems. In C. B. McGuire, R. Radner, & K. J. Arrow (Eds.) *Decision and organization: A volume in honor of Jacob Marschak* (pp. 297–336). Amsterdam: North-Holland.

Johnson, J. W., Pratt, J. W., & Zeckhauser, R. J. (1990). Efficiency despite mutually payoff-relevant private information: The finite case. *Econometrica, 58*, 873–900.

Kosenok, G., & Severinov, S. (2008). Individually rational, budget-balanced mechanisms and allocation of surplus. *Journal of Economic Theory, 140*, 126–161.

Makowski, L., & Mezzetti, C. (1994). Bayesian and weakly robust first best mechanisms: Characterizations. *Journal of Economic Theory, 64*, 500–519.

Maskin, E. S. (1986). Optimal Bayesian mechanisms. In W. P. Heller, R. M. Starr, & D. A. Starrett (Eds.), *Essays in honor of Kenneth J. Arrow: Volume 3, uncertainty, information, and communication* (pp. 229–238). Cambridge: Cambridge University Press.

Matsushima, H. (1991). Incentive compatible mechanisms with full transferability. *Journal of Economic Theory, 54*, 198–203.

Matsushima, H. (2007). Mechanism design with side payments: Individual rationality and iterative dominance. *Journal of Economic Theory, 133*, 1–30.

Myerson, R., & Satterthwaite, M. (1983). Efficient mechanisms for bilateral trading. *Journal of Economic Theory, 29*, 265–281.

Szalay, D. (2008). Contracts with endogenous information. *Games and Economic Behaviour, 65*, 586–625.

Vickrey, W. (1961), Counterspeculation, auctions and competitive sealed tenders. *Journal of Finance, 72*, 44–61.

Walker, M. (1978). A note on the characterization of mechanisms for the revelation of preferences. *Econometrica, 46*, 147–152.

Walker, M. (1980). On the nonexistence of a dominant strategy mechanism for making optimal public decisions. *Econometrica, 48*, 1521–1540.

Feasible Nash Implementation of Social Choice Rules When the Designer Does Not Know Endowments

Leonid Hurwicz, Eric Maskin, and Andrew Postlewaite

Preface by E. Maskin

Leo Hurwicz quite literally changed my life. I was a math major at Harvard and had only a vague idea of what economics is all about. Then, one term I wandered almost by accident into course on information taught by Leo's old friend, Kenneth Arrow. A major part of the course was devoted to Leo's work on mechanism design.

This was the early 1970s, and mechanism design was just getting started and still on the periphery of economics. But it was soon apparent to me that this was great stuff. It had the precision and power and sometimes even the beauty of mathematics. And it could be used to answer some of the big questions of the economic world: What does decentralization mean? When does a free market perform better than a planned one? Which economic system uses information most efficiently? Leo was even able to show in a now famous theorem that in a 2-person economy it is impossible to implement an efficient, individually rational allocation in dominant strategies. It doesn't get better than that! So on the basis of Leo's work, I decided I would change directions and try to do mechanism design myself.

This paper is a revised, shortened version of a book contribution originally published together with the late Leonid Hurwicz in a festschrift for Stan Reiter.
Hurwicz L., Maskin E., Postlewaite A. (1995) Feasible Nash Implementation of Social Choice: Rules When the Designer Does not Know Endowments or Production Sets.
In: Ledyard J.O. (ed) The Economics of Informational Decentralization: Complexity, Efficiency, and Stability. Springer, Boston, MA © Springer Science+Business Media New York 1995.
The preface for this version was added by Eric Maskin.

L. Hurwicz
University of Minnesota, Minnesota, MN, USA

E. Maskin (✉)
Harvard University, Cambridge, MA, USA
e-mail: emaskin@fas.harvard.edu

A. Postlewaite
University of Pennsylvania, Philadelphia, PA, USA
e-mail: apostlew@econ.upenn.edu

W. Trockel (ed.), *Social Design*, Studies in Economic Design,
https://doi.org/10.1007/978-3-319-93809-7_7

A couple of years later, Ken introduced me to Leo in person at the theory workshop then held every summer at Stanford, and I learned that mechanism design had a sense of humor—although a peculiar one. "Why do most economists prefer French fries to hash browns?", Leo asked me. It's because fries are potato optimal.

Andy Postlewaite was also at Stanford that summer, and he had discovered a puzzling phenomenon: it appeared that Walrasian outcomes on the boundary of the feasible set are not implementable in Nash equilibrium—contrary to what people had previously thought. Well, Leo, Andy and I thought about that for a while and soon got to the bottom of it. And we wrote up a short manuscript of 8 pages or so—suitable for publication as a note in the *Journal of Economic Theory*.

But would Leo actually submit the paper to the journal? "Let me put it this way," said Leo. "Wouldn't you first like to know what happens if agents can destroy their endowments?" And of course we did want to know.

So, a year later, we had answered that question and now had a manuscript of 30 pages, appropriate for a regular article in *Econometrica*. But was Leo now ready to actually send it in? "Let me put it this way," he said. "Before publishing the paper we ought to find out what happens if production can occur." And we had to admit he was right.

Six years later, when we had actually done the finding out, we had a gargantuan manuscript of 80 pages that was too long for any journal. So we thought we should turn the paper into a monograph. But was Leo prepared to do this? "Let me put it this way: No. After all, the proofs and exposition still need refinement." So over the next 11 years, at erratic intervals, Leo would send Andy and me updated versions of the manuscript in which a lemma here or a definition there would be improved.

I'm pretty sure things would have continued that way indefinitely if Stan Reiter had not been gracious enough to reach the age when it became appropriate to present him with a festschrift. And so—a full 21 years after we had started work on it—the paper was finally published in Stan's festschrift. But Leo was able to put all this in perspective. It just goes to show, he said, that when writing a paper, the first 20 years are always the hardest.

I am happy that this volume in honor of Leo will contain a somewhat abridged version of our three-way paper.

1 Introduction

The aim of the present paper is to analyze the problem of assuring the feasibility[1] of a mechanism (game form), implementing in Nash equilibrium[2] a given social choice rule abbreviated as (*SCR*) when the mechanism is constrained as to the way

[1]Earlier models of tatonnement and of proposed mechanisms designed to implement social choice rules (e.g., Walras or Lindahl) were criticized for not guaranteeing the feasibility at disequilibrium points. Some, like the Walrasian auctioneer, were not balanced (1), others failed to assure individual feasibility. (See Wilson 1976.)

[2]From now on "implementation" is to be understood in the sense of Nash non-cooperative equilibria. Let n be the number of players, Z the outcome space (the space of allocations), S the joint strategy space, i.e., $S = S^1 \times \ldots \times S^n$, where S^i is the strategy domain of the ith player, and

in which it is permitted to depend on endowments. A social choice rule is a correspondence specifying outcomes considered to be desirable in a given economy (environment). A mechanism is defined by (a) an outcome function and (b) a strategy domain prescribed for each player. Our outcome functions are not permitted to depend at all on the initial endowments. As to strategy domains, the ith agent's strategy domain S^i is only permitted to depend on that agent's endowment, but not on the endowments of other agents. (For earlier results concerning endoment manipulation, see Postlewaite (1979) and Sertel (1994).)

A possible (but not necessary) interpretation is that those formulating the rules[3] of the game have no knowledge of the endowments; they may have no way of preventing the players from either understating or even destroying their own endowments, but they may formulate rules making an overstatement of their own endowments impossible, for instance, by requiring the players to "place the claimed endowments on the table." In that case, an agent's strategy domain is limited by his/her (true) endowment. As for the final allocations, these are determined by a formula based only on the agents' claims and hence are not directly dependent on the true values of the endowments.[4]

In a pure exchange economy, whether or not the designer knows the individual endowments (as well as the traders' admissible consumption sets), suppose it is required that the outcome function be informationally decentralized, in the sense defined in Sect. 2. It is then seen, from Proposition 1 in Sect. 2, that feasibility out of equilibrium makes it unavoidable that each unit's strategic domain would depend on its initial endowment. It is furthermore to be noted that this result applies to all informationally decentralized mechanisms, regardless of the equilibrium concept[5] used. A stronger conclusion, at the expense of a stronger assumption is obtained in

---

let h: $S \to Z$ be the outcome function. An *SCR*, denoted by F, is a correspondence from the space E of environments into Z, specifying for each environment (economy) e in E a nonempty set in the outcome space Z. An environment (economy) is defined as an n-tuple of characteristics $e^i = (C^i, \omega^i, R^i)$, where, for the ith agent, C^i is the admissible consumption set, w^i the initial endowment, and R^i the (weak) preference relation. I.e, $e = (e^1, \ldots, e^n)$ and E is the class of a priori admissible environments. A possible interpretation is that the designer believes (correctly) that an environment (economy) outside of E will not occur.

We say that a mechanism (S, h) Nash *implements* an *SCR F over* a class of environments E if it is the case that, for every e in E, (1) the set of Nash equilibrium outcomes $N_{S,h}(e)$ generated by the mechanism (S, h) is nonempty, and (2) this set $N_{S,h}(e)$ is a subset of $F(e)$. (The term sometimes used in the literature for this concept is "weakly implements.") The mechanism (S, h) is said to *fully* implement F over E if, for every e in E, $N_{S,h}(e) = F(e)$. In most of the present paper we actually deal with a singleton-valued correspondence F, i.e., one equivalent to a function. In that case the two concepts of implementation coincide and we simply say that (S, h) implements the social choice function f, abbreviated *SCF*, the function equivalent to the singleton-valued correspondence F. (A method for extending our results to correspondences is illustrated in the Appendix to Sect. II.A.1, in Hurwicz et al. 1995.)

[3]Those formulating the rules are often collectively referred to as "the designer." Hence the title of this paper.

[4]Of course, because of the non-exaggeration requirement, an agent's claim as to his/her own endowment provides partial information as to the true endowment, namely that the true endowment is at least as high as that claimed.

[5]For example, maximin, Nash, etc.

Proposition 2 of Sect. 2. We obtain: (i) certain conditions on the nature of game forms necessary for the implementability of SCRs; (ii) certain conditions that must be satisfied by an SCR in order that it be implementable; (iii) sufficient conditions for the implementability of an SCR, established by constructing an implementing game form.

When a mechanism is said to be feasible, all values of the outcome function, rather than only the equilibrium values, lie in the set of feasible outcomes. We shall denote by $A(e)$ the set of outcomes feasible in the environment e. This defines a correspondence $A(\cdot)$ from the space of environments (economies) into the space Z of outcomes (allocations).

Let us illustrate this in the situation of pure exchange private goods economies without free disposal with n traders. (Section 3 of the paper is devoted to this case.) Here, the ith trader's *characteristic* e^i is defined by his/her consumption set C^i, initial endowment ω^i, and preference relation R^i, written $e^i = (C^i, \omega^i, R^i)$.[6] The *environment* e is defined as the list of characteristics, i.e., $e = (e^1, \ldots, e^n)$. The space of feasible outcomes in this economy consists of all net trade lists $x = (x^1, \ldots, x^n)$, each x^i an element of the commodity space $\mathbb{R}^m$, satisfying the following two conditions: (a) *Individual feasibility*—every agent remains within his/her consumption set, i.e., $\omega^i - x^i \in C^i$; (b) *compatibility* or *balance*—the sum of all net trades is the null vector of the commodity space, written $\Sigma x^i = 0$.

In earlier mechanism design literature, the balance condition was observed, but not the individual feasibility. This contrasts with the conventional Walrasian auctioneer scenario where the reverse is the case. In the present paper, the emphasis is on mechanisms satisfying both conditions.

Looking at the problem of constructing a feasible game form implementing a given SCR over a class E of environments, we must distinguish situations in which the designer knows the feasible set $A(e)$ for each e in E, i.e., the feasibility correspondence $A(\cdot)$, from those in which the designer has no such information. Maskin's algorithm[7] (1977) for constructing a mechanism implementing a given SCR postulates a class E of environments with a common set A of feasible outcomes known to the designer.[8] In this paper we are interested in the situation where the feasible set is not known to the designer. Since the balance condition does not contain any unknown parameters, we are dealing in our illustrative example with a situation where the designer does not know the traders' initial endowments.

Section 3 is devoted to pure exchange economies without public goods. In Sect. 3, to gain insight into the problem, we start with the case where the designer does know preferences, but not the endowments. We then construct two types of endowment revelation games (involving, respectively, the withholding and destruction of endowments), each analogous to Maskin's algorithm for unknown preferences. The strategy space for each trader consists of n-tuples of claimed

---

[6]Preferences do not affect feasibility.

[7]Maskin's construction is an algorithm in the sense that it is a 'recipe' for constructing implementing mechanisms for a class of SCR's (by inserting the SCR F in an outcome function schema), rather than a single mechanism.

[8]On the other hand, the designer does not know which preference profile (from a known family of profiles) will prevail.

endowments. Thus, the ith trader claims that the vector of players' endowments is $w^i = (w_1^i, \ldots, w_n^i)$ where w_j^i is j's endowment according to i's claim. It is assumed that i knows his/her true endowment ω^i. An important restriction imposed on the nature of the strategy space is that a trader cannot exaggerate his/her own endowment; i.e., $w_i^i \leqslant \omega^i$. This means that the individual strategy domains depend on the true endowments. In Sect. 2 of the paper, it is shown that some such restriction is unavoidable.[9]

Two variants of an endowment game are considered: withholding (Sect. 3.1), and destruction (Sect. 3.2). When a trader is *withholding* a part of the endowment, he/she (falsely) claims some $w_i^i \leqslant \omega^i$ (so that $w_i^i \neq \omega^i$) as own endowment, but—in addition to the commodity bundle allocated by the outcome function—he/she can also consume the difference $\omega^i - w_i^i$. By contrast, when a trader is *destroying* the part $\omega^i - w_i^i$ of the endowment, this part is not available for consumption. In Sect. II.C of Hurwicz et al. (1995), we consider a mixture of withholding and destruction. Implementation under withholding is called W-implementation, that under destruction as D-implementation. When preferences are assumed known to the designer, they are dealt with respectively in Theorem 1 (Sect. 3.1.1) and Theorem 3 (Sect. 3.2). Under withholding, we assume individual rationality, under destruction, "non-confiscatoriness" of the social choice function.

The more interesting case is, of course, when the designer knows neither endowments nor preferences. Under withholding, this is referred to as W-R-implementation and is dealt with in Theorem 2, Sect. 3.1.2. It is shown there, for the case of withholding, how to deal with this situation. The proof involves combining the game form for the withholding game, for known preferences constructed in Sect. 3.1.1, with a Maskin type game form, for situations where endowments but not preferences are assumed known to the designer (see Maskin 1977; Saijo 1988; Hurwicz 1986).

As in the Groves–Ledyard (1977) treatment of public goods and in Maskin's 1977 algorithm, all our constructions assume that there are at least three agents ($n > 2$). Subsequent to the circulation of earlier versions of this paper, feasible game forms have been constructed for exchange economies with two agents in economies with free disposal[10] (see, in particular, Nakamura 1989, 1990).

The mechanisms used in our existence proofs are far from informationally efficient. In fact, Page (1989) and Hong and Page (1994) show how the size of the message space can be substantially reduced. In the next section of this introduction, we provide a few additional comments concerning the contents of this paper.

While Theorem 1 only deals with social choice *functions*, it is indicated in the appendix to Sect. II.A.1 of Hurwicz et al. (1995) how, for the endowment withholding game with preferences known to the designer, the result can be extended to the implementation of social choice *correspondences*. Analogous extensions from SCF's to SCR's (correspondences) seem to be possible for our other cases, but are not dealt with in the paper.

[9]When the goods are physical their existence (and ownership) might have to be shown. Similarly, proof might be required for claimed rights or entitlements, or ever claimed skills. See discussion in Hong and Page (1994).

[10]I.e., where the balance condition is in the form of a weak inequality rather than equality (called "weak balance").

2 The Dependence of Strategy Domains on Initial Endowments

In what follows, we show that, when the outcome function is privacy preserving with respect to endowments (but possibly "parametric" in the sense of Hurwicz (1972, pp. 310–313), the strategy domain of each person in a pure exchange economy must vary with that person's initial endowment. These results apply to noncooperative games in general and not merely to Nash equilibria. Proposition 1 and the corollary are valid whether or not the designer knows the initial endowments.

We consider a class E of pure exchange economies with the set of goods $L=\{1,\ldots,l\}$. The set of agents is denoted by $N=\{1,\ldots,n\}$. The ith person's true initial endowment is $\mathring{\omega}^i$, but sometimes the circle above ω is omitted. We write $\underset{\to}{\omega}=(\omega^1,\ldots,\omega^n)$ and $\omega^i=(\omega_1^i,\ldots,\omega_l^i)$ for each i in N. Each person's consumption set is contained in the nonnegative orthant $\mathbb{R}_+^l$.

Let $E=E^1\times\ldots\times E^n$, with the generic element of E^i denoted by $e^i=(\omega^i,R^i)$, $\omega^i\in\mathbb{R}_{+0}^l$; here R^i denotes the ith agent's (weak) preference relation, assumed to be reflexive, transitive, and total.

Assumption 1 We assume that for every i in N, every r in L, and every positive number ε, there is $e^i\in E^i$, $e^i=(\omega^i,R^i)$, such that $0<\omega_r^i<\varepsilon$.

Restricting ourselves, for the sake of simplicity, to single-valued social choice rules (performance correspondences), we denote a social choice function (performance function) by $f\colon E\to\mathbb{R}^{ln}$. The values of f specify net trades. Feasibility requirements are: for all $e\in E$ and all $r\in L$,

$$\textit{balance:}\quad \sum_{i\in N} f_r^i(e)=0 \tag{1}$$

$$\textit{individual feasibility:}\quad f_r^i(e)\geqslant-\omega_r^i \quad \text{for all} \quad i\in N. \tag{2}$$

where f_r^i denotes the net allocation of the rth good to the ith person, and ω_r^i the initial endowment of the ith person in the rth good.

To avoid triviality, we assume that there is at least one person $i\in N$, a good $r\in L$, and an economy $\bar{e}\in E$, such that, for a social choice rule f implementable on $\bar{e}$,

$$f_r^i(\bar{e})\neq 0. \tag{3*}$$

From feasibility, it follows that there is at least one person $j\in N$, a good $r\in L$, and an economy $\bar{e}\in E$, such that

$$f_r^j(\bar{e})<0. \tag{3}$$

We shall write

$$f_r^j(\bar{e})=-a,\quad a>0. \tag{3'}$$

We now define a noncooperative game with the ith strategy domain denoted by S_i. Since the question is whether, or in what way, this domain depends on the initial endowments, we write $S_i = S_i(e^i) = S_i(\omega^i, R^i)$. (That is, the S_i may be "parametric," but must not depend on the characteristics of other agents.) This, of course, does not a priori preclude the possibility that $S_i(\cdot)$ is constant, i.e., that, for any two environments $\bar{e}$, $\bar{\bar{e}}$, we would have $S_i(\bar{e}^i) = S_i(\bar{\bar{e}}^i)$. However, the following proposition shows that, in fact, at least some persons' domains do vary with their own endowments.

Write $S = S(e) = S_1(e^1) \times \ldots \times S_n(e^n)$.

We shall permit the outcome functions to be "parametric," i.e., to depend on the initial endowments, but in a privacy-preserving way. That is, the ith individual's net allocation z^i is given by

$$z^i = h^i(s, e^i), \quad s \in S(e), \quad i \in N.$$

One could, of course, confine oneself to "nonparametric" outcome functions where $z^i = h^i(s)$. By permitting the dependence of h^i on $\mathring{\omega}^i$ (perhaps even on e^i), however, we strengthen the result.

We impose on the outcome functions the following feasibility restrictions for all $r \in L$, all $s \in S$, and all $e \in E$:

$$\textit{balance:} \quad \sum_{i \in N} h_r^i(s, e^i) = 0 \tag{1*}$$

$$\textit{individual feasibility:} \quad h_r^i(s, e^i) \geqslant -\mathring{\omega}_r^i \quad \text{for all} \quad i \in N. \tag{2*}$$

We assume that the game form $(h, S(\cdot))$ implements f on E. By definition, this implies that for every e in E, there exists s^* in $S(e)$, such that for every i in N, and for every r in L, $h_r^i(s^*, e^i) = f_r^i(e)$.

Proposition 1 *Assume Assumption* 1 *holds, let* $e^* \in E$ *and let* f *satisfying* (3*) *be implementable on* e^*. *Let further* j, r, e^* *and* a *be those specified in* (3′), *with* $e^* = (\omega^{*i}, R^{*i})_{i \in N}$. *Then there exists a strategy* n-*tuple* $s = (s_i)_{i \in N}$ *and an economy* $e^{**} = (w^{**i}, R^{**i})_{i \in N}$, *with* $\omega_r^{**j} = \omega_r^{*j}$, *while* $\omega^{**k} = \omega^{*k}$ *for all* $k \in N/\{j\}$, *such that* $s_j \in S_j(e^{*j})$ *but* $s_j \notin S_j(e^{**j})$.

Proof Since $(h, S(\cdot))$ implements f on E, there exists s in $S(e^*)$, $s = (s_1, \ldots, s_n)$, $s_i \in S_i(e^{*i})$ for all i in N, and such that, for some j,

$$h_r^j(s, e^{*j}) = f_r^j(e^*) = -a, \quad a > 0$$

and $s_j \in S_j(e^{*j})$. By Assumption (1), there is an environment e^{**} in E, such that

$$0 < \omega_r^{**j} < a,$$

while

$$\omega^{**k} = \omega^{*k} \quad \text{for all} \quad k \in N \backslash \{j\}.$$

By showing that $s_j \notin S_j(e^{**j})$, we shall complete the proof. Suppose, to the contrary, that s_j does belong to $S_j(e^{**j})$. Since the characteristics of others remain unchanged, it follows that $s \in S(e^{**})$. Using the individual feasibility requirement (2*) and previously established relations we obtain

$$h_r^j(s, e^{**j}) \gtreqless -\omega^{**j}_r > -\mathrm{a} = h_r^j(s, e_j^*),$$

while

$$\Sigma_{k \neq j} h_r^k(s,\ e^{**k}) = \Sigma_{k \neq j} h_r^k(s,\ e^{*k}).$$

Adding, we find that

$$\Sigma_{i \in N} h_r^i(s,\ e^{**i}) > \Sigma_{i \in N} h_r^i(s,\ e^{*i}),$$

which contradicts the balance requirement in (1*). Q.E.D.

Remark 1 Thus s_j depends on e^j. s_j need not depend on ω^j, but if it does not vary with ω^j, then it must vary with R^j.

Corollary 1 *If for* every *person $j \in N$, there exists a good $r \in L$ and an economy $\bar{e} \in E$, such that*

$$f_r^j(\bar{e}) \neq 0,$$

then, for every $j \in N$, the domain correspondence $S_j(e^j)$ is non-constant; more specifically, there exists $s^* = (s_i^*)_{i \in N}$ and an economy $\bar{\bar{e}} = (\bar{\bar{\omega}}^i, \bar{\bar{R}}^i)_{i \in N'}$ with $\bar{\bar{\omega}}^j_r \neq \bar{\omega}^j_r$ while $\bar{\bar{\omega}}^k = \bar{\omega}^k$ for all $k \in N \backslash \{j\}$, such that $s_j^* \in S_j(\bar{e}^j)$ but $s_j^* \notin S_j(\bar{\bar{e}}^j)$.

Proof Follows immediately from the preceding proposition.

Assume now that an agent's strategy is independent of preferences but may depend on his/her endowment, so that i's strategy domain can be written as $S_i(\omega^i)$. We shall next show that, under Assumption 2 on the social choice function (stated below), if, in environment e^* agent i has a greater endowment of a particular good than in environment e^{**}, while the other agents' endowments of all goods are the same, then i's strategy domain $S_i(\omega^{*i})$ must contain elements not present in $S_i(\omega^{**i})$.

To state (2), we first introduce a class of environments. We shall denote by $E/\underset{\rightarrow}{\omega}$ the class of all environments in E whose endowment profile equals $\underset{\rightarrow}{\omega}$, while preferences vary.

Hence, $f_r^i(E/\underset{\rightarrow}{\omega})$ is the set of net allocations in the rth good to the ith agent produced by the performance function f, as environments trace out the class $E/\underset{\rightarrow}{\omega}$. The additional assumption is as follows:

Assumption 2

$$\forall i \in N, \quad r \in L, \quad \varpi_r^i \geqslant 0,$$
$$\inf f_r^i(E/\varpi) = -\varpi_r^i.$$

Remark 2 It appears that, when the postulated class of environments is sufficiently rich, Assumption 2 is satisfied for social choice functions which always yield allocations that are Pareto optimal and individually rational.

Proposition 2 *Assume Assumption 2 holds, and let e^*, e^{**} be two environments such that, for some agent i and a good r, $\omega^{*i}_r > \omega^{**i}_r$, while $w^{*j} = w^{**j}$ for all j not equal to i. Then there exists a strategy available to i in e^* but not in e^{**}.*

Proof By Assumption 2, there is a sequence $\{e^{*k}\}$, $k = 1, 2, \ldots$ of environments[11] such that each e^{*k} belongs to the class E/ω^*, so that for each e^{*k} the endowment profile is ω^*, by individual feasibility $f_r^i(e^{*k}) \gtrless -\omega^{*i}_r$, and, by Assumption 2, $\lim f_r^i(e^{*k}) = -\omega^{*i}_k$ as k tends to infinity.

Write $c = \omega^{*i}_r - \omega^{**i}_r$. By hypothesis, $c > 0$. Then there exists a number c', with $0 \lessgtr c' < c$, such that, for a sufficiently large integer K, we have

$$f_r^i(e^K) = -\omega^{*i}_r + c'.$$

Write $)i(= (1, \ldots, i-1, i+1, \ldots, n)$. Since h implements f, there exists a strategy n-tuple $s^{*K} = \langle s^{*Ki}, s^{*K,)i(} \rangle$ such that (supressing in our notation the possible dependence of h^i on e^i) $h(s^{*K}) = f(e^{*K})$, and hence

$$h_r^i(s^{*Ki}, s^{*K,)i(}) = -\omega^{*i}_r + c'.$$

Hence,

$$s^{*Ki} \in S_i(\omega^{*i}).$$

But, since $c' < c$, it follows from the definition of c that

$$-\omega^{*i}_r + c' < -\omega^{**i}_r,$$

and hence $h_r^i(s^{*K}) < -\omega_r^i$, which violates the individual feasibility requirement for agent i in the environment e^{**}. Since s^{*K}, $)i($ was available members of $)i($ in e^*, and $S_j(\omega^{**j}) = S_j(\omega^{*j})$ for j in $)i($ (since, by hypothesis, $\omega^{**j} = \omega^{*j}$ for j not equal to i), we conclude that

[11]The environments e^{ik} has the endowment profile ϖ but may differ with respect to preferences.

$$s^{*Ki} \notin S_i(\omega^{**i}).$$

Q.E.D.

In what follows we, sketch the construction used in Theorem 1 (where endowments may be withheld, but preference profiles are known).

3 Pure Exchange in Private Goods

3.1 Withholding

3.1.1 The Endowment Game (with Endowments Unknown but Preferences Known)

Notation and Assumptions

(i) Vectors

Let m be a positive integer. Then

$$\mathbb{R}^m = \{x | x = (x^1, \ldots, x^m), x^r \text{ a real number for all } 1 \leqslant r \leqslant m\}.$$

Let $x, y \in \mathbb{R}^m$. Then $x \geqslant y$ means $x^r \geqslant y^r$ for all $1 \leqslant r \leqslant m$; $x \geq y$ means $x \geqslant y$, but $x \neq y$; and $x > y$ means $x^r > y^r$ for all $1 \leqslant r \leqslant m$. $\mathbb{R}^m_+ = \{x \in \mathbb{R}^m | x \geqslant 0\}$; $\mathbb{R}^m_{++} = \{x \in \mathbb{R}^m | x > 0\}$; $\mathbb{R}^l_{+0} = \mathbb{R}^l_+ \setminus \{0\}$, so that $x \in \mathbb{R}^l_{+0}$ means $x \geq 0$; $\mathbb{R}^{ln}_{+0} = \mathbb{R}^l_{+0} \times \ldots \times \mathbb{R}^l_{+0}$ (n times). For $a, b \in \mathbb{R}^m$, $[a, b] = \{x \in \mathbb{R}^m | a \leqslant x \leqslant b\}$, $(a, b] = \{x \in \mathbb{R}^m | a \leqslant x \leqslant b, x \neq a\}$.

(ii) Environment

$N = \{1, \ldots, n\}$ = the set of agents; $n \geqslant 3$.
$L = \{1, \ldots, l)$ = the set of goods.
$\mathring{\omega}_i$ = the true endowment of agent i; $\underset{\to}{\mathring{\omega}}_i \in \mathbb{R}^l_{+0}$ for all i.
$\underset{\to}{\mathring{\omega}} = (\mathring{\omega}_1, \ldots, \mathring{\omega}_n)$ = the endowment profile.
$\mathbb{R}^l_+$ is assumed to be the individually feasible consumption set for every agent.
$\mathring{R}_i$ = the true preference relation of agent i on $\mathbb{R}^l_+ \times \mathbb{R}^l_+$.
$\mathring{P}_i$ = the true strict preference of agent i (i.e., $x\mathring{P}_i y$ iff $x\mathring{R}_i y$ but not $y\mathring{R}_i x$).
$\mathring{R}_i$ is reflexive, transitive, and convex on $\mathbb{R}^l_+ \times \mathbb{R}^l_+$ (i.e., preferences are selfish); $\mathring{R}_i$ is assumed strictly increasing in all goods for all agents (i.e., $x \geq y$ implies $x\mathring{P}_i y$).

(iii) Performance

$Z = \{\underset{\sim}{z} \in \mathbb{R}^{ln} | \underset{\sim}{z} = (z_1, \ldots, z_n);\ z_i \in \mathbb{R}^l,\ \forall i \in N;\ \Sigma_{i \in N} z_i = 0\}$ = the set of balanced *net* trades[12]. Given a configuration $z = (z_1, \ldots, z_n)$ of net trades, agent i's final (total) holdings are $\mathring{\omega}_i + z_i$.

f = the performance function[13] (social choice rule).

$f\colon \mathbb{R}^{ln}_{+0} \rightarrow Z$.

Let $\underset{\sim}{v} = (v^1, \ldots, v^n) \in \mathbb{R}^{ln}_{+0};\ v^i \in \mathbb{R}^l_{+0}, \forall i \in N$.

$f = (f_1, \ldots, f_n)$; if $z = (z_1, \ldots, z_n) = f(\underset{\sim}{v})$, then $z_i = f_i(\underset{\sim}{v})$; so, $f_i\colon \mathbb{R}^{ln}_{+0} \rightarrow \mathbb{R}^l$.

$f(\underset{\sim}{\mathring{\omega}})$ is interpreted as the optimal[14] net trade configuration when the true endowment profile is $\underset{\sim}{\mathring{\omega}}$; $f_i(\underset{\sim}{\mathring{\omega}})$ is agent i's optimal net trade for the profile $\underset{\sim}{\mathring{\omega}}$.

It is assumed that $v^i + f_i(\underset{\sim}{v}) \geqq 0$, for all i and all $\underset{\sim}{v} \in \mathbb{R}^{ln}_{+0}$.

(iv) Strategies and Outcome Functions

For each $i \in N$, let T_i be an arbitrary nonempty set. It is assumed that the strategy space S_i of agent i is of the form

$$S_i = (0, \mathring{\omega}_i] \times T_i,$$

where T_i is independent of $\underset{\sim}{\mathring{\omega}}$.

We also define $S = S_1 \times \ldots \times S_n$.

Generically, we write for the corresponding elements

$$s_i = (w^i_i, t_i),\ s = (s_1, \ldots, s_n),\ \text{and}^{15}\ s = (s_i;\ s_{)i(}),$$

where $t_i \in T_i$, $0 \leqslant w^i_i \lesseqgtr \mathring{\omega}_i$, $s_i \in S_i$, $s_{)i(} \in \prod_{j \neq i} S_j$[16], $s \in S$.

If we interpret the component w^i_i of $s_i = (w^i_i, t_i)$ as a profession of agent i's endowment, the inequality $0 \leqslant w^i_i \lesseqgtr \mathring{\omega}_i$ means that the agent cannot overstate his own endowment; on the other hand, the endowment can be understated (in one or more commodity components), but the claimed endowment w^i_i (like the true endowment

---

[12]The amount received by i is a positive component z_i.

[13]To simplify exposition, we confine ourselves in this section to single-valued social choice rules; subsequently, we shall extend our treatment to correspondences.

[14]The term "optimal" is always used in the sense of the given performance function f.

[15]We use, here and elsewhere, the somewhat imprecise notation which identifies $(S_i, S_1, \ldots, S_{i-1}, S_{i+1}, \ldots, S_n)$ with $(S_1, \ldots, S_{i-1}, S_i, S_{i+1}, \ldots, S_n)$.

[16]$\prod_{j \neq i} S_j = S_1 \times \ldots \times S_{i-1} \times S_{i+1} \times \ldots \times S_n$.

$\mathring{\omega}_i$) must be semi-positive (i.e., different from the null vector and nonnegative in all commodity components).[17]

h = the outcome function (game form).
h: $S \rightarrow Z$.
$h = (h_1, \ldots, h_n)$; if $z = (z_1, \ldots, z_n) = h(s)$, then $h_i(s) = z_i$; so, h_i: $S \rightarrow \mathbb{R}^l$.
$h(s)$ = then net trade configuration resulting from the strategic configuration s.
$h_i(s)$ = agent i's net trade resulting from the strategic configuration s.

Given s, agent i's final (total) holdings are

$$\mathring{\omega}_i + h_i(s).$$

For net trades z_i', $z_i'' \in z$, we shall sometimes write $z_i' \mathring{R}_i z_i''$ to mean $(\mathring{\omega}_i + z_i')\mathring{R}_i(\mathring{\omega}_i + z_i'')$, etc.

It will be assumed that, for all $i, s = (s_i, s_{)i(}), s_i = (w_i^i, t_i)$,

$$w_i^i + h_i(s) \geqq 0.$$

That is, the outcome function will never deprive the agent of goods in excess of his claimed endowment.

Since $w_i^i \leqq \mathring{\omega}_i$, a fortiori, the outcome function will never require the agent to give up more of any good than there was in the true initial endowment. Thus, individual feasibility is assured.

Furthermore, since h takes its values in Z, we have $\Sigma_{i \in N} h_i(s) = 0$ for all $s \in S$; hence, balance is also assured. Thus, feasibility is preserved at all points of the strategy space, out of equilibrium as well as at equilibrium.

On the other hand, since $w_i^i \leqslant \mathring{\omega}_i$ is permitted, the agent is able to *withhold* a part of the true endowment. Complete withholding is ruled out by the requirement $w_i^i \geqslant 0$.

We shall say that the outcome function h W-implements[18] (in Nash equilibrium (NE)) the performance function f for $\mathring{\underset{\sim}{R}}$ of true preference profiles, if: for any true endowment profile $\mathring{\underset{\sim}{\omega}}$, (1) an NE exists, and, further, (2) for any NE configuration s^* of strategies, $\mathring{\underset{\sim}{\omega}} + h(s^*) = \mathring{\underset{\sim}{\omega}} + f(\mathring{\underset{\sim}{\omega}})$; i.e., every Nash outcome is f-optimal.

---

[17] It would be possible to relax our assumptions by replacing the requirement $\mathring{\omega}_i \geqslant 0$ by $\mathring{\omega}_i \geqq 0$ and, at the same time weaken $w_i^i \geqslant 0$ to: $w_i^i \geqslant 0$ if $\mathring{\omega}_i \geqslant 0$. But we cannot permit an agent to claim $w_i^i = 0$ when $\mathring{\omega}_i \geqslant 0$. For let all agents claim zero endowments while in fact $\sum_{i \in N} \mathring{\omega}_i \geqslant 0$. Then, since the possibility of withholding means that $w_i^i + h_i(s) \geqq 0$ for all $i \in N$, the net Nash allocation would have to be 0 for everyone, and this might be non-optimal.

If the assumptions were relaxed along the indicated lines, a minor modification would have to be made in the outcome function.

[18] Here W- is mnemonic for withholding, as distinct from strategies to be labeled D-, in which an agent may not withhold but only destroy his endowment, and from those labeled WD-, where the agent may do both.

Definition 1 f is individually rational (IR) if, for all i in N, and all $\underset{\rightarrow}{\mathring{\omega}} \in \mathbb{R}^{ln}_{+0}$, $(\mathring{\omega}_i + f_i(\underset{\rightarrow}{\mathring{\omega}}))\mathring{R}_i\mathring{\omega}_i$.

Proposition 3 *If preferences $\underset{\rightarrow}{\mathring{R}}$ are continuous and nondecreasing, and if f is W-implementable (in NE) for $\underset{\rightarrow}{\mathring{R}}$, then f is individually rational (IR). ("W-implementable" stands for "withholding-implementable.")*

Proof Suppose f is implementable by h: $S \rightarrow \mathbb{R}^{ln}$, but is not IR. Then there exist $\underset{\rightarrow}{\mathring{\omega}} \in \mathbb{R}^{ln}_{+0}$ and $i \in N$ such that[19] $0\ \mathring{P}_i f_i(\underset{\rightarrow}{\mathring{\omega}})$. Since h implements f, there exists an NE $s^* = (s_1^*, \ldots, s_n^*) \in S$ for $(\underset{\rightarrow}{\mathring{\omega}}, \underset{\rightarrow}{\mathring{R}})$, such that $h_i(s^*) = f_i(\underset{\rightarrow}{\mathring{\omega}})$. Hence $0\ \mathring{P}_i h_i(s^*)$.

Then, by the assumed continuity of R_i, the semi-positivity of $\mathring{\omega}_i$, and the non-decreasing preferences, there exists a real number $\varepsilon > 0$ and an i-feasible net trade $b = (b_1, \ldots, b_l)$, where[20] $b \leqslant 0$, $\| b \| = \in$, and, furthermore,

$$b\mathring{P}_i h_i(s^*).$$

But, for any $t_i \in T_i$ and[21] $s_i = (-b, t_i)$, we have $h_i(s_i, s^*_{)i(}) \gneqq b$, since $w_i^i + h_i(s') \gneqq 0$ for all s'. Hence, $h_i(s_i, s^*_{)i(})\mathring{R}_i b \mathring{P}_i h_i(s^*)$, which contradicts the supposition that s_i^* is an NE strategy.

Q.E.D.

Definition 2[22] f is *non-confiscatory* (NC) if $\forall i \in N$, $\forall\, \underset{\rightarrow}{\mathring{\omega}} \in \mathbb{R}^{ln}_{+0}$, $\mathring{\omega}_i + f_i(\underset{\rightarrow}{\mathring{\omega}}) \geqslant 0$.

Remark 3 It may be noted that, when $\mathring{\omega}_i \geqslant 0$ and preferences are strictly increasing, IR implies NC. Clearly, however, f may be NC and not IR.

Theorem 1 *(1) If f is IR, and if the assumptions (including strictly increasing[23] preferences) preceding the above proposition are satisfied, then f is W-implementable (in NE). ("W-implementable" stands for "withholding-implementable.")*

(2) If preferences are continuous[24] and strictly increasing, f is W-implementable if and only if it is IR (individually rational).

Proof The proof of (2) follows from (1) and the preceding proposition. To establish (1), we construct an outcome function h, which W-implements f.

For $i \in N$, let the strategy space of the ith agent be

$$S_i = \{(w_i^1, \ldots, w_i^n) \in \mathbb{R}^{ln}_{+0} | w_i^j \in \mathbb{R}^l_{+0}, 0 \leqslant w_i^i \lneqq \mathring{\omega}_i, i, j \in N\}.$$

[19] Here 0 is a net trade (the l-dimensional null vector), strictly preferred by i to the net trade $f_i(\underset{\rightarrow}{\mathring{\omega}})$.

[20] With $\| x \|$ denoting the norm of the vector x; any norm can be used.

[21] That is, $w_i^i = -b$.

[22] When the requirement $\mathring{\omega}_i \in \mathbb{R}^l_{+0}$ is relaxed to $\mathring{\omega}_i \in \mathbb{R}^l_+$, the above definition is generalized as follows: f is non-confiscatory (NC) if $\forall i \in N$, $\forall \underset{\rightarrow}{\mathring{\omega}} \in \mathbb{R}^{ln}_+$, $\mathring{\omega}_i \geqslant 0$ implies $\mathring{\omega}_i + f_i(\underset{\rightarrow}{\mathring{\omega}}) \geqslant 0$.

[23] But not necessarily continuous.

[24] Note that the continuity of preferences is only needed for the necessity part of Theorm 1.2.

For $s_i \in S_i$, we shall sometimes write

$$s_i = w_i = (w_i^i, w_i^{)i(}),$$

where

$$w_i^{)i(} = (w_i^1, \ldots, w_i^{i+1}, w_i^{i+1}, \ldots, w_i^n)$$

and

$$w_k = (w_k^1, \ldots, w_k^n), \quad \text{with} \quad w_k^r \in \mathbb{R}^l_{+0} \quad \text{for all} \quad k \in N, r \in N.$$

We interpret w_i^j as agent i's statement about j's endowment. For all $i, j \in N$, it is assumed that $w_i^j \geqslant 0$; i.e., each agents's statement attributes to everybody, including himself, positive holdings of some commodity. In the spirit of informational decentralization (privacy-preserving property of the mechanism), it is assumed that an agent has no useable information about the other agents' endowments. Therefore, for $j \neq$ i, there is no upper bound on w_i^j. By contrast, an agent is assumed to know his own endowment. While he may conceal or destroy a part of it, he is not permitted to exaggerate it; hence, the requirement that $w_i^i \lesseqgtr \mathring{\omega}_i$ for all $i \in N$. (We might, for instance, imagine that the rules of the game require that the agent "put on the table" the reported amount w_i^i.)

Notice that this S_i has the structure of the strategy space $S_i = (0, \mathring{\omega}_i] \times T_i$, introduced in the previous section. In $s_i = (w_i^i, w_i^{)i(})$, the component $w_i^{)i(}$ corresponds to t_i in $s_i = (w_i^i, t_i)$.

We will define the outcome function $h(w_1, \ldots, w_n)$, with $w_i \in \mathbb{R}^{ln}_{+0}$ for each $i \in N$, by the following rules:

(a) (*The case of unanimity*)

If, for some $\underset{\sim}{v} \in \mathbb{R}^{ln}_{+0}, s = (s_1, \ldots, s_n) \in S, s_i = \underset{\sim}{v}$ for all $i \in N$, then

$$h(s) = f(\underset{\sim}{v}).$$

To state rules (b) and (c), we use the following notation.

Let $s = (s_1, \ldots, s_n) \in S, s_j = w_j = (w_j^1, \ldots, w_j^n),\ w_j^k \in \mathbb{R}^l_{+0}, k, j \in N$. We define

$$M(s) = \{i \in N | w_i^i \geqslant w_j^i, \forall j \neq i,\ j \in N\};$$

$$w(s) = \sum_{i \in N} w_i^i;$$

$$\beta_i(s) = \sum_{\substack{j \in N \\ j \neq i}} \sum_{\substack{k \in N \\ k \neq i}} \| w_j^j - w_k^j \|, \ i \in N.$$

(When there is no danger of confusion, we suppress the argument s and write, respectively M, w, β_i.)

The second rule is then as follows:

(b) If $M(s) = \emptyset$, but there is no $\underset{\sim}{y}$ such that $s_i = \underset{\sim}{y}$ for all $i \in N$, then $\Sigma_{j \in N} \beta_j(s) > 0$, and we set

$$h_i(s) = \left[\frac{\beta_i(s)}{\sum\limits_{j \in N} \beta_j(s)} \right] w(s) - w_i^i, \; i \in N.$$

The third rule is:

(c) If $M(s) \neq \emptyset$, we set

$$h_i(s) = \begin{cases} \dfrac{1}{\#M(s)} w(s) - w_i^i & \text{for} \quad i \in M(s) \\ -w_i^i & \text{for} \quad i \notin M(s). \end{cases}$$

We shall now prove three claims which together imply that this outcome function h does W-implement (in NE) the performance function f.

These claims are: (1) the unanimous announcement of the true endowment profile by all agents is a Nash equilibrium; (2) the unanimous announcement of a false endowment profile is not a Nash equilibrium; and (3) in the absence of unanimity, there is no Nash equilibrium.

Claim 1 The unanimous announcement of the true endowment profile by all agents is an NE. That is,

$$\begin{aligned} &\text{if} \qquad s_i = \underset{\to}{\mathring{\omega}}, \quad \forall i \in N, \\ &\text{then} \qquad \mathrm{s} = (\mathrm{s}_1, \ldots, \mathrm{s}_\mathrm{n}) \;\text{ is a NE for } \underset{\to}{\mathring{\omega}}. \end{aligned}$$

Proof of Claim 1 For such a unanimous announcement s of the true endowment profile $\underset{\to}{\mathring{\omega}}$, by rule (a),

$$h(s) = f(\underset{\to}{\mathring{\omega}}).$$

Suppose s is not an NE. Then there is an agent j and some $\tilde{s}_j$, such that

$$(+++) \qquad h_j(\tilde{s}_j, s_{)j()}) \mathring{P}_j h_j(s_j, s_{)j()})$$

where $(s_j, s_{)j()}) = s$. Necessarily, $\tilde{s}_j \neq \underset{\to}{\mathring{\omega}}$, and also, for $\tilde{s}_j = (\widetilde{w}_j^j, \widetilde{w}_{)j()})$, by the non-exaggeration rule,

$$\tilde{w}_j^j \lesseqgtr \mathring{\omega}_j.$$

Writing $\tilde{s} = (\tilde{s}_j, \tilde{s}_{)j(}) = (\tilde{s}_j, s_{)j(})$, so that $\tilde{s}_r = s_r$ for all $r \neq j$, it follows[25] that

$$j \notin M(\tilde{s}).$$

Let k be any agent other than j; i.e., $k \neq j$. Since[26] $n \gtreqless 3$, there exists a third agent m, with $m \neq j$ and $m \neq k$.

Now $\tilde{s}_r = s_r = \underset{\rightarrow}{\mathring{\omega}}$ for all $r \neq j$. Hence,

$$\tilde{s}_m = \underset{\rightarrow}{\mathring{\omega}} \quad \text{and} \quad \tilde{s}_k = \underset{\rightarrow}{\mathring{\omega}}.$$

Since $\tilde{s}_r = (\mathring{\omega}_1, \ldots, \mathring{\omega}_n) = (w_r^1, \ldots, w_r^n), \ \forall\, r \neq j$, we have

$$w_k^k = w_m^k,$$

and hence,

$$k \notin M(\tilde{s}).$$

Since k was an arbitrary agent other than j, it follows that no agent other than j is in $M(\tilde{s})$, and we have seen above that j is not in $M(\tilde{s})$. So,

$$M(\tilde{s}) = \emptyset.$$

Thus, $M(\tilde{s}) = \emptyset$ but $\tilde{s}$ is not unanimous, so rule (b) applies to $\tilde{s}$.

Since $s_i = \tilde{s}_i$ for all $i \neq j$, we have

$$\beta_j(\tilde{s}) = \sum_{k \neq j} \sum_{i \neq j} \| w_k^k - w_i^k \| = 0,$$

and so

$$h_j(\tilde{s}) = 0 \cdot w(\tilde{s}) - \tilde{w}_j^j = -\tilde{w}_j^j.$$

Since f is IR,

$$h_j(s) = h_j(\underset{\rightarrow}{\mathring{\omega}}, \ldots, \underset{\rightarrow}{\mathring{\omega}}) = f_j(\underset{\rightarrow}{\mathring{\omega}}) \mathring{R}_j 0.$$

[25]Because $\tilde{w}_i^j = w_i^j = \mathring{\omega}_j$ for all $i \neq j$.

[26]By assumption, $\#N = n \gtreqless 3$.

Because preferences are strictly increasing and $\widetilde{w}_j^j \geqslant 0$,

$$0\,\mathring{P}_j(-\widetilde{w}_j^j).$$

Therefore,

$$h_j(s)\mathring{P}_j(-\widetilde{w}_j^j),$$

and so

$$h_j(s)\mathring{P}_j h_j(\tilde{s}),$$

which contradicts the above supposition that $h_j(\tilde{s})\mathring{P}_j h_j(s)$. Hence s is an NE.

Q.E.D.

Claim 2 The unanimous announcement of a false endowment profile is not an NE. That is, if $s=(\underset{\to}{v},\ldots,\underset{\to}{v})$, $\underset{\to}{v}\in\mathbb{R}^{ln}_{+0}$ with $\underset{\to}{v}\neq\underset{\to}{\mathring{\omega}}$, then s is not an NE.

Proof of Claim 2 Since s is unaminous, rule (a) again applies, and so

$$h(s)=f(\underset{\to}{v}).$$

Suppose s is a Nash equilibrium.

Since $\underset{\to}{v}$ is not the true endowment profile, and agents cannot overstate their endowments, then there must be an agent i such that

$$w_r^i \leqslant \mathring{\omega}_i, \quad \forall r\in N.$$

(We have $\underset{\to}{v}=(v^1,\ldots,v^n)$, $v^k=w_r^k$, $\forall k, r\in N$. Since $\underset{\to}{v}\neq\underset{\to}{\mathring{\omega}}$, it must be that, for some i, $v^i\neq\mathring{\omega}_i$. $\therefore w_i^i\neq\mathring{\omega}_i$, $\therefore w_i^i\leqslant\mathring{\omega}_i$. But $w_r^i=v^i=w_i^i$, $\forall r\in N$. Therefore, $w_r^i\leqslant\mathring{\omega}_i, \forall r\in N$.)

Consider $\tilde{s}=(\tilde{s}_i,\tilde{s}_{)i(})$ such that

$$\begin{aligned} \tilde{s}_{)i(} &= s_{)i(}, \\ \tilde{s}_i^k &= s_i^k \qquad \text{for all} \quad k\neq i \end{aligned}$$

while

$$\tilde{s}_i^i=\mathring{\omega}_i.$$

(That is, $\tilde{s}_i^i\neq s_i^i$ since $s_i^i=w_i^i\leqslant\mathring{\omega}_i$.)

Then

$$M(\tilde{s})=\{i\},$$

and, by rule (c)

$$h_i(\tilde{s}) = w(\tilde{s}) - \tilde{w}_i^i = \sum_{k \neq i} w_k^k = \sum_{j \neq i} v^j$$

We shall show below that

$$(+) \qquad h_i(\tilde{s}) \geqslant h_i(s).$$

Since preferences are strictly increasing, the inequality (+) implies

$$h_i(\tilde{s}) \mathring{P}_i h_i(s).$$

Therefore, when (+) holds, agent i has an incentive to deviate from s_i, and so s is not an NE. That is, Claim 2 follows.

To establish (+), we note that, since, for our outcome function $h(\cdot)$, $h_j(s') \geqq -w_j^j, \forall j$, and $\Sigma_{k \in N} h_k(s') = 0, \forall s' \in S$,[27] we have[28]

$$h_i(s') \leqq \sum_{j \neq i} w_j^j, \quad \forall s' \in S.$$

But $s = (\underset{\sim}{v}, \underset{\sim}{v}, \ldots, \underset{\sim}{v})$, $\underset{\sim}{v} = (v^1, \ldots, v^n)$ implies $v^k = w_k^k$ for all $k \in N$; hence

$$h_i(s) \leqq \sum_{j \neq i} v^j.$$

Suppose

$$(++) \qquad h_i(s) = \sum_{j \neq i} v^j.$$

We shall show that (++) cannot be true. Then, from the inequality in the preceding line, it will follow that

$$h_i(s) \leqslant \sum_{j \neq i} v^j.$$

But we have already shown above that $h_i(\tilde{s}) = \Sigma_{j \neq i} v^j$. Hence, $h_i(s) \leqslant h_i(\tilde{s})$, which is the inequality (+) above. It remains to show that (++) yields a contradiction.

---

[27]These properties of h(·) can be verified directly.

[28]*Proof*: (omitting reference to s'):

$$h_j \geqq -w_j^j \text{ implies } \sum_{j \neq i} h_j \geqq -\sum_{j \neq i} w_j^j.$$

But balance implies $\Sigma_{j \neq i} h_j = -h_i$. Hence, the previous inequality can be written as $-h_i \geqq -\Sigma_{j \neq i} w_j^j$ which is equivalent to $h_i \leqq \Sigma_{j \neq i} w_j^j$.

Writing

$$h_j(s) = x_j, \quad j \neq i,$$

the balance requirement then yields

$$\sum_{j \neq i} x_j + \sum_{j \neq i} v^j = 0.$$

But, $x_j \gtreqqless -v^j$, so that

$$x_j = -v^j + \varepsilon_j \quad j \neq i$$
$$\varepsilon_j \gtreqqless 0.$$

Hence, the balance equation can be written as

$$\sum_{j \neq i} (-v^j + \varepsilon_j) + \sum_{j \neq i} v^j = 0,$$

and this implies $\varepsilon_j = 0, \forall j \neq i$; hence,

$$x_j = -v^j, \quad \forall j \neq i.$$

That is, if $h_i(s) = \sum_{j \neq i} v^j$,

then

(*) $$h_j(s) = -v^j, \quad \forall j \neq i.$$

As noted in Remark 3, before Theorem 1, under our assumption, IR (individual rationality) implies NC (non-confiscatority), so that

$$v^j + f_j(\underset{\sim}{v}) \geqslant 0.$$

But here

$$h_j(s) = f_j(\underset{\sim}{v}).$$

Hence

$$v^j + h_j(s) \geqslant 0$$

which contradicts (*).

Q.E.D.

Claim 3 In the absence of unanimity there is no NE. That is, if for some $i, j \in N$, $s_i \neq s_j$, then $s=(s_1,\ldots,s_n)$ is not an NE.

Proof of Claim 3 Let $s=(s_1,\ldots,s_n)=(w_1,\ldots,w_n)$ with $s_i \neq s_j$ for some $i,\ j \in N$. We consider three cases: (i) $M(s)=N$; (ii) $M(s)\neq \emptyset$, $M(s)\neq N$; (iii) $M(s)=\emptyset$.

(i) Suppose first that $M(s)=N$. Then consider $\tilde{s}$ with

$$\begin{aligned} \tilde{s}_k &= s_k \quad \text{for all} \quad k \neq 1, \\ \tilde{s}_1^q &= s_q^q \quad \text{for all} \quad q \in N. \end{aligned}$$

(We shall sometimes write $\tilde{s}_p = \widetilde{w}_p,\ p \in N$.)

For any agent $r \neq 1$,

$$r \notin M(\tilde{s}),$$

since $\tilde{s}_1^r = \tilde{s}_r^r$.

On the other hand, we shall show that

$$1 \in M(\tilde{s}).$$

Notice that $1 \in M(s)$, since $N=M(s)$ by hypothesis. Hence, by definition of $M(\cdot)$,

$$s_1^1 \geqslant s_r^1, \quad \forall r \neq 1.$$

Thus, by the construction of $\tilde{s}$,

$$\tilde{s}_1^1 \geqslant \tilde{s}_r^1,$$

and so

$$1 \in M(\tilde{s});$$

therefore, rule (c) applies to $\tilde{s}$.

Since it was shown previously that nobody else belongs to $M(\tilde{s})$, we have now established that

$$M(\tilde{s}) = \{1\}.$$

Rule (c) implies therefore

$$h_1(\tilde{s}) = 1 \cdot w(\tilde{s}) - \widetilde{w}_1^1 = w(s) - w_1^1 = \sum_{k \neq 1} w_k^k.$$

But $h_1(s) \leqslant \Sigma_{k\neq 1} w_k^k$ because $M(s)=N$, so that, under s, part of $\Sigma_{k\neq 1} w_k^k$ was allocated to persons other than 1. [That is, $\beta_k(s) > 0,\ \forall k \neq 1$.]

Therefore,

$$h_1(\tilde{s}) \geqslant h_1(s)$$

and consequently, because of strictly increasing preferences,

$$h_1(\tilde{s})\, \mathring{P}_1\, h_1(s).$$

Hence, in case (i), s is not an NE.

(ii) Suppose now that $M(s) \neq \emptyset$, $M(s) \neq N$. Since $M(s) \neq \emptyset$, rule (c) applies to s. Because $M(s) \neq N$, there is an agent $j \notin M(s)$ who, by rule (c), gets

$$h_j(s) = -w_j^j.$$

Now consider $\tilde{s}$ where, for all k and all $i \neq j$,

$$\tilde{s}_i^k = s_i^k,$$
$$\tilde{s}_j^j = s_j^j,$$

and

$$\tilde{s}_j^k = s_k^k.$$

For any $r \neq j$, we have

$$\tilde{s}_j^r = \tilde{s}_r^r,$$

and so, by definition of $M(\cdot)$,

$$r \notin M(\tilde{s}) \quad \text{for all} \quad r \neq j.$$

Furthermore, since (by construction) $j \notin M(s)$ and $\tilde{w}_i^j = w_i^j$ for all $i \neq j$, we have $j \notin M(\tilde{s})$.

Thus,

$$M(\tilde{s}) = \emptyset,$$

and so either rule (a) or rule (b) applies to $\tilde{s}$. But rule (a) cannot be applicable because unanimity in $\tilde{s}$ is impossible: since $n \geqslant 3$ and $M(s) \neq \emptyset$, there is a person $k \in M(s)$, $k \neq j$, and a person $i \neq j$, $i \neq k$ such that,

$$w_k^k \geqslant w_i^k;$$

hence $s_k \neq s_i$. But, w_k^k and w_i^k are unchanged in $\tilde{s}$, and so $\tilde{s}_k \neq \tilde{s}_i$. Hence, there is no unanimity in $\tilde{s}$ and rule (a) does not apply to $\tilde{s}$. Hence, rule (b) applies to $\tilde{s}$.

For agents j, k, and i just referred to, we have

$$\beta_j(\tilde{s}) \geqq \|w_k^k - w_i^k\| > 0.$$

Since $w(s) \geqslant 0$, it follows that

$$h_j(\tilde{s}) \geqslant -w_j^j = h_j(s),$$

and so, by the assumption of strictly increasing preferences,

$$h_j(\tilde{s}) \mathring{P}_j h_j(s).$$

Hence, in case (ii), s is not an NE.

(iii) Finally, suppose that $M(s) = \emptyset$ and *s is not unanimous*. Since, by the hypothesis of Claim 3, not all announced profiles are the same, there exist agents i and j, $i \neq j$, with

$$w_i^i \neq w_j^i,$$

We now distinguish two subcases, according to whether $\beta_j(s) = 0$ or $\beta_j(s) > 0$.

Subcase (iii.1) $\beta_j(s) = 0$.

Consider $\tilde{\tilde{s}}$ defined by

$$\begin{aligned} \tilde{\tilde{s}}_k &= s_k \quad \text{for all} \quad k \neq j \\ \tilde{\tilde{w}}_j^j &= \frac{1}{2} w_j^j \\ \tilde{\tilde{w}}_j^r &= w_j^r \quad \text{for all} \quad r \neq j. \end{aligned}$$

We note that, since s is not unanimous and $M(s)$ is empty, Rule (b) applies to s, and hence

$$h_j(s) = -w_j^j.$$

But, also, $\tilde{\tilde{s}}$ is not unanimous, because $\tilde{\tilde{w}}_i^i = w_i^i$, $\tilde{\tilde{w}}_j^i = w_j^i$ by construction,[29] and $w_i^i \neq w_j^i$ by the above hypothesis.

[29]Since $i \neq j$.

Also, $M(\tilde{\tilde{s}})$ is empty because $M(s)$ is empty, and the change from w_j^j to $\tilde{\tilde{w}}_j^j = \frac{1}{2}w_j^j$ (while $\tilde{\tilde{w}}_r^j = w_r^j$ for $r \neq j$) does not enlarge the set M. Hence, Rule (b) also applies to $\tilde{\tilde{s}}$. Now, since $\tilde{\tilde{s}}_k = s_k$ for $k \neq j$, $\beta_j(s) = 0$ implies $\beta_j(\tilde{\tilde{s}}) = 0$. Therefore,

$$h_j(\tilde{\tilde{s}}) = -\tilde{\tilde{w}}_j^j.$$

But,

$$-\tilde{\tilde{w}}_j^j = -\frac{1}{2}w_j^j \geqslant -w_j^j,$$

because, by assumptions on messages, $w_j^j \geqslant 0$. Hence, by the assumption of strictly increasing preferences,

$$h_j(\tilde{\tilde{s}})\mathring{P}_j h_j(s).$$

So, $\tilde{\tilde{s}}$ is better than s for agent j, and hence s is not a Nash equilibrium.

Subcase (iii.2) $\beta_j(s) > 0$.

In this situation, consider $\tilde{s}$, such that

$$\tilde{s}_k = s_k \quad \text{for all} \quad k \neq j$$

and

$$\tilde{s}_j^r = s_r^r \quad \text{for all} \quad r.$$

By construction, $\beta_j(\tilde{s}) = \beta_j(s) > 0$ and $\Sigma_{k \neq j}\beta_k(\tilde{s}) < \Sigma_{k \neq j}\beta_k(s)$. Also, $M(\tilde{s}) = M(s) = \emptyset$, so Rule (b) applies to both $\tilde{s}$ and s. Therefore, $h_j(\tilde{s}) \geqslant h_j(s)$. And so, again by the assumption of strictly increasing preferences, s is not a Nash equilibrium.

Q.E.D.

3.1.2 The Game with Both Preferences and Endowments Unknown to the Designer

Notation and Assumptions

Here the performance correspondence (SCR) f associates elements of $\mathbb{R}^{ln}$ (net trades) with ordered pairs $(\underset{\rightarrow}{\omega}, \underset{\rightarrow}{R})$ consisting of endowment and preference profiles. The set of these elements is denoted by $f(\underset{\rightarrow}{\omega}, \underset{\rightarrow}{R})$. It is assumed that $f(\underset{\rightarrow}{\omega}, \underset{\rightarrow}{R})$ is non-empty for all $(\underset{\rightarrow}{\omega}, \underset{\rightarrow}{R})$ in its domain.

For the sake of simplicity, we shall assume in what follows that this correspondence is single-valued, i.e., a function. Subsequently, we shall indicate the modifications required to extend the results to the general case of correspondences.

We shall consider two games. The *main game*, in which both the endowments and preferences are unknown, and withholding (but not destruction) is permitted, is called the W-R game. In such a game, for any $i \in N$, a generic element of the ith strategy space S_i is denoted by s_i, with

$$s_i = (w_i, d_i).$$

$w_i \in \mathbb{R}^{ln}_{+0}$ as before,[30] $w_i = (w_i^1, \ldots, w_i^n)$, $w_i^j \in \mathbb{R}^l_{+0}$. $d_i \in D_i$ where D_i is an arbitrary set (the ith domain). The outcome function of this game is $h: S_1 \times \ldots \times S_n \to \mathbb{R}^{ln}$.

We shall also consider an *auxiliary game*, designed for situations where the endowment is given (though perhaps incorrectly) while preferences are unknown. Let the given endowment profile be $v = (v^1, \ldots, v^n)$, $v^i \in \mathbb{R}^l_{+0}$, $i \in N$. We denote by $A(v)$ the set of feasible net allocations in a pure exchange economy when v is the initial endowment profile and each consumption set is the nonegative orthant; i.e., $A(v) = \{(z^1, \ldots, z^n) \in \mathbb{R}^{ln}: z^i \in \mathbb{R}^l, \Sigma_{i \in N} z^i = 0, z^i \geqq -v^i, i \in N\}$.

We denote by g^v an outcome function, $g^v: D_1 \times \ldots \times D_n \to \mathbb{R}^{ln}$, for an auxiliary game when the set of feasible allocations is $A(v)$ and the strategic domains are D_i, $i \in N$. The mapping associating the outcome function g^v with the profile v is called the *auxiliary game form g*.

The set of Nash equilibria of this game (a subset of $D_1 \times \ldots \times D_n$) for the preference profile $\underset{\rightarrow}{R}$ is denoted by $\upsilon_{g^v}(\underset{\rightarrow}{R})$, and the corresponding set of Nash allocations (a subset of $\mathbb{R}^{ln}_{+0}$) by $N_{g^v}(\underset{\rightarrow}{R})$.

Definition 3 f is R-implementable through the auxiliary game form g if, for every $v \in \mathbb{R}^{ln}_{+0}$, there exist domains $D_1, \ldots, D_n$ and an auxiliary outcome function $g^v: D_1 \times \ldots \times D_n \to \mathbb{R}^{ln}$, such that

$$N_{g^v}(\underset{\rightarrow}{R}) = f(v, \underset{\rightarrow}{R}) \quad \text{for all} \quad (v, \underset{\rightarrow}{R}).$$

(That is, every Nash allocation generated by the auxiliary game is f-optimal for v and $\underset{\rightarrow}{R}$, and every f-optimal allocation for v and $\underset{\rightarrow}{R}$ is attainable as a Nash allocation of the auxiliary game.)

Definition 4 For each $i \in N$, let the ith person's strategy set be of the form

$$S_i = S_i(\mathring{\omega}_i) \subset \mathbb{R}^l_{+0} \times T_i,$$

where T_i is an arbitrary set. A generic element of S_i is denoted by $s_i = (w_i^i, t_i)$.[31] Write $S = S(\underset{\rightarrow}{\mathring{\omega}}) = S_1 \times \ldots \times S_n$, and[32]

[30] $\mathbb{R}^m_{+0} = \{x \in \mathbb{R}^m: x \geqq 0, x \neq 0\}$.

[31] The w_i^i component can be interpreted as the i-th agent's claim concerning his own initial endowment.

[32] That is, $A(w_1^1, \ldots, w_n^n)$ would be the set of feasible net allocations if $(w_1^1, \ldots, w_n^n)$ were the true endowment profile.

$$A(w_1^1, \ldots, w_n^n) = \{(z^1, \ldots, z^n) \in \mathbb{R}^{ln}: z^i \in \mathbb{R}^l, \sum_{i \in N} z^i = 0,$$
$$z^i \geqq -w_i^i, \forall i \in N\}.$$

An outcome function h: $S \to \mathbb{R}^{ln}$ is said to be $\underset{\to}{\mathring{\omega}}$-feasible if

$$h(s) \in A(\mathring{\omega}_1, \ldots, \mathring{\omega}_n) \quad \text{for all} \quad s \in S,$$

where $s = (w_i^i, t_i)_{i \in N}$.

Definition 5 A SCR (performance correspondence) f is W-R-implementable (in NE) if, for every $\underset{\to}{\mathring{\omega}} \in \mathbb{R}^{ln}_{+0}$, and for every $i \in N$, there exist strategic domains

$$S_i = S_i(\mathring{\omega}_i) \subset \mathbb{R}^l_{+0} \times T_i,$$

where T_i is an arbitrary set, and an $\underset{\to}{\mathring{\omega}}$-feasible outcome function

$$h: \prod_{i \in N} S_i \to \mathbb{R}^{ln},$$

such that:

$$\forall \underset{\to}{\mathring{R}} \in R,$$

there is an NE s for $\underset{\to}{\mathring{\omega}}$ and $\underset{\to}{\mathring{R}}$ (i.e., $s \in \upsilon_h(\underset{\to}{\mathring{\omega}}, \underset{\to}{\mathring{R}})$)

such that

$$h(s) \in f(\underset{\to}{\mathring{\omega}}, \underset{\to}{\mathring{R}}).$$

Remark 4 In our applications, $S_i(\mathring{\omega}_i) = (0, \mathring{\omega}_i) \times T_i$ and $T_i = \mathbb{R}^{l(n-1)}_{+0} \times D_i$ where D_i is an arbitrary set.

Theorem 2.A *Let f be an IR social choice rule (performance function) which is R-implementable (in NE) through an auxiliary form g: $v \to g^v$. Then f is W-R-implementable in NE (by a "combination" of g with the endowment game of Sect.* 3.1.1*).*

Proof We will construct an outcome function h as follows.

Let $\mathring{\omega}_i$ and the corresponding strategy spaces $S_i(\mathring{\omega}_i)$, $i \in N$, be given. By construction, $s_i = (w_i, d_i)$, $w_i = (w_i^1, \ldots, w_i^i, \ldots, w_i^n)$, and $w_i^i \leqq \mathring{\omega}_i$.

Now we distinguish two types of situations according as to whether there exists $\underset{\to}{v} \in \mathbb{R}^{ln}_{+0}$ such that $\underset{\to}{v} = w_i$ for all $i \in N$.

If such $\underset{\rightarrow}{\nu}$ does not exist, we follow rules (b) and (c) above and conclude that s is not an NE (see Claim 3′).

On the other hand, suppose that $\underset{\rightarrow}{\nu}$ does exist. Then the outcome is dictated by the outcome function $g^{\underset{\rightarrow}{\nu}}$ generated through the mapping g for this $\underset{\rightarrow}{\nu}$. It then turns out (see Claims 1′ and 2′) that an NE obtains only if $\underset{\rightarrow}{\nu}$ coincides with the true endowment profile $\underset{\rightarrow}{\mathring{\omega}}$. But then, by the assumption on g, it follows that $N_h(\underset{\rightarrow}{\mathring{\omega}}, \underset{\rightarrow}{\mathring{R}}) = f(\underset{\rightarrow}{\mathring{\omega}}, \underset{\rightarrow}{\mathring{R}})$.

Formally, the rule (a) of the endowment game (W-game) described in the previous section is replaced by the following Rule (a′): if for some $\underset{\rightarrow}{\nu}$ such that, for all $i \in N$,

$$s_i = (\underset{\rightarrow}{\nu}, d_i)$$

for some $(d_1, \ldots, d_n) \equiv \underset{\rightarrow}{d}$, then, for $s = (s_1, \ldots, s_n)$, we set

$$h(s) = g^{\underset{\rightarrow}{\nu}}(\underset{\rightarrow}{d}).$$

The rules governing cases where there is no unanimity as to endowments are unchanged. The right hand sides of the definitions of $M(s)$ and $w(s)$ remain the same as in the W-game, although now $s_i = (w_i, d_i)$ rather than $s_i = w_i$. The two other rules ((b′) and (c′)) are the same as rules (b) and (c) for the W-game, again with $s_i = (w_i, d_i)$.

Theorem 2.B.B[33] *Let* $n \geqslant 3$, *let endowments be semi-positive* ($\omega^i \geqslant 0$), *and preferences continuous and strictly increasing. Then, a social choice function f is W-R-implementable in NE if and only if it is monotone and individually rational (IR).*

Proof (i) Sufficiency. For $n \geqslant 3$ and monotone f, Theorem 5 in Maskin (1977)[34] shows that there exists a function g which R-implements f in NE.[35] Hence, by Theorem 2.A, the individually rational social choice function f is W-R-implementable.

(ii) Necessity. If f is R-implementable, it is monotone by Theorem 2 of Maskin (1977). If f is W-implementable, it is IR by Proposition 3 in Sect. 3.1.1.

Claim 1′ Correct unanimity with regard to endowments yields an NE.

Let $s^* = (s_1^*, \ldots, s_n^*)$, and, for all $i \in n$, $s_i^* = (\underset{\rightarrow}{\mathring{\omega}}, d_i^*)$, such that $d^* = (d_1^*, \ldots, d_n^*)$ is an NE for $g^{\underset{\rightarrow}{\mathring{\omega}}}$ given $\underset{\rightarrow}{\mathring{R}}$, i.e., $d^* \in \nu_{g^{\mathring{\omega}}}(\underset{\rightarrow}{\mathring{R}})$. Then s^* is an NE for h given $(\underset{\rightarrow}{\mathring{\omega}}, \underset{\rightarrow}{\mathring{R}})$; i.e., $s^* \in \nu_h(\underset{\rightarrow}{\mathring{\omega}}, \underset{\rightarrow}{\mathring{R}})$.

[33]Note that the continuity of preference is only needed for the necessity part of this theorem.

[34]See also the theorem in Saijo (1988, p. 698), and theorem M^1 in Hurwicz (1986, p. 86); in the latter the assumptions of transitivity and completeness are dispensed with. The latter paper follows Maskin's original schema, with Lemmas 1 (p. 88) and 2 (p. 90) corresponding to Maskin's Theorems 4 and 5, respectively.

[35]This is so because, for $n \geqslant 3$, in a pure exchange economy with strictly increasing preferences, the "no veto power" (NVP) requirement in Maskin's Theorem 5 is necessarily satisfied.

Proof Suppose s^* is not an NE. By the assumption concerning d^*, for any agent i, it would not help to depart from d_i^* while retaining $w_i = \underset{\rightarrow}{\mathring{\omega}}$.

Consider therefore $\tilde{s} = (\tilde{s}_l, \ldots, \tilde{s}_n)$, such that $\tilde{s}_j = s_j^*$ for all $j \neq i$, while $\tilde{s}_i = (\tilde{w}_i, \tilde{d}_i)$ with $\tilde{w}_i \neq \underset{\rightarrow}{\mathring{\omega}}$. ($\tilde{d}_i$ may or may not equal d_i^*.) Since, by the outcome rules, $\tilde{w}_i^i \lneqq \mathring{\omega}_i$, it follows that $M(\tilde{s}) = \varnothing$ and so rule (b′) applies.

But

$$\beta_i(\tilde{s}) = 0,$$

since other agents remain unanimous with regard to endowments. Hence, rule (b′) prescribes

$$h_i(\tilde{s}) = -\tilde{w}_i^i.$$

By our assumptions on the auxiliary game form g and d^*,

$$h(s^*) = f(\underset{\rightarrow}{\mathring{\omega}}, \underset{\rightarrow}{\mathring{R}}).$$

Since f is assumed to be IR,

$$f_i(\underset{\rightarrow}{\mathring{\omega}}, \underset{\rightarrow}{\mathring{R}}) \mathring{R}_i 0,$$

hence

$$h_i(s^*) \mathring{R}_i 0,$$

and therefore

$$-\tilde{w}_i^i \mathring{P}_i 0,$$

which contradicts the requirement of semi-positivity for endowment messages and strictly increasing preferences. Hence s^* is an NE for $(\underset{\rightarrow}{\mathring{\omega}}, \underset{\rightarrow}{\mathring{R}})$.

Claim 2′ Incorrect unanimity concerning endowments does not yield an NE.

Let $s = (s_l, \ldots, s_n)$, $s_i = (\underset{\rightarrow}{v}, d_i) \forall i \in N$, $\underset{\rightarrow}{v} = (v^l, \ldots, v^n)$, $v^i \in \mathbb{R}_{+0}^{ln}$, $\underset{\rightarrow}{v} \neq \underset{\rightarrow}{\mathring{\omega}}$. Then s is not an NE for $(\underset{\rightarrow}{\mathring{\omega}}, \underset{\rightarrow}{\mathring{R}})$.

Proof Suppose that s is an NE for $(\underset{\rightarrow}{\mathring{\omega}}, \underset{\rightarrow}{\mathring{R}})$.

By the outcome rules, $\underset{\rightarrow}{v} \lneqq \underset{\rightarrow}{\mathring{\omega}}$, and (since $\underset{\rightarrow}{v} \neq \underset{\rightarrow}{\mathring{\omega}}$ by hypothesis), $v^i \leqslant \mathring{\omega}_i$ for some i by virtue of the non-exaggeration requirement.

Then, by reasoning exactly like that in the proof of Claim 2, we show that Claim 2′ will have been established if [with $\tilde{s}_i = (\tilde{w}_i, d_i), \tilde{w}_i = (\tilde{w}^i_i, w_{)i(}), \tilde{w}^i_i = \mathring{\omega}_i$, and $\tilde{s}_j = s_j \, \forall j \neq i$]

$$(+) \qquad h_i(\tilde{s}) \geqslant h_i(s),$$

and that if (+) fails then

$$(0) \qquad h_i(s) = \sum_{j \neq i} v^j$$

and

$$(*) \qquad h_j(s) = -v^j \; \forall \, j \neq i.$$

It will therefore suffice to show that the last two equalities yield a contradiction. To get this contradiction, we shall first prove the following:

Auxiliary Proposition *If s is an NE for $(\mathring{\underset{\sim}{\omega}}, \mathring{\underset{\sim}{R}})$, then s is also an NE for $(\underset{\sim}{v}, \mathring{\underset{\sim}{R}})$.*

Proof (1) Consider agent i. We know that our rules never give to an agent more than the others have "put on the table." That is, for all s'_i,

$$h_i(s'_i, s_{)i(}) \lesseqgtr \sum_{j \neq i} w^j_j = \sum_{j \neq i} v^j.$$

But, by (0) above,

$$h_i(s) = \sum_{j \neq i} v^j.$$

Hence

$$h_i(s'_i, s_{)i(}) \lesseqgtr h_i(s) \quad \text{for all } s'_i,$$

and so, by the monotonicity of preferences, s_i is a Nash equilibrium strategy for agent i.

(2) Now consider any agent j other than i. Suppose s_j is not a Nash equilibrium strategy for j in the economy $(\underset{\sim}{v}, \mathring{\underset{\sim}{R}})$.

Then there must exist a strategy s'_j for j with the characteristic $(v^j, \mathring{R}_j)$ such that

$$(\alpha) \qquad h_j(s'_j, s_{)j(}) \; \mathring{P}_j h_j(s).$$

Now, since by the rules of the game $h_j(s^*) \geqq -w_j^j$ always, we have in particular

$$(\beta) \qquad h_j(s_j', s_{)j()}) \geqq -v^j = h_j(s)$$

where the last equality follows from (∗) above.

Since replacing $\geqq$ by = in (β) would contradict (α), if follows that $\geqq$ in (β) can be replaced by $\geqslant$, and (β) becomes

$$h_j(s_j', s_{)j()}) \geqslant h_j(s).$$

In view of the assumed strict monotonicity of preferences, the latter inequality implies

$$h_j(s_j', s_{)j()})\, \mathring{P}_j h_j(s)$$

where j's characteristics are $(\mathring{\omega}_j, \mathring{R}_j)$, and so s is not an NE in the economy$(\mathring{\underset{\sim}{\omega}}, \mathring{\underset{\sim}{R}})$. This contradiction of our initial hypothesis completes the proof of the Auxiliary Proposition.

We now return to the proof of Claim 2′. By Rule (a′), since s is unanimous as to endowments, we have

$$h(s) = g^{\underset{\sim}{v}}(\underset{\sim}{d}),$$

and

$$(\gamma) \qquad h_j(s) = g_j^{\underset{\sim}{v}}(\underset{\sim}{d}) \;\; \forall j \in N.$$

Now, by the Auxiliary Proposition, $\underset{\sim}{d}$ constitutes an NE in the game $g^{\underset{\sim}{v}}$ for $\mathring{\underset{\sim}{R}}$, and, by hypothesis, $g^{\underset{\sim}{v}}$ R-implements f. Therefore

$$g^{\underset{\sim}{v}}(\underset{\sim}{d}) = f(\underset{\sim}{v}, \mathring{\underset{\sim}{R}}),$$

and so

$$(\delta) \qquad g_j^{\underset{\sim}{v}}(\underset{\sim}{d}) = f_j(\underset{\sim}{v}, \mathring{\underset{\sim}{R}}).$$

Using in turn (δ), (γ), and (∗), we obtain

$$f_j(\underset{\sim}{v}, \mathring{\underset{\sim}{R}}) = g_j^{\underset{\sim}{v}}(\underset{\sim}{d}) = h_j(s) = -v^j, \quad \forall j \neq i;$$

hence,

$$f_j(\underset{\to}{v}, \underset{\to}{\mathring{R}}) = -v^j, \quad \forall j \neq i.$$

But this contradicts the hypothesis that f is NC, i.e., that

$$f_j(\underset{\to}{v}, \underset{\to}{\mathring{R}}) \geqslant -v^j, \quad \forall j \in N.$$

This contradiction implies that (+) holds, and hence that, by the strict monotonicity of preferences,

$$h_i(\tilde{s})\mathring{P}_i h_i(s).$$

So s is not an NE for $(\underset{\to}{\mathring{\omega}}, \underset{\to}{\mathring{R}})$. This completes the proof of Claim 2′.

Claim 3′ If there is no unanimity as to endowments, then there is no NE.

Proof We proceed as in the proof of Claim 3 except for (iii), which is replaced by the following:

(iii)′ Finally, suppose that s is not unanimous as to endowments and $M(s) = \emptyset$. Since, by the hypothesis of Claim 3′, not all announcements in s are the same, there exist agents i and j, $i \neq j$, with

$$w_i^i \neq w_j^i,$$

We now distinguish two subcases according to whether $\beta_j(s) = 0$ or $\beta_j(s) > 0$.

Subcase (iii.1)′ $\beta_j(s) = 0$.

Consider $\tilde{\tilde{s}}$ defined by

$$\begin{aligned} \tilde{\tilde{s}}_k &= s_k && \text{for all} \quad k \neq j \\ \tilde{\tilde{w}}_j^j &= \tfrac{1}{2} w_j^i && \\ \tilde{\tilde{w}}_j^r &= \tilde{\tilde{w}}_j^r && \text{for all} \quad r \neq j, \end{aligned}$$

and the second component of $\tilde{\tilde{s}}_j$ arbitrary (e.g., $\tilde{\tilde{d}}_j = d_j$).

We note that since s is not unanimous as to endowments and $M(s)$ is empty, rule (b)′ applies to s, and hence

$$h_j(s) = -w_j^j.$$

But, also, $\tilde{\tilde{s}}$ is not unanimous as to endowments because $\tilde{\tilde{w}}_i^i = w_i^i$, $\tilde{\tilde{w}}_j^i = w_j^i$ by construction,[36] and $w_i^i \neq w_j^i$ by hypothesis.

[36]Since $i \neq j$.

Also, $M(\tilde{\tilde{s}})$ is empty because $M(s)$ is empty, and the change from w_j^j to $\tilde{\tilde{w}}_j^j = \frac{1}{2}w_j^j$ (while $\tilde{\tilde{w}}_r^j = w_r^j$ for $r \neq j$) does not enlarge the set M. Hence Rule (b)′ also applies to $\tilde{\tilde{s}}$. Now, since $\tilde{\tilde{s}}_k = s_k$ for $k \neq j$, $\beta_j(s) = 0$ implies $\beta_j(\tilde{\tilde{s}}) = 0$. Therefore,

$$h_j(\tilde{\tilde{s}}) = -\tilde{\tilde{w}}_j^j.$$

But,

$$-\tilde{\tilde{w}}_j^j = -\frac{1}{2}w_j^j \geqslant -w_j^j,$$

because, by assumptions on messages, $w_j^j \geqslant 0$. Hence, by the assumption of strictly increasing preferences,

$$h_j(\tilde{\tilde{s}})\mathring{P}_j h_j(s).$$

So, $\tilde{\tilde{s}}$ is better than s for agent j, and hence s is not a Nash equilibrium.

Q. E. D.

Subcase (iii.2)′ $\beta_j(s) > 0$.

In this situation consider $\tilde{s}$, such that

$$\tilde{s}_k = s_k \quad \text{for all} \quad k \neq j$$

and

$$\tilde{s}_j^r = s_r^r \quad \text{for all} \quad r.$$

By construction, $\beta_j(\tilde{s}) = \beta_j(s) > 0$ and $\Sigma_{k \neq j}\beta_k(\tilde{s}) < \Sigma_{k \neq j}\beta_k(s)$. Also, $M(\tilde{s}) = M(s) = \emptyset$, so rule $(b)'$ applies to both $\tilde{s}$ and s. Therefore, $h_j(\tilde{s}) \geqslant h_j(s)$. And so, again by the assumption of strictly increasing preferences, s is not a Nash equilibrium.

Q.E.D.

3.2 *Destruction of Endowments*

In this section, we consider an alternative game, in which the agents may destroy a part of their endowment but are not able to withhold (conceal) any of it. D-implementability is defined analogously to W-implementability, with destruction replacing the withholding of endowments. We again assume pure exchange, with semi-positive initial endowments ($\mathring{\omega}_i \geqslant 0$) and strictly increasing preferences.

It then turns out that the outcome function introduced in Sect. 3.1.1, with the modification indicated under Claim 3,[37] D-implements any non-confiscatory (NC)[38] performance function when preferences are known to the designer.[39] Similarly, when f is monotone as well as NC, outcome functions of the type considered in Sect. 3.1.2 implement f when neither endowments nor preferences are known to the designer.

In what follows we state the result for the case of known preferences and indicate the modifications in the proof for W-implementation needed to make it valid for D-implementation. The theorem on D-implementability when both endowments and preferences are unknown is the same as part (1) of the theorem on W-implementability, with NC replacing IR.

The notation for strategies remains the same as in Sect. 3.1 but the interpretation differs. In particular, given s, agent i's final (total) holdings $H^i(s)$ equal $w_i^i + h_i(s)$ where w_i^i denotes i's (true) endowment after destruction. Similarly, for $i \neq j$, w_i^j denotes i's estimate of j's endowment after destruction. It is still assumed that $w_i^k \geqslant 0$ (i.e., $w_i^k \in \mathbb{R}^l_{+0}$) for all i, k in N. Hence, an agent cannot destroy all of his endowment.

The result for the case of known preferences is given by the following:

Theorem 3 *f is D-implementable (in NE) for $\mathring{\underset{\sim}{R}}$ if it is non-confiscatory (NC).*

Proof The proof is very much the same as that for W-implementability. In particular, in the former proof we used the fact (see Remark 3 in Sect. 3.1.1 that IR implies NC, while here only NC is assumed. We shall therefore only spell out those parts of the proof of D-implementability which differ significantly from the proof of W-implementability, with page references to the former proof.[40]

First, for the destruction game, we replace rule (b) by the following rule (b*), consisting of two parts, (b_1^*) and (b_2^*).[41]

In order to state these rules we must define numbers t_i ($i = 1, \ldots, n$) as follows. Consider $s = (s_1, \ldots, s_n)$ where $s_i = (s_i^1, \ldots, s_i^n) = (w_i^1, \ldots, w_i^n)$, with w_i^j—as before—denoting the value of j's endowment claimed by i (called i's *estimate* of j's endowment). Denote by $t^i(s)$ the number of distinct commodity space points among the elements $w_1^i, \ldots, w_n^i$, to be called the *number of estimates* (*in* s) *of* i*'s endowment*, and define $t(s) = \max\{t^1(s), \ldots, t^n(s)\}$. We shall call $t(s)$ the *number of estimates in* s.

[37] It may be that this same modification would also work in Sect. 3.1.1.

[38] f is *non-confiscatory* (NC) if $\forall i \in N$, $\forall \underset{\sim}{\mathring{\omega}} \in \mathbb{R}^{ln}_{+0}$, $\mathring{\omega}_i + f_i(\underset{\sim}{\mathring{\omega}}) \geqslant 0$.

[39] NC is however, not a necessary condition for D-implementability.

[40] Note, however, that for purposes of this section $z_i' \; R_i \; z_i''$ should be interpreted as $(w_i^i + z_i') \; R_i \; (w_i^i + z_i'')$.

[41] I.e., the formula of rule (b) for W-implementation applies.

The rule (b^*) then reads as follows

If $M(s) = \emptyset$, and $t(s) = 2$, then (b_1^*)

$$h_i(s) = [\beta_i(s)/\Sigma_{j \in N}\beta_j(s)] \cdot w(s) - w_i^i, \quad i \in N. \tag{\#}$$

If $M(s) = \emptyset$, and $t(s) > 2$, then (b_2^*)

$$h_i(s) = [\beta_i^*(s)/\Sigma_{j \in N}\beta_j^*(s)] \cdot w(s) - w_i^i, \quad i \in N, \tag{\#\#}$$

where

$$\beta_k^*(s) = 1 + \beta_k(s), \; k \in N.$$

The changes in the proof of the three claims, here labeled respectively with double primes, are indicated below.

Claim 1″ Here we must replace the part of the W-proof using the IR property of *f* by an argument using the NC property only. We therefore substitute for the last ten lines of the proof of Theorem 1[42,43] the following paragraph:

Since *f* is NC, and preferences are strictly increasing,

$$\mathring{\omega}_j + f_j(\underset{\rightarrow}{\mathring{\omega}})\mathring{P}_j 0.$$

But here

$$\mathring{\omega}_j + h_j(s) = \mathring{\omega}_j + f_j(\underset{\rightarrow}{\mathring{\omega}})$$

and

$$\widetilde{w}_j^i + h_j(\tilde{s}_j, s_{)j(}) = \widetilde{w}_j^j - \widetilde{w}_j^j = 0.$$

Hence,

$$(\mathring{\omega}_j + h_j(s))\mathring{P}_j(\widetilde{w}_j^j + h_j(\tilde{s}_j, s_{)j(}))$$

which contradicts our supposition (+ + +) in the proof of Claim 1 and in the proof of Theorem 1.

Remark 5 This argument would not be valid for withholding where, under $\tilde{s}$, the total final holdings equal $\mathring{\omega}_j - \widetilde{w}_j^j$ rather than 0.

---

[42]The paragraph starting with the words “Since *f* is IR ...”

[43]Ending with “Hence *s* is an NE.”

Claim 2″ Replace the sentence after (+) in the proof of Claim 2 in the proof of Theorem 1 with:

Since preferences are strictly increasing and $\tilde{w}_i^i \geqslant w_i^i$, the inequality (+) implies

$$(\tilde{w}_i^i + h_i(\tilde{s}))\mathring{P}_i(w_i^i + h_i(s)).$$

Claim 3″ In the absence of unanimity there is no NE.

Proof We consider three cases:

$$\text{(i)}''\ M(s) = N; \quad \text{(ii)}''\ M(s) \neq \emptyset,\ M(s) \neq N; \quad \text{(iii)}''\ M(s) = \emptyset.$$

(i)″ Suppose first that $M(s) = N$. Then consider $\tilde{s}$ with

$$\begin{aligned} \tilde{s}_k &= s_k \quad \text{for all} \quad k \neq 1, \\ \tilde{s}_1^q &= s_q^q \quad \text{for all} \quad q \in N. \end{aligned}$$

(That is, agent one accepts everyone's self-evaluation.)

Then

$$M(\tilde{s}) = \{1\}.$$

(This is proved exactly as in Theorem 1, Claim 3(*i*).)

Since $M(\tilde{s}) \neq \emptyset$, rule (c) applies. Therefore,

$$h_1(\tilde{s}) = 1 \cdot w(\tilde{s}) - \tilde{w}_1^1 = w(s) - w_1^1 = \sum_{i=1}^{n} w_i^i - w_1^1.$$

On the other hand, since $M(s) = N$, rule (c) also applies to s and yields

$$h_1(s) = \frac{1}{n} \sum_{i=1}^{n} w_i^i - w_1^1.$$

Since $\sum_{i=1}^{n} w_i^i \geqslant 0$ (by the rule $w_i \geqslant 0$), and $n > 1$, it follows that

$$h_1(\tilde{s}) \geqslant h_1(s).$$

Hence, since $\tilde{w}_1^1 = w_1^1$,

$$H_1(\tilde{s}) \geqslant H_1(s),$$

and, by strictly increasing preferences, $H_1(\tilde{s})\mathring{P}_1 H_1(s)$. So s is not an NE in case (i)″.

(ii)″ $M(s) \neq \emptyset$, $M(s) \neq N$.

Since $M(s) \neq \emptyset$ and there is no unanimity, rule (c) applies to s. Because $M(s) \neq N$, there is an agent $j \notin M(s)$ who, by rule (c), gets

$$h_j(s) = -w_j^j.$$

(Since this is the case of destruction, $H_j(s) = w_j^j + h_j(s) = w_j^j - w_j^j = 0$.)

Now suppose that agent j accepts everyone's self-evaluation. Thus

$$\tilde{s}_r = s_r \quad \text{for all} \quad r \neq j$$

and

$$\tilde{s}_j^q = s_q^q \quad \text{for all} \quad q.$$

Then (by the argument in Theorem 1)

$$M(\tilde{s}) = \emptyset.$$

Hence rule (c) does not apply. But neither does rule (a) because $\tilde{s}$ is not unanimous. (This is seen as follows: since $n \gtreqless 3$ and $M(s) \neq \emptyset$, there is a person $k \in M(s)$, $k \neq j$, and a person i, with $i \neq j$, $i \neq k$, such that

$$w_k^k \geqslant w_i^k;$$ [44]

hence $s_k \neq s_i$. But since $k \neq j$ and $i \neq j$, we have $\tilde{s}_k = s_k$ and $\tilde{s}_i = s_i$ by construction. Hence $\tilde{s}_k \neq \tilde{s}_i$, and so $\tilde{s}$ is not unanimous.)

Since $\tilde{s}$ is not unanimous and $M(\tilde{s}) \neq \emptyset$, rule (b*) applies to $\tilde{s}$.

For agents j, k, and i referred to above, we have

$$\beta_j(\tilde{s}) \gtreqless \|w_k^k - w_i^k\| > 0,$$

since $w_k^k \geqslant w_i^k$.

From $w(\tilde{s}) = w(s) \geqslant 0$, it follows that $h_j(\tilde{s}) = \dfrac{\beta_j(\tilde{s})}{\Sigma \beta_r(\tilde{s})} w(\tilde{s}) \geqslant 0$. On the other hand, $\beta_q^*(\tilde{s}) > 0$ by construction for all $q \in N$ and all $\tilde{s} \in s$, so that $\dfrac{\beta_j^*(\tilde{s})}{\Sigma \beta_r^*(\tilde{s})} w(\tilde{s}) \geqslant 0$. Hence, whether rule ($b_1^*$) or rule ($b_2^*$) applies, we have

[44] In fact $k \in M(s)$ means that $w_k^k \geqslant w_r^k$ for all r in $N/\{k\}$.

$$h_j(\tilde{s}) \geqslant -w_j^j = h_j(s).$$

(The last equality was exhibited above.)

But $\tilde{w}_j^j = w_j^j$, so $H_j(\tilde{s}) \geqslant H_j(s)$, and, by strictly increasing preferences, $H_j(\tilde{s})\mathring{P}_j H_j(s)$. Therefore, s is not an NE.

(iii)″ Finally, suppose there is no unanimity in s; hence the number of estimates $t(s)$ is at least 2, and $M(s) = \emptyset$. We distinguish two cases: case 1: The number $t(s)$ of estimates is 2; case 2: the number of estimates is at least three.

Consider first case 1 where the number of estimates is two, i.e., $t(s) = 2$. In this case, we distinguish two subcases, 1a, where all $\beta_k(s) > 0$, $k \in N$, and 1b, where not all $\beta_k(s)$ are positive (i.e., some are zero).

Subcase 1a Here $t(s) = 2$, and $\beta_k(s) > 0$ for all k in N. Since there is no unanimity, there are agents i and j such that, in s, $w_i^j \neq w_j^j$. Let i change his strategy from s_i to $\tilde{s}_i$, so that, in $\tilde{s}_i$, $\tilde{w}_i^j = w_j^j$, while other components of $\tilde{s}_i$ are the same as in s_i. Then $\beta_i(\tilde{s}) = \beta_i(s) > 0$, where the equality follows from the definition of $\beta_i(.)$ and the inequality holds by the hypothesis of case A. Also, $\beta_j(\tilde{s}) = \beta_j(s)$. But, since our theorem assumes n > 2, there is at least one agent r other than i or j, and for all such agents $\beta_r(\tilde{s}) < \beta_r(s)$. Clearly, $t(\tilde{s}) = t(s) = 2$, so rule (b_1^*) applies. It follows from the above properties of the β's that $h_i(\tilde{s}) > h_i(s)$, and hence s is not an NE.

Subcase 1b Here, still, $t(s) = 2$, but there exists some agent i such that $\beta_i(s) = 0$. Here the argument depends on whether i has a strategy $\tilde{s}_i$ such that $t(\tilde{s}) > 2$, with $\tilde{s}$ non-unanimous and leaving the set $M(\tilde{s})$ empty.

Consider first the *sub-subcase 1b′* where such a strategy $\tilde{s}$ is available to agent i. The situation with $\tilde{s}$ qualifies then under rule (b_2^*). Now since $\beta_i(s) = 0$, it follows from (#) that $H_i(s) = 0$. On the other hand, since $\beta_i^*(\tilde{s}) > 0$ by construction, it follows from (##) that $H_i(\tilde{s}) \geqslant 0$. Again, s is not an NE.

But suppose (*sub-subcase 1b*), that i has no strategy $\tilde{s}_i$ qualifying under rule (b_2^*). This can only happen if, under s, all agents other than i ("the crowd") are announcing identical profiles but different from that announced by i (the only "dissident").[45]

Here again there are two possibilities:

(i) The dissident and the crowd agree about i's endowment; i.e., $w_i^i = w_j^i$ for all $j \neq i$. Then i can adopt the strategy $\tilde{s}_i$ with $\tilde{w}_i^i = w_i^i$ and $\tilde{w}_i^j = w_j^j$ for all $j \neq i$. With others retaining their strategies from s, this will result in a unanimous $\tilde{s}$, so that

[45] For suppose that among agents other than i there are present at least two distinct profiles, say for agents j and k. If j and k disagree as to i's endowment, so that $w_j^i \neq w_k^i$, then i can choose $\tilde{w}_i^i \geqslant 0$, so that $\tilde{w}_i^i$ is simultaneously different from w_j^i and w_k^i and not higher than ω^i. On the other hand if j and k agree about i's endowment, then they must disagree about the endowment of some agent r other than i (since, by hypothesis, they are in disagreement). In that case agent i can choose $\tilde{w}_i^r$ that is different both from w_j^r and w_k^r (without removing any existing disagreements). In either case, the result is that $(\tilde{s}) > 2$, contrary to the hypothesis of 1.B".

$h_j(\tilde{s}) = f(\tilde{s})$. Since f is, by assumption in the Theorem, NC (non-confiscatory), it follows that $\tilde{w} + h_i(\tilde{s}) \geqslant 0$. On the other hand, since $t(s) = 2$, so that (b_1^*) applies, and $\beta_i(s) = 0$, formula (#) yields $w_i^i + h_i(s) = w_i^i + (-w_i^i) = 0$. Hence $\tilde{s}$ yields to agent i a bigger outcome, i.e., $\tilde{w}_i^i + h_i(\tilde{s}) \geqslant 0 = w_i^i + h_i(s)$, so—by the assumed monotonicity of preferences—s is not an NE.

(ii) The dissident and the crowd disagree about i's endowment; i.e., $w_i^i \neq w_j^i$ for all $j \neq i$. For any j in the crowd, $\beta_j(s) > 0$, $j \neq i$. Then any member of the crowd r (with $r \neq i$) can change from s_r to $\tilde{s}_r$ such that $\tilde{w}_r^i = w_i^i$, while other components of s_i remain unchanged. This does not change the number of disagreements, so $t(\tilde{s}) = 2$, continues to hold, $\tilde{s}$ is not unanimous, and $M(\tilde{s})$ is still empty. Hence formula (#) in (b_1^*) applies. Now $\beta_i(\tilde{s}) = \beta_i(s) = 0$ and $\beta_r(\tilde{s}) = \beta_r(s) > 0$.[46] But for any agent k other than i or r (i.e., any member of the crowd other than r) $\beta_k(\tilde{s}) < \beta_k(s)$. Thus for agent r, in the expression for $h_r(\tilde{s})$ in (#) the numerator is positive and the same as in $h_r(s)$ while the denominator is smaller; also, $w(\tilde{s}) = w(s)$. Hence $h_r(\tilde{s}) > h_r(s)$ and so s is not an NE.

We now proceed to case 2, with $t(s) > 2$, i.e., where the number of estimates in s is three or more. Hence formula (##) in rule (b_2^*) defines the outcomes under s.

Since $t(s) = \max\{t^1(s), \ldots, t^n(s)\} > 2$, there exist three agents i, j, and k such that among the three estimates w_i^i, w_j^i, and w_k^i no two are equal. Let now agent j change the endowment estimate profile from s_j to $\tilde{s}_j$ so that $\tilde{w}_j^p = w_j^p$ for all $p \neq i$, and $\tilde{w}_j^i$ such that $\tilde{w}_j^i$ is closer (in norm) to w_i^i than w_j^i was, while still $\tilde{w}_j^i \neq w_i^i$, and $\tilde{w}_j^i \neq w_k^i$. Hence formula (##) in rule (b_2^*) applies to $\tilde{s}$ as well as to s. (All components of $\tilde{s} = (\tilde{s}_1, \ldots, \tilde{s}_n)$, except $\tilde{s}_j$, are the same as those of s.)

Note that, since the components of $\tilde{s}$ other than $\tilde{s}_j$ are unchanged, we have $\beta_j(\tilde{s}) = \beta_j(s)$. Also, $\beta_i(\tilde{s}) = \beta_i(s)$. However, for r other than i or j, it is the case that $\beta_r(\tilde{s}) < \beta_r(s)$. The same relations hold respectively for the β^*'s. Hence in the quotient of formula (##) for $h_j(\tilde{s})$, the numerator is the same as for $h_j(s)$ and positive, while the denominator is smaller. It follows that $h_j(\tilde{s}) > h_j(s)$, and therefore s is not an NE. This completes the proof of Theorem 3.

Remark 6 If rule (b*) had not been substituted for rule (b), Claim 3″ section (iii)″, would no longer be true (when $M(s) = \emptyset$). This is shown by the following counterexample:

$$n = 3;\ l = 1;\ s = \begin{pmatrix} s_1 \\ s_2 \\ s_3 \end{pmatrix} = \begin{pmatrix} w_1^1 & w_1^2 & w_1^3 \\ w_2^1 & w_2^2 & w_2^3 \\ w_3^1 & w_3^2 & w_3^3 \end{pmatrix} = \begin{pmatrix} 1 & 5 & 4 \\ 2 & 3 & 4 \\ 1 & 3 & 4 \end{pmatrix}.$$

Assume that $\mathring{\omega}_1 = 1$, $\mathring{\omega}_2 = 3$, $\mathring{\omega}_3 = 4$. (So $w_i^i = \mathring{\omega}_i$, for $i = 1, 2, 3$.)

---

[46] $\beta_r(\tilde{s}) = \beta_r(s)$ because $\beta_r(\cdot)$ does not depend on r's statements concerning the others' endowments.

This s is not unanimous, and $M(s) = \emptyset$. If the mechanism were generally rules (a), (b), and (c), then rule (b) would apply here to s. Contrary to Claim 3″, this s is a Nash equilibrium.

Proof

(1) No $\tilde{s}$ can be unanimous (because if one player changes, the other two still disagree). So rule (a) will not apply to $\tilde{s}$.
(2) For every $\tilde{s}$, we have $M(\tilde{s}) = \emptyset$. This is so because, by hypothesis, every agent is already telling the truth about himself (i.e., he is destroying nothing), so he cannot raise his w_i^i; therefore $M(s) = \emptyset$ implies $M(\tilde{s}) = \emptyset$, So rule (c) will not apply to $\tilde{s}$.
(3) Hence rule (b) applies to any $\tilde{s}$ (as well as to s).
(4) We have $\beta_1(s) = \beta_2(s) = 0$ and $\beta_3(s) > 0$. By rule (b), agent 3 gets everything (i.e., $H_3(s) = w_1^1 + w_2^2 + w_3^3$), while the other two agents get nothing (i.e., $H_1(s) = H_2(s) = 0$). Certainly, therefore, agent 3 cannot do any better under any change of his strategy $\tilde{s}_3$.

As for agent 2, $H_2(\tilde{s}) = 0$ for any change of his strategy $\tilde{s}_2$, because $\tilde{s}_2$ does not enter $\beta_2(\cdot)$, so that $\beta_2(s_1, \tilde{s}_2, s_3) = 0$ for all $\tilde{s}_2$. Hence, agent 2 cannot do any better under any change of his strategy $\tilde{s}_2$.

Agent 1 is in exactly the same situation as agent 2.

So, no agent can do any better by unilateral strategy change, and hence s is a Nash equilibrium.

It is of some interest to see why and how the situation differs in the withholding game, in contrast to the destruction game being considered here.

In the withholding game, comments (1), (2), and (3) of the above proof remain valid. It also remains true that $\beta_1(s) = \beta_2(s) = 0$ and $\beta_3(s) > 0$. It is still true that agent 3 cannot improve his situation, but either of the other two agents can. Thus, in the W-game, let agent 2 choose $\tilde{w}_2^2 = \frac{1}{2} w_2^2$. (Recall that $w_2^2 = \mathring{\omega}_2$.) Then[47] $H_2^W(\tilde{s}) = \mathring{\omega}_2 - \tilde{w}_2^2 = \mathring{\omega}_2 - \frac{1}{2}\mathring{\omega}_2 = \frac{1}{2}\mathring{\omega}_2 \geqslant 0$, which is better than $H_2^W(s) = 0$. On the other hand, $H_2^D(\tilde{s}) = \tilde{w}_2^2 - \tilde{w}_2^2 = 0$, which is no improvement.

Remark 7 If rule (b) must be modified (as seen in Remark 1), it is natural to ask why it cannot be replaced by rule (b_2^*), rather than the more complex rule (b^*), which distinguishes between disagreement situations depending on whether there are more than two distinct strategy profiles. The answer is that rule (b_2^*) would be inappropriate in the proof of Claim 1″, while rule (b_1^*) does work.

Remark 8 We may note that we need not distinguish the cases $\beta_j(s) = 0$ from $\beta_j(s) > 0$ when rule (b_2^*) applies, since in both cases $\beta_j^*(s) > 0$, and the derived conclusion is due to changes in the denominator of $\beta_j^*(s) / \Sigma \beta_j^*(s)$, while the positive numerator remains constant. On the other hand, as in Theorems 1 and 2, we must distinguish these two cases when rule (b_1^*), which is identical with rule (b), does apply.

[47] The superscript refers to the game (W or D).

References

Groves, T., & Ledyard, J. (1977). Optimal allocation of public goods a solution to the 'free rider' problem. *Econometrica, 45*, 783–811.

Hong, L., & Page, S. (1994). Reducing informational costs in endowment mechanisms. *Economic Design, 1*(1), 103–117.

Hurwicz, L. (1972). On informationally decentralized systems. In C. B. McGuire & R. Radner (Eds.), *Decision and organization*. Amsterdam: North-Holland.

Hurwicz, L. (1986). On the implementation of social choice rules in irrational societies. In W. P. Heller, R. M. Starr, & D. A. Starrett (Eds.), *Essays in Honor of Kenneth J. Arrow* (Vol. 1). Cambridge: Cambridge University Press.

Hurwicz, L., Maskin, E., & Postlewaite, A. (1995). Feasible Nash implementation of social choice: Rules when the designer does not know endowments or production sets. In J. O. Ledyard (Ed.), *The economics of informational decentralization: Complexity, efficiency, and stability*. Boston, MA: Springer.

Maskin, E. (1977). *Nash equilibrium and welfare optimality*. Published in *Review of Economic Studies* 66: 1999, 23–38.

Nakamura, S. (1989). *Efficient feasible Nash mechanisms with production and externalities*. Doctoral dissertation, University of Minnesota.

Nakamura, S. (1990). A feasible Nash implementation of Walrasian equilibria in the two-agent economy. *Economics Letters, 34*, 5–9.

Page, S. (1989). *Reducing the dimension of the message space in HMP*. Unpublished note, dated July 1989.

Postlewaite, A. (1979). Manipulation via endowments. *Review of Economic Studies, 46*, 255–262.

Saijo, T. (1988). Strategy space reduction in Maskin's theorem: Sufficient conditions for Nash implementation. *Econometrica, 56*(3), 693–700.

Sertel, M. (1994). Manipulating Lindahl equilibrium via endowments. *Economics Letters, 34*, 5–9.

Wilson, R. (1976). *Competitive processes of price formation: A survey of several models*. Mimeo: Stanford University (IMSSS).

Design of Tradable Permit Programs Under Imprecise Measurement

John O. Ledyard

1 Introduction

The formal approach to mechanism design began with Hurwicz's 1960 paper. He recognized that the information about the economic environment, such as technological possibilities, preferences, and endowments, is dispersed among economic agents and that the "informational tasks entailed by the mechanism imply costs in real resources used to operate the mechanism" (Hurwicz and Reiter, 2006, p.1). He then focused on those informational tasks and searched for mechanisms that produced an efficient resource allocation and were informationally efficient in the sense that the messages were smaller than those of other possible mechanisms that produced an efficient resource allocation. In Hurwicz's 1972 paper, he took mechanism design to the next level by introducing the concept of incentive compatibility. Soon after, in 1973, Gibbard introduced the Revelation Principle which led theorists to focus on direct mechanisms, mechanisms whose messages are everything an agent knows about the environment, and to ignore the "informational tasks" entailed by such a mechanism. I believe more attention needs to be put on the costs of acquiring the information necessary to attain a measure of resource efficiency and the trade-off between the two types of efficiency. In this paper, I take a very small step in that direction.

I look at one of the informational tasks involved in cap and trade programs. Such programs possess a degree of informationally efficiency since they employ the price

J. O. Ledyard (✉)
Division of the Humanities and Social Sciences, California Institute of Technology, Pasadena, CA, USA
e-mail: jledyard@caltech.edu

W. Trockel (ed.), *Social Design*, Studies in Economic Design,
https://doi.org/10.1007/978-3-319-93809-7_8

mechanism to communicate needed information about the environment.[1] But there is another, usually ignored, part of the problem—identifying the size of the output decisions of producers in a commons. In order to enforce the rules of a cap and trade program, it is necessary to compare that output to the permits held by the agents. But it is often the case in practice that it is difficult or impossible to achieve valid measurement of that output. Instead imprecise measurement is possible and less expensive. In this paper I pursue what is lost, if anything, by relying on that imprecise measurement when using cap and trade to manage a commons.

The tragedy of the commons is well-known: an unmanaged, common resource will be over used and the benefits from its use will be lower than would be possible under a benevolent dictator. It is also well-known that, if use can be accurately and precisely measured, it is possible to design a tradable permit program such that, under a fairly general set of conditions, the permit market equilibrium allocation is efficient for the given aggregate permit level and everyone is better off after the permit program than before. I will refer to such a program as satisfactory.

In practice the implementation of a tradable permit program is often postponed or never undertaken because there does not exist an inexpensive technology able to measure violations accurately and precisely. However, sometimes there is an inexpensive technology available that can measure violations imprecisely. That is, there is random measurement error. Examples of imprecise measurement in commons problems are easy to find. I look at only two: fisheries and ground water.

In the management of fisheries, it is the catch that is often permitted. The easiest and most direct way to measure the catch is at the landing where the catch is unloaded and can easily be weighed. But that is actually only an inaccurate and imprecise measure of what has really been caught. For example, high grading, keeping the best and tossing the rest, would mean more fish are caught than measured. The best place to measure the actual catch is at the point of catch, but that is remote. Nevertheless, a variety of methods have been employed to try to get such measurements. Remote sensing through satellites, onboard human observers, and/or surveillance cameras are some of the technologies used. While these seem to improve measurement, there are still errors. The measurement is imprecise.

In the management of ground water, it is important to know the amount of water pumped from an aquifer. The obvious way to measure it is to place a meter at the pump. But in many situations, meters do not exist. Meters are costly and the manpower to read them is expensive. There have been several alternative measurement schemes proposed to be used in place of direct metering. One is to measure the electricity used by the pumps. The Turlock Irrigation District in California uses this method.[2] Another method is to use remote sensing by satellite or unmanned drones, using evapotranspiration as an indirect measure of water use.[3]

---

[1] I will side-step adverse selection incentive problems by assuming competitive behavior on the part of the economic agents.

[2] See also Zekri (2009) for a deeper discussion of this method.

[3] See Water Education Foundation (2015) for a discussion of this method.

Of course, both these technologies have errors in measurement. The measurement is imprecise.

To the best of my knowledge, the implications of imprecise measurement for permit markets have not been studied before. In the literature, measurement is always perfect—precise and accurate. Some studies have considered the case when, although the measurement is perfect, it is only done with some probability—thereby lowering the costs. This is often called random or imperfect monitoring.[4] The main conclusion of this research is that, with risk neutral producers, if the penalty rate per unit violation is high enough then the equilibrium allocations in a competitive permit market will be exactly the same as under perfect measurement. As long as the expected cost of a violation is high enough, producers in a commons will chose to hold permits exactly equal to their planned output. Because of that, the combination of perfect measurement and random monitoring can lead to a satisfactory program.

In this paper, I study whether imprecise measurement could be the basis for a satisfactory permit program. Unfortunately and perhaps not surprisingly, the answer is no. It is useful to understand why. With imprecise measurement, producers will buy more permits than their actual output in order to insure against the potential that measurement will indicate higher production than is actually the case. This leads to two types of resource inefficiency. One occurs because aggregate output will be less than the aggregate amount of permits. The second occurs if the errors in measurement are biased. If errors are not proportional to production, then aggregate output will not be efficiently allocated. Imprecise measurement can also make it difficult or impossible to guarantee that all producers will be at least as well off as they were before the program was put into place. This is because the aggregate cost of holding permits to insure against mis-measurement can be greater than the increase in aggregate benefits from better management of the commons. This can cause political problems.

In spite of these negative findings, there are some positive results. First, if measurement errors are proportional to use, then it is possible to design a tradable permit program with imprecise measurement such that the equilibrium aggregate output is efficiently allocated. Second, there are easily calculated subsidies that allow the design of a tradable permit program that guarantees that producers will be better off than they were before the program. Third, as the precision of the measurement approaches perfect accuracy, the equilibrium of the market with imprecise measurement approaches what it would be with perfect measurement and the size of the subsidies necessary to guarantee a Pareto-improvement decline to zero.

[4]See Malik (1990), Malik (1992), Stranlund and Dhanda (1999), Stranlund (2007), Murphy and Stranlund (2007), and Stranlund et al. (2008).

2 The Commons

A collection of producers, named $i = 1, \ldots, N$, are involved in a commons. Each chooses a level of production, q_i.[5] A producer's economic profits are $b_i(q_i, Q)$ where $Q = \sum_i q_i$. The fact that aggregate output affects an individual's profits creates the externality that is at the heart of the commons problem. There is a set of standard assumptions that guarantees this model of the commons is well-behaved.

Assumption 1 (Regular Commons)

(i) $b_i(q_i, Q)$ *is strictly concave and increasing in* q_i*, decreasing in* Q*, and continuously differentiable in* q_i *and* Q*.*
(ii) $\lim_{x \to 0} b_{iq}(x, Q) = \infty$.[6]
(iii) $b_{iqQ} \leq 0, \forall i$.

Assumption 1(i) is standard. Assumption 1(ii) ensures that no producer will ever want to drop out. This is not necessary for many of the results in this paper but does make them a little cleaner. Assumption 1(iii) is similar to a single crossing condition. It ensures that aggregate demand is downward sloping.

I assume throughout that producers behave competitively in the commons. That is, each producer acts as if their individual choice of production level will not affect aggregate production.

In the absence of collective management, autarky reigns.

Definition 1 (Autarkic Equilibrium) An Autarkic Equilibrium is (q^a, Q^a) where (i) q_i^a solves $\max_q b_i(q, Q^a), \forall i$ and (ii) $Q^a = \sum_i q_i^a$.

Economic efficiency is a standard benchmark and a desirable target for public policy.

Definition 2 (Efficient Allocation) An allocation $\tilde{q}$ is Efficient if and only if $(\tilde{q}, \tilde{Q})$ solves $\max_{q,Q} \sum_i b_i(q_i, Q)$ subject to $Q = \sum_i q_i$.

It is well known that Autarkic equilibria are generally not efficient. Thus, there is the opportunity to design a policy that will guide the producers on the commons to higher aggregate profits. A tradable permit program can be one such policy.

In Sect. 3, I provide some basic results for a permit market with accurate measurement. This material is reasonably well known and is provided to introduce the reader to the notation, concepts and prior results that I will later refer to. Section 4 contains the main results of this paper.

---

[5]The model would be essentially the same if the producers were choosing an input.

[6]I use the following notation for derivatives of functions: $f_x = \partial f(x, y, z)/\partial x$ and $f_{xy} = \partial^2 f(x, y, z)/\partial x \partial y$. The index i is the name of a producer and is not a variable.

3 Tradable Permits with Valid Measurement

A tradable permit program specifies an aggregate level of permits, L, allocates it to the producers, requires the producers to keep their production level to no more than the permits that they hold, and allows trading of those permits. The initial allocation of permits is l^o, where $\sum l_i^o = L$.

A valid measurement technology is both accurate and precise. With valid measurement of q, a tradable permit program can be enforced by monitoring and imposing a penalty for producing more than the permits held. I assume that the penalty is a function only of the level of violation $v_i = q_i - l_i$. This is standard in the literature.[7] The penalty to be paid by i is indicated by $P(v_i)$. One typical and simple form of the penalty function has $P(v_i) = \max\{0, av_i\}$ where a is a positive constant. I allow a wider range of possibilities.

I assume producers behave competitively in the permit market.[8] Let p be the market price of permits.

Definition 3 (Market Equilibrium under Valid Measurement) A Permit Market Equilibrium under Valid Measurement is (q^*, v^*, p^*, Q^*) such that

(i) each producer i chooses (q_i^*, v_i^*) to solve

$$\max_{(q_i, v_i)} b_i(q_i, Q^*) - p^*(q_i - v_i - l_i^o) - P(v_i),$$

(ii) $Q^* = \sum_i q_i^*$, and
(iii) $\sum_i (q_i^* - v_i^*) = L$.

In Definition 3, (q, Q) and v can be solved for independently since they are separable. We use this fact to define a supply of output and a demand for violations.

Definition 4 (Supply and Demand under Valid Measurement)

(a) Supply under valid measurement is $[q^V(p), Q^V(p)]$ where
$q_i^V(p) \in \arg\max_{q_i} b_i(q_i, Q^V(p)) - pq_i$ and $Q^V(p) = \sum_{i=1}^N q_i^V(p)$.
(b) Demand under valid measurement is $[v^V(p), V^V(p)]$ where
$v_i^V(p) \in \arg\max_v pv_i - P(v_i)$ and $V^V(p) = \sum_{i=1}^N v_i^V(p)$.

The market equilibrium price satisfies $Q^V(p^*) - V^V(p^*) = L$. To ensure the market is well-behaved, I impose a regularity assumption.

[7] Of course, P could depend on more than just v_i. For example, P could depend on the percentage violation so that the penalty is $P(q_i/l_i)$. But when $\partial P/\partial q_i + \partial P/\partial l_i \neq 0$, permit market equilibria will generally not be efficient. Thus, the design choice is usually $P(v_i)$.

[8] This is rarely true in practice. Even if there is a large number of producers, most extant permit markets are disorganized and thinly traded. They tend to violate the Law of One Price and, therefore, traders' behaviors are not really competitive. There are ways to design a trading mechanism to avoid this, but that rarely happens. I leave the design of those markets to another paper.

Assumption 2 (Regular Accurate Measurement)

(i) $P(v_i) = 0$ *on* $v_i \leq 0$. $P(v_i)$ *is convex, increasing, and continuously differentiable on* $v_i \geq 0$.
(ii) $L > Q^V(P_v(0^+))$.[9]

Assumption 2 (i) is a slight generalization of the linear penalty, $\max\{0, av_i\}$.

Assumption 2 (ii) is needed to ensure that the penalty is strong enough so that producers do not want violations in equilibrium.

Result 1 *Under Assumptions 1 and 2,* $Q_p^V(p) < 0$ *and* $V_p^V(p) \geq 0$.[10]

The geometry of this market is displayed in Fig. 1 in Sect. 6.1.

3.1 Efficiency and Political Viability

Permit markets mimic competitive markets which, under the appropriate conditions, produce allocations that are efficient. But permit markets do not necessarily produce efficient allocations unless the supply of permits, L, is exactly right. Instead, permit markets produce allocations that are efficient given the total amount of permits, L.

Definition 5 (Efficiency Given X) The allocation $\tilde{q}$ is Efficient given X if and only if $\tilde{q}$ solves $\max_q \sum_i b_i(q_i, X)$ subject to $\sum_i q_i = X$.

If the total number of permits, L, is chosen appropriately, aggregate profits after the program is implemented will be higher than before. But producers won't receive those increases unless the program is actually adopted and implemented. A satisfactory program must be politically viable.

If the permit program can be designed in a way that all producers are better off after the introduction of the permit program than they are in the autarkic equilibrium, then the program will be more likely to be adopted. In the language of mechanism design, such a property of the design is called voluntary participation or individual rationality. The well-known result is that all producers will be better off at the permit market equilibrium than in autarky as long as the distribution of the initial permits, l^o, is such that all producers are at least as well off at l_i^o as they are in autarky.[11]

[9] $P_v(0^+) = lim_{x \to 0^+} P_v(x)$.

[10] The proofs are omitted since they are standard and straightforward. For details see, e.g., Ledyard (2018).

[11] Since it is rare that such a distribution is unique, there may still be serious political bargaining over the allocation of L.

Result 2 *Let (q^*, v^*, p^*, Q^*) be a permit market equilibrium. Under Assumptions 1 and 2:*

A) Efficiency

(i) q^* is efficient given L, $Q^* = L$, and $v^* = 0$.[12]
(ii) The permit market equilibrium is independent of the initial distribution, l^o, and only depends on L.

B) Voluntary Participation

(i) For any initial allocation of permits, l^o, such that $b_i(l_i^o, L) > b_i(q^a, Q^a), \forall i$, it will be true that $b_i(q_i^*, Q^*) - p^*(q_i^* - v_i^* - l_i^o) - P(v_i^*) > b_i(q_i^a, Q^a), \forall i$.
(ii) Let q^e be efficient given L and $Q^e = \sum_i q_i^e$. There exists an l^o satisfying (i) if and only if $\sum_i b_i(q_i^e, Q^e) > \sum_i b_i(q_i^a, Q^a)$.

The proofs are omitted since they are standard and straightforward.[13]

Result B(i) holds because the producer has been put into an initial position at least as good as autarky and, by choosing $q_i = l_i^o$, she can protect that position. Anything she then decides to do will be at least as good. If the commons effect of Q is large enough, as in fisheries, then letting l^o be based on autarkic output, so that $l_i^o = \frac{q_i^a}{Q^a} L$, will often be sufficient.

Result B(ii) holds because of the quasi-linearity of profits. Locally, if $L < Q^a$, there exists an l^o satisfying (i).

To summarize, with a valid measurement technology it is possible to design a tradable permit program such that, under a fairly general set of conditions, the market equilibrium is efficient for the given permit total and everyone is better off than with autarky.

4 Tradable Permits with Imprecise Measurement

As shown in the previous section, under ideal conditions, the benefits of using a tradable permit system to manage an over-used commons are increased aggregate profits and political viability. But, often implementation of such a system is postponed or never undertaken because the conditions are not ideal. One reason for this can be the absence of an inexpensive technology able to provide valid measurements of violations. However, there often is an inexpensive technology which is approximately valid. In this section, I study the possibilities for the design

[12]If the penalty is weak so that $L < Q^V(P_v(0^+))$, then $v_i^* > 0$ and $Q^* > L$. In this case, q^* is not efficient given L. But q^* will be efficient given Q^*. See, e.g., Malik (1990).

[13]For details, see, e.g., Ledyard (2018).

of a tradable permit system when the measurement technology involves an indirect measure of q_i that contains statistical uncertainty. Such a measurement is imprecise.

I model imprecise measurement technologies as follows. The indirect measure of producer i's output is $w_i = q_i + \epsilon$ where ϵ is the measurement error. I assume ϵ is a random variable with density $f(\epsilon, q_i)d\epsilon$. Further, I assume that $\mathcal{E}(\epsilon|q_i) = \int \epsilon f(\epsilon, q_i)d\epsilon = 0$; that is, the measurement technology is accurate. This is pretty much without loss of generality since, if $\mathcal{E}(\epsilon|q_i)$ is not accurate but is a known function of q_i, then one can adjust the penalty function to account for the inaccuracy.[14] Finally, I assume that the measurement error is the same for every producer. This is not entirely without loss of generality.[15]

The measured violation is $w_i - l_i = \epsilon + q_i = v_i + \epsilon$. The penalty to be paid by the producer, based on the measured violation, is $\rho(w_i - l_i)$. When the producer chooses her output, she faces an *expected penalty payment* of $\mathcal{P}(v_i, q_i) = \int \rho(v_i + \epsilon) f(\epsilon, q_i)d\epsilon$.

With imprecise measurement, the definition of market equilibrium in Sect. 3 needs to be slightly altered to allow for the fact that the expected penalty now depends on q_i.

Definition 6 (Market Equilibrium Under Imprecise Measurement) A permit market equilibrium under imprecise measurement is (q^*, v^*, p^*, Q^*) such that

(i) each producer i, chooses (q_i^*, v_i^*) to solve

$$\max_{(q_i, v_i)} b_i(q_i, Q^*) - p^*(q_i - v_i - l_i^o) - \mathcal{P}(v_i, q_i)$$

(ii) $Q^* = \sum_i q_i^*$,
and
(iii) $\sum_i q_i^* - \sum_i v_i^* = L$.

(q, Q) and v are no longer independent as they were under valid measurement. But we can still define demand and supply functions.

Definition 7 (Supply and Demand Under Imprecise Measurement) Supply under imprecise measurement is $[q^I(p), Q^I(p)]$ and Demand under imprecise

[14] See Ledyard (2018).

[15] Differences across producers might occur in water markets for different crops or different irrigation technologies, and in fishing markets for different gear types. These differences would affect the efficiency results below to some extent. I leave it to the reader to work out those implications.

measurement is $[v^I(p), V^I(p)]$ where

$$q_i^I(p) \in \arg\max_{q_i} b_i(q_i, Q^I(p)) - pq_i - \mathcal{P}(v_i^I(p), q_i) \text{ and } Q^I(p) = \sum_{i=1}^{N} q_i^I(p).$$

$$v_i^I(p) \in \arg\max_{v_i} pv^i - \mathcal{P}(v_i, q_i^I(p)) \text{ and } V^I(p) = \sum_{i=1}^{N} v_i^I(p).$$

The market equilibrium price satisfies $Q^I(p^*) - V^I(p^*) = L$.

To ensure the market is well-behaved, I impose a regularity assumption.

Assumption 3 (Regular Imprecise Measurement)

A. *Errors*

(i) *The indirect measure of output is* $w_i = q_i + \delta h(q_i), \forall i$, *where* δ *is a random variable with density* $g(\delta)$ *and* $\mathcal{E}[\delta] = \int \delta g(\delta) d\delta = 0$.
(ii) *There is a* $\underline{\delta} > 0$ *such that* $\delta \geq -\underline{\delta}$ *and* $\frac{q}{h(q)} \geq \underline{\delta}$ *for all* $q \geq 0$.[16]
(iii) *The size of the errors,* $h(q)$, *is positive, non-decreasing, continuously differentiable, and convex in* q. *That is,* $h(q_i) > 0$, $h_q(q_i) \geq 0$ *and* $h_{qq}(q_i) \geq 0$, *for all* $q_i > 0$. *Also* $h(0) = 0$.

B. *Enforcement*

(i) $\rho(m_i) = \max\{0, am_i\}$ *with* $a > 0$, *where* $m_i = w_i - l_i$ *is the measured violation.*
(ii) $L > Q^I(P_v(0+)) = Q^I[a(1 - G(0))]$.

Assumption 3A provides a structure that is like a single crossing property. It keeps the densities of the errors under control when q changes. Increasing $h(q)$ is then like applying a mean-preserving spread to the error.

Assumption 3B is similar to Assumption 2. Assumption 3B(ii) ensures that penalties are strong enough that producers will not want violations in equilibrium.

Under Assumption 3, the expected penalty function becomes

$$\mathcal{P}(v_i, q_i) = a \int_{-\frac{v_i}{h(q_i)}} [v_i + \delta h(q_i)] g(\delta) d\delta. \tag{1}$$

The geometry of a permit market under imprecise measurement satisfying Assumption 3 is illustrated in Fig. 2 in Sect. 6.2. The crucial fact is that $Q^I(p)$ is downward sloping.

[16]This ensures that $w_i \geq 0$. If errors are proportional to output with $h(q) = \tau q$ for some $\tau > 0$, then this will be true if $\tau \underline{\delta} \leq 1$.

Result 3 *Under Assumptions 1 and 3, $Q^I_p(p) < 0$.*

Proof Let $k_i = -\frac{v_i}{h(q_i)}$, and $R(k_i) = \int_{k_i} \delta g(\delta) d\delta$. The first order conditions for a solution to Definition 6(i) and (ii) for a given p are:

$$b_{iq}(q_i, Q) - p - h_q(q_i) a R(k) = 0 \tag{2}$$

$$p - a(1 - G(k)) = 0 \tag{3}$$

$$v_i + h(q_i)k = 0 \tag{4}$$

$$\sum_i q_i - Q = 0 \tag{5}$$

$$\sum_i q_i - \sum_i v_i = L \tag{6}$$

To solve for the partial equilibrium comparative statics, differentiate (2)–(5) with respect to p to get:

$$b_{iqq} q^I_{ip} + b_{iqQ} Q^I_p - 1 - h_{qq} a R q^I_{ip} - h_q a R_k k^I_p = 0 \tag{7}$$

$$1 + agk^I_p = 0 \tag{8}$$

$$v^I_{ip} + h_q k q^I_{ip} + hk^I_p = 0 \tag{9}$$

$$[\sum_i q^I_{ip}] - Q^I_p = 0. \tag{10}$$

From (8), $k^I_p(p) = -\frac{1}{ag(k)} < 0$. By definition $R_k = -k^I g$. Since $v^I_i < 0, k^I > 0$. Solving (7) for q^I_{ip} yields

$$q^I_{ip} = \frac{1 + h_q k - b_{iqQ} Q^I_p}{b_{iqq} - h_{qq} a R} = \alpha_i(p) - \beta_i(p) Q^I_p(p).$$

From (10)

$$Q^I_p = \sum_i \alpha_i - (\sum_i \beta_i) Q_p.$$

Solving gives

$$Q^I_p = \frac{\sum_i \alpha_i}{1 + \sum_i \beta_i}.$$

Since $\alpha_i < 0$ and $\beta_i \geq 0$, the result follows. □

4.1 *Efficiency and Political Viability: Impossibility*

Although the equilibrium equations for valid measurement and imprecise measurement look very similar, the latter create serious problems for both the efficiency and the political viability of permit programs.

4.1.1 Efficiency

Efficiency given L requires that, at a permit market equilibrium (q^*, v^*, p^*, Q^*), $v_i^* = 0$ for all producers. With imprecise measurement, under Assumptions 1 and 3, this will not be true. Instead, $v_i^* < 0$ for all i and, therefore, $Q^* < L$ in equilibrium.

Result 4 *Under Assumptions 1 and 3, at a permit market equilibrium* (q^*, v^*, p^*, Q^*), $0 < p^* < a[1 - G(0)]$, $v_i^* < 0, \forall i$, *and* $Q^* < L$.

Proof ($p^* < a[1 - G(0)]$) Assume the contrary. Then $v_i^* > 0$ and, so, $Q^* > L$. But $Q(p^*) \leq Q(a[1 - G(0)]) < L$ which is a contradiction.

($v_i^* < 0$) From profit maximization $v_i^* \in \arg\max_{v_i} p^* v_i - \int_{\frac{-v_i}{h(q_i)}} a(v_i + \delta h(q_i))g(\delta)d\delta$. The first order condition for this is $p^* = a\int_{-\frac{v_i}{h(q_i)}} g(\delta)d\delta = a[1 - G(-\frac{v_i}{h(q_i)})]$. Since $p^* < a[1 - G(0)]$, it follows that $v_i^* < 0$. This implies $Q^* < L$. □

With imprecise measurement, producers will pay a penalty even if $q = l$. They can reduce that expected penalty cost by producing less than the permits they hold. Because of this, if regulators issue a total of permits L equal to the desired aggregate output target, Q, actual output will be less than Q and, therefore, not efficient given Q.

Knowing that equilibrium output is less than L, one might ask whether regulators could change L to some other $\hat{L}$ so as to move $Q^I(p^*)$ to L. For the regular case, the answer is yes. Consider Fig. 2. The amount of permits that works is $\hat{L} = Q^I(p^V) - V^I(p^V) > L$, where p^V is the equilibrium price under perfect measurement. There must be enough permits to allow the producer to use the extras to insure herself against a faulty measurement. That is, an additional amount $\hat{L} - L^* = -V^I(p^V)$ must be added to L^*. Of course, to compute $\hat{L}$ the regulator must know, prior to the implementation of the program, the functions $Q^I(p)$ and $V^I(p)$ and the price p^V, which they do not.

4.1.2 Political Viability

With valid measurement, by Result 2B, there are distributions of the initial permit, l^o, such that all producers will be at least as well off in the permit market equilibrium as they would have been without the program. With imprecise measurement, this may not be true because a producer's expected penalties are positive even if she chooses to buy licenses equal in number to her production levels. That is, $\mathcal{P}(0, q_i) = ah(q_i) \int_0 \delta g(\delta) d\delta > 0$.

To see the effect of imprecise measurement on political viability, consider the following result which adapts Result 2B to imprecise measurement.

Result 5 *Under Assumptions 1 and 3,*

(i) for any initial allocation of permits, l^o*, such that* $b_i(l_i^o, L) - \mathcal{P}(0, l_i^o) > b_i(q_i^a, Q^a)$*, at the market equilibrium* (q^*, v^*, p^*, Q^*)

$$b_i(q_i^*, Q^*) - p^*(q_i^* - v_i^* - l_i^o) - \mathcal{P}(v_i^*, q_i^*) > b_i(q_i^a, Q^a), \forall i,$$

and

(ii) there exists an l^o *satisfying (i) if and only if there is a* $\hat{q}$ *such that* $\sum_i \hat{q} = L$ *and* $\sum_i b_i(\hat{q}, L) - \sum_i b_i(q^a, Q^a) > \sum_i \mathcal{P}(0, \hat{q}_i)$*.*

Proof (i) Let (q^*, v^*, p^*, Q^*) be the permit market equilibrium. Then $b_i(q_i^*, Q^*) - p^*(q_i^* - v_i^* - l_i^o) - \mathcal{P}(v_i^*, q_i^*) \geq b_i(l_i^o, Q^*) - p^*(l_i^o - 0 - l_i^o) - \mathcal{P}(0, l_i^o) \geq b_i(l_i^o, L) - \mathcal{P}(0, l_i^o) > b_i(q_i^a, Q^a)$. The first inequality follows from profit maximization. The second follows because $L > Q(a[1 - G(0)])$ implies $Q^* \leq L$. The last follows from the assumption on l^o. □

Comparing this to Result 2 under valid measurement, it is easy to see that imprecise measurement introduces a potential barrier to voluntary participation. If either the penalty rate or the errors are large, the expected penalty with no violations, $ah(q_i) \int_0 \delta g(\delta) d\delta$, is large. Then, if the gains from improving the management of the commons are not very large, it may be difficult or impossible to find an appropriate l_i^o.

4.2 Efficiency and Political Viability: Possibility

In spite of the difficulties described in the previous section, some positive results can be found.

4.2.1 Efficiency

Although there is no general efficiency result when measurement is imprecise, if measurement errors are proportional to output, then even though q^* is not efficient given L, q^* is efficient given Q^*. That is, production will be organized efficiently given the aggregate output level.

Assumption 4 (Errors Are Proportional to Output)

$$h_{qq}(q_i) = 0, \forall q, \forall i.$$

This assumption along with Assumption 3 A(iii) imply that $h(q) = \tau q$ for some $\tau > 0$.

Result 6 *Under Assumptions 1, 3, and 4, q^* is efficient given Q^*.*

Proof Under Assumptions 1 and 3, the FOC for a permit market equilibrium are found in (2)–(6). From (3) it follows that k^* is the same for all i. Therefore, from (2) and (4), q^* is efficient given Q^* if and only if the $h_q(q_i^*)$ are equal for each i. This is true under Assumption 4 since $h_{qq}(q_i) = 0$. □

Without Assumption 4, the equilibrium will not be efficient given Q^*. It is easy to see why. If $h_{qq} > 0$, then $h_q(\hat{q}) > h_q(\tilde{q})$ iff $\hat{q} > \tilde{q}$. Let $\bar{h}_q = \frac{\sum h_q(q_i^*)}{N}$ be the average value of h_q in equilibrium. If $h_q(q_i^*) > \bar{h}_q$, then in equilibrium q_i^* is relatively smaller than desired for efficiency. If $h_q(q_i^*) < \bar{h}_q$, then in equilibrium q_i^* is relatively larger than desired for efficiency. The fact that the imprecise measurement errors are getting worse as q gets larger means that those who produce a large amount will have more incentive to cut back on their output to avoid the penalties from mis-measurement. Non-proportional measurement errors interfere with efficiency given Q^*.

4.2.2 Political Viability

There is a way to design around this problem by using the same insights employed for valid measurement. Put the producer in an initial position that is at least as good as autarky and that she can protect. If the regulator gives each producer an initial subsidy of $P_i^o = \mathcal{P}(0, l_i^o)$, and if the producer then chooses $(q_i, v_i) = (l_i^o, 0)$, she can guarantee that her expected penalty less the subsidy will be zero. That plus the appropriate initial permit allocation guarantees voluntary participation.

Result 7 *Under Assumptions 1 and 3, for any initial allocation of permit, l^o, such that $b_i(l_i^o, L) > b_i(q_i^a, Q^a)$ for all i, there are lump-sum payments $P_i^o = \mathcal{P}(0, l_i^o)$ such that the market equilibrium (q^*, v^*, p^*, Q^*) satisfies*

$$b_i(q_i^*, Q^*) - p^*(q_i^* - v_i^* - l_i^o) - \mathcal{P}(v_i^*, q_i^*) + P_i^o > b_i(q_i^a, Q^a), \forall i.$$

Proof Let (q^*, v^*, p^*, Q^*) be the permit market equilibrium. Then $b_i(q_i^*, Q^*) - p^*(q_i^* - v_i^* - l_i^o) - \mathcal{P}(v_i^*, q_i^*) + P_i^o \geq b_i(l_i^o, Q^*) - p^*(l_i^o - 0 - l_i^o) - \mathcal{P}(0, l_i^o) + P_i^o \geq b_i(l_i^o, L) > b_i(q_i^a, Q^a)$. The first inequality follows from profit maximization. The second follows because $L > Q(a[1 - G(0)])$ implies $Q^* \leq L$. The last follows from the assumption on l^o and the fact that $\mathcal{P}(0, l_i^o) = P_i^o$. □

If the measurement technology is well understood, then $P_i^o = ah(l_i^o) \int_0 \delta g(\delta) d\delta$ is easy to calculate.

Although the subsidies in Result 7 help solve the problem of generating voluntary participation, they create a new problem for the designers. If the subsidies are deployed and equilibrium output is less than or equal to L, then producers will be receiving a positive net aggregate subsidy. The permit program will not be self-financing in expected value.

Result 8 *Under Assumptions 1, 3, and 4*, $\sum_i \mathcal{P}(0, l_i^o) > \sum_i \mathcal{P}(v_i^*, q_i^*)$.

Proof Under Assumptions 3 and 4, in equilibrium $\sum_i[\mathcal{P}(0, l_i^o) - \mathcal{P}(v_i^*, q_i^*)] = a\tau l_i^o \int_0 \delta g(\delta)d\delta - a\tau q_i^* \int_{\frac{-v^*}{h(q_i^*)}} \delta g(\delta)d\delta$. By (3), $\int_{\frac{-v^*}{h(q_i^*)}} \delta g(\delta)d\delta = \frac{p^*}{a}$ is the same for every i. Therefore the aggregate net subsidy is $S = \sum_i[\mathcal{P}(0, l_i^o) - \mathcal{P}(v_i^*, q_i^*)] = a\tau[\sum_i l_i^o] \int_0 \delta g(\delta)d\delta - a\tau[\sum_i q_i^*] \int_{\frac{-v^*}{h(q_i^*)}} \delta g(\delta)d\delta = a\tau[L \int_0 \delta g(\delta)d\delta - Q^* \int_{\frac{-v^*}{h(q_i^*)}} \delta g(\delta)d\delta]$.

Under Assumption 3, $v_i^* < 0$ which implies that $\int_0 \delta g(\delta)d\delta > \int_{\frac{-v^*}{h(q_i^*)}} \delta g(\delta)d\delta$. Also $L > Q^*$. Therefore, $S > 0$. □

The voluntary participation of the producers has been bought with funding from outside of the market—presumably from taxpayers. This creates a new friction against adoption. Nevertheless, such subsidies may be justified. The rationalization of the management of the commons can lead to gains, not only for producers, but also for the consumers of the products of the commons. This is certainly true for fisheries, and probably true for many other situations. Using some of these gains to ease the transition to permit markets might be a very good bargain for all concerned.

4.3 Precision

In this section, I explore what happens to efficiency and political viability as the measurements become more precise. The easiest way to do that is to introduce a precision parameter, η and replace $h(q_i)$ with $\eta h(q_i)$ in the measurement model. Thus, $w_i = q_i + \eta h(q_i)\delta$. I will assume $\eta \leq 1$. With this small change, the measurement error is $w_i - q_i = zh(q_i)$ where $z = \delta\eta$. The expected value of z is 0 and the variance of z is $\eta Var(\delta)$. As η decreases, precision increases.

The penalty that a producer now faces, given output q_i and actual violation $v_i = q_i - l_i$, is

$$\mathcal{P}^\eta(v_i, q_i) = a \int_{-\frac{v_i}{\eta h(q_i)}} [v_i + \delta h(q_i)\eta]g(\delta)d\delta. \tag{11}$$

As one might expect, increased precision improves efficiency and eases political viability. In the limit as $\eta \to 0$, the equilibrium under imprecise measurement approaches the equilibrium under valid measurement. The permit market equilibrium becomes efficient given L and no subsidies are required for voluntary participation.

Result 9 *Let* $(q(\eta), v(\eta), p(\eta), Q(\eta))$ *be the market equilibrium under imprecise measurement with the expected penalty* $\mathcal{P}^\eta(v_i, q_i) = a \int_{-\frac{v_i}{\eta h(q_i)}} [v_i + \delta h(q_i)\eta]g(\delta)d\delta$

and let (q^*, v^*, p^*, Q^*) *be the market equilibrium under valid measurement with the penalty* $P(v_i) = \max\{av, 0\}$. *Under Assumptions 1 and 3,*

(i) $\lim_{\eta\to 0}(q(\eta), v(\eta), p(\eta), Q(\eta)) = (q^*, v^*, p^*, Q^*)$ *and*
(ii) $\lim_{\eta\to 0}\mathcal{P}^\eta(0, l_i^o) = 0$.

Proof First note that, as precision increases, the penalty function under imprecise measurement approaches the penalty function under valid measurement. For any q_i such that $0 < q_i < \infty$,

$$\lim_{\eta\to 0}\mathcal{P}^\eta(v_i, q_i) = av \text{ if } v \geq 0 \tag{12}$$

$$\lim_{\eta\to 0}\mathcal{P}^\eta(v_i, q_i) = 0 \text{ if } v \leq 0. \tag{13}$$

The first order conditions for a market equilibrium under imprecise measurement are:

$$b_{iq}(q_i, Q) - p - \mathcal{P}^\eta_{q_i}(v_i, q_i) = 0 \tag{14}$$

$$p - \mathcal{P}^\eta_{v_i}(v_i, q_i) = 0 \tag{15}$$

$$\sum_i q_i - Q = 0 \tag{16}$$

$$\sum_i q_i - \sum_i v_i = L \tag{17}$$

It is easy to show that $\mathcal{P}^\eta_{q_i}(v_i, q_i) \to 0$ as $\eta \to 0$. Also $\mathcal{P}^\eta_{v_i}(v_i, q_i) \to a$ as $\eta \to 0$ if $v > 0$ and $\mathcal{P}^\eta_{v_i}(v_i, q_i) \to 0$ as $\eta \to 0$ if $v^i < 0$. Result 9(i) follows from the implicit function theorem.

$\mathcal{P}^\eta(0, l_i^o) = \int_0 \delta h(l_i^o)\eta g(\delta)d\delta = \eta h(l_i^o)\int_0 \delta g(\delta)d\delta = \eta\mathcal{P}(0, l_i^o)$. Result 9(ii) follows. □

5 Final Comments

In this paper I have explored the geometry, efficiency, and political viability of permit market equilibria when enforcement can only use imprecise measurement. I provide a set of conditions (Assumption 3) such that the geometry corresponds to most economists' intuitions. These conditions are similar in spirit to those needed under accurate measurement (Assumption 2).

Unlike the standard results when there is accurate and precise measurement of use, when measurement is imprecise permits will not be efficiently allocated. There are two sources of the inefficiency: aggregate output is less than the number of permits as producers overbuy to insure against mis-measurement and there can be bias in the efficient allocation of output as mis-measurement can have different marginal effects on producers depending on their size. Further, when measurement is imprecise, it may not be possible to find an initial allocation of permits such that all firms are better off than they were before implementation of the program. The reason is that all firms will face positive expected penalty payments, even if they have negative actual violations. This could seriously affect the political viability of the program, even if measurement were free.

But there are also some positive results. First, if the errors are proportional to output, then aggregate output will be efficiently allocated among firms. The only inefficiency will be that aggregate output is less than the target, the number of permits. This can be compensated for by issuing more permits than the target output. Second, there are easily calculated individual subsidies that will make it possible to find an initial allocation of permits to guarantee voluntary participation. This will mean that taxpayers must subsidize the implementation of the program. But if the benefits from the program are large enough and some of these benefits accrue to consumers, then the program may still be politically viable.

Finally, if the measurement errors are small, inefficiencies will be small and the subsidies required for voluntary participation will be small. In this case, it may well be reasonable to proceed with the inexpensive, imprecise measurement system as if it were accurate.

6 Figures and Comments

6.1 Valid Measurement

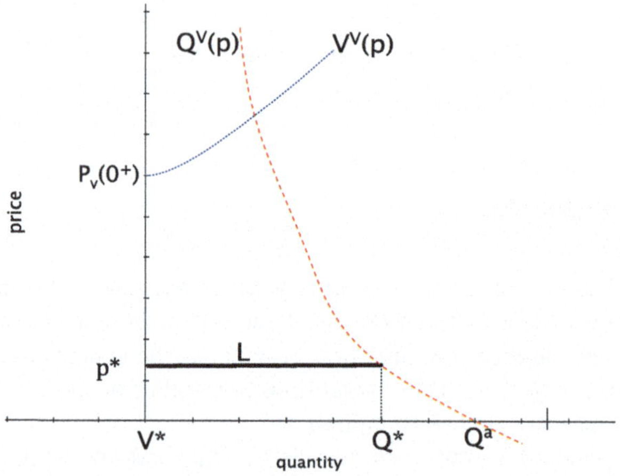

Fig. 1 Market equilibrium with valid measurement

Figure 1 displays the standard situation for a permit market equilibrium under valid measurement. The equilibrium under autarky is Q^a. $P_v(0^+) = \lim_{v \to 0^+} dP(v)/dv$, the right-hand derivative of the penalty function, is the intercept of $V^V(p)$.

The dashed red line, denoted $Q^V(p)$, is the partial equilibrium aggregate supply of output of the commons. The dashed blue line, denoted $V^V(p) = NP_v^{-1}(p)$, is the partial equilibrium aggregate demand for violations.

The equilibrium price of the permit market is p^*. Since $L \geq Q^V(P_v(0^+))$, equilibrium demand is $V^* = V^V(p^*) = 0$ and equilibrium supply is $Q^* = Q^V(p^*) = L$.

6.2 *Imprecise Measurement*

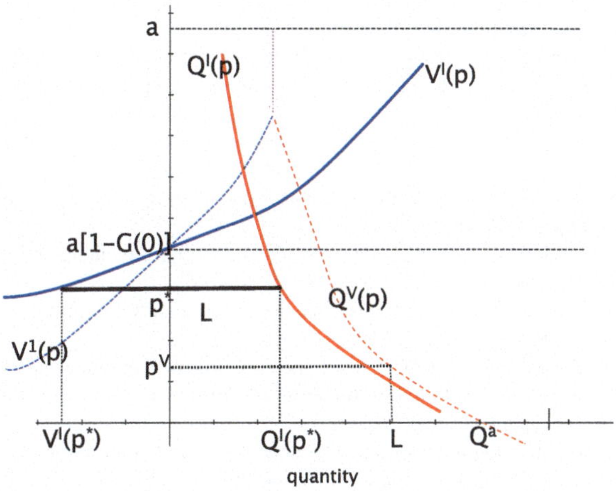

Fig. 2 Market equilibrium with imprecise measurement

The dashed red line, $Q^V(p)$, comes from Fig. 1 and is the aggregate supply under valid measurement. p^V is the equilibrium price under valid measurement. The dashed blue line, $V^1(p)$, is the partial equilibrium aggregate demand for violations if measurement errors are independent of q_i. That is $V^1(p) = \sum_i v_i^1(p)$ where $v_i^1(p) = \arg\max_{v_i} pv_i - a\int_{-v_i}(v_i + \delta)g(\delta)d\delta$.

The solid blue and red lines, $Q^I(p)$ and $V^I(p)$, are the partial equilibrium supply and demand from Definition 7. Because a penalty is assessed for measured violations larger than actual violations, $Q^I(p)$ lies to the left of aggregate demand under accurate measurement. Because $V^I(p) = -[\sum_i h(q_i)]G^{-1}(1 - \frac{p}{a})$, $V^I(p)$ rotates clockwise from $V^1(p)$.[17]

p^* is the equilibrium price under imprecise measurement. The equilibrium aggregate supply of output is $Q^* = Q^I(p^*) < L$. The equilibrium aggregate demand for violations is $V^* = V^I(p^*) < 0$ because excess permits are held in equilibrium. $V^* + L^* = Q^*$.

Acknowledgements I thank the Max Factor Family Foundation in partnership with the Jewish Community Foundation of Los Angeles for its financial support of this project.

References

Hurwicz, L. (1960). Optimality and informational efficiency in resource allocation processes. In *Proceedings of a Symposium on Mathematical Methods in the Social Sciences, 1959* (pp. 27–46). Palo Alto: Stanford University Press.

Hurwicz, L. (1972). On informationally decentralized systems. In R. Radner & C. B. McGuire (Eds.), *Decisions and organizations: A volume in honor of Jacob Marschak* (pp. 297–336). Amsterdam: North Holland.

Hurwicz, L., & Reiter, S. (2006). *Designing economic mechanisms*. Cambridge: Cambridge University Press.

Ledyard, J. (2018). Imprecise measurement and tradable permits: Geometry and comparative statics, Working Paper, Caltech.

Malik, A. S. (1990). Markets for pollution control when firms are noncompliant. *Journal of Environmental Economics and Management, 18*(2), 97–106.

Malik, A. S. (1992). Enforcement costs and the choice of policy instruments for controlling pollution. *Economic Inquiry, 30*(4), 714–721.

Murphy, J. J., & Stranlund, J. K. (2007). A laboratory investigation of compliance behavior under tradable emissions rights: Implications for targeted enforcement. *Journal of Environmental Economics and Management, 53*(2), 196–212.

Stranlund, J. K. (2007). The regulatory choice of noncompliance in emissions trading programs. *Environmental and Resource Economics, 38*(1), 99–117.

Stranlund, J. K., & Dhanda, K. K. (1999). Endogenous monitoring and enforcement of a transferable emissions permit system. *Journal of Environmental Economics and Management, 38*(3), 267–282.

Stranlund, J. K., Murphy, J. J., Spraggon, J. M. (2008). Imperfect enforcement of emissions trading and industry welfare: A laboratory investigation, working paper

Water Education Foundation. (2015). The view from above: The promise of remote sensing, http://www.watereducation.org/western-water-excerpt/view-above-promise-remote-sensing.

Zekri, S. (2009). Controlling groundwater pumping online. *Journal of Environmental Management, 90*(11), 3581–3588.

[17]This is drawn for the case that $\frac{[\sum_i h(q_i)]}{N} > 1$. If that is not true, then $V(p)$ rotates in a counter-clockwise direction.

Second Thoughts of Social Dilemma in Mechanism Design

Tatsuyoshi Saijo

1 Introduction

Why have we been using second thoughts? The second thoughts here refer to giving a player a chance to change the strategy in decision-making after observing the strategies of others in a sequential game. We will show that second thoughts are not an innocent device in our daily life but are human wisdom that plays an important role in resolving problems such as social dilemmas.

Consider a social dilemma game such as a prisoner's dilemma game. Saijo et al. (2018) designed a two stage mechanism called the approval mechanism to implement Pareto efficient outcome when the number of players is two.[1] After having played a usual prisoner's dilemma, players can approve or reject the other's choice of cooperation (*C*) or defection (*D*) in the next approval stage. If both players approve the other's choice, the outcome is the result of the chosen strategies. However, if either rejects the other's choice, the outcome is the same as if they had mutually defected from the prisoner's dilemma. In theory, such an approval mechanism implements cooperation in elimination of weakly dominated strategies (*EWDS*), although this is not the case in the subgame perfect *Nash* equilibrium. They then showed that it works well even from early periods experimentally. They also

[1] Saijo and Shen (2018) showed that the approval (or mate choice) mechanism works well in a class of quasi-dilemma games including prisoner's dilemma games. See also Masuda et al. (2014) for public good provision problems.

T. Saijo (✉)
Research Institute for Humanity and Nature, Kyoto, Japan

Research Institute for Future Design, Kochi University of Technology, Eikokuji, Kochi, Japan

Tokyo Foundation of Policy Research, Roppongi, Tokyo, Japan

W. Trockel (ed.), *Social Design*, Studies in Economic Design,
https://doi.org/10.1007/978-3-319-93809-7_9

found that subjects understood the subgame perfection part well and used *EWDS* instead of *Nash* equilibria at each subgame.

In implementation literature such as Hurwicz and Schmeidler (1978), Hurwicz (1979) and Maskin (1999), they used Nash equilibria as the basic equilibrium concept, and then Moore and Repullo (1988), in their path breaking and influential paper, constructed mechanisms and found conditions on social goals to implement them in subgame perfect *Nash* equilibrium. However, as Fehr et al. (2015) recently show, the experimental performance of mechanisms implementing social goals with subgame perfect Nash equilibria is rather limited. On the other hand, Saijo et al. (2018) found an *affinity* among the mechanism, subgame perfection, and *EWDS*, but not Nash.[2] A basic question is whether this affinity works well beyond two players.

Huang et al. (2017) extended the idea of the approval mechanism to include the cases with more than two players in a social dilemma game. In the first stage, each player chooses either *C* or *D*. Knowing the choices in the first stage, all *C* players can change from *C* to *D* in the second stage unless all choose *C*. If a player chooses *D* in the first stage, then the other *C* players will change to *D* in the second stage. Once players understand this logic, no player would take *D* in the first stage, and hence the mechanism implements cooperation. They conducted a series of experiments and found that the performance of the mechanism is limited in early rounds if the number of players is at least three. In order to overcome this problem, we introduce second thoughts, as a new tool in implementation theory, avoiding complication of the mechanism.

Although second thoughts, allowing players to reconsider their decisions after observing them, have been widely used in our daily life, no theoretic analysis has been done. In the Huang et al. mechanism, we add one stage called the second thought stage between the social dilemma stage and the approval stage. All *D* players have a chance to change from *D* to *C* in the second thought stage unless all choose *D* in the first stage. After a player chose *D* in the first stage, the player notices that the other *C* players will change to *D* later. Understanding this logic, the *D* player changes to *C* in the second thought stage. What we find is that second thoughts in social dilemma work very well theoretically. First, second thoughts change the payoff structure of the game in favor of cooperation. Second, second thoughts make mechanisms robust even when players deviate from *EWDS*.

In the following, we show the two-player case in Sect. 2, the three-player case in Sect. 2, and then the general case in Sect. 3. Section 4 is for further research.

[2]Varian (1994) constructed a simple mechanism called the compensation mechanism that implements a social goal in subgame perfect Nash equilibrium, but the experimental performance is limited as Andreoni and Varian (1999) showed.

2 The Simplified Approval Mechanism with Second Thoughts for $n = 2$

Let $n \geq 2$ be the number of players, and each player has endowment $w > 0$. Each player must choose to contribute either the entire w for the production of a public good y or nothing. The production function of y is linear, namely, $y = \alpha m w$ where $1 > \alpha > 1/n$ and m is the number of players who choose cooperation (C) (i.e., those that contribute the entire w). Hence, the payoff of a player who chooses no contribution (defection or D) is $\alpha m w + w = (\alpha m + 1)w$, while the contributor's payoff is $\alpha m w$. We term a player who chooses C (D) as a C (D) player, respectively.

We consider a mechanism that has a new stage after the *PD* game, due to Huang et al. (2017). If all participants are either C players or D players, the game ends. The payoff of a player in the former case is $\alpha n w$ and that in the latter case is w. If the number of C players is at least one and at most $n - 1$, then only C players can proceed to the second stage, in which they have the opportunity to change their decisions from C to D. This mechanism is called the *simplified approval mechanism* or the *SAM* in short. A natural behavioral procedure found in previous experiments on approval mechanisms is *subgame perfect elimination of weakly dominated strategies* (*SPEWDS*), which is also adopted, for example, in Kalai (1981). This requires two properties: (1) subgame perfection and (2) that players do not choose weakly dominated strategies in each subgame and in the reduced normal form game.

Figure 1 illustrates the case of $n = 2$, $\alpha = 0.7$, and $w = 10$. Players 1 and 2 face a prisoner's dilemma game in the first stage. Knowing that player 2 chose D in the first stage in subgame a, player 1 proceeds to the second stage and faces a choice between C and D. Player 1 chooses D in subgame a since 10 dominates 7, or $10 > 7$. Similarly, player 2 chooses D in subgame b. Then, as the reduced normal form game on the right-hand side of Fig. 1 shows, player 1 chooses C after eliminating weakly dominated strategy D. Similarly, player 2 also chooses C in the first stage and hence, (C,C) is the outcome. Hereafter, a strategy profile with parentheses, such as (C,C), represents the choice in the reduced normal form game, and a sequence

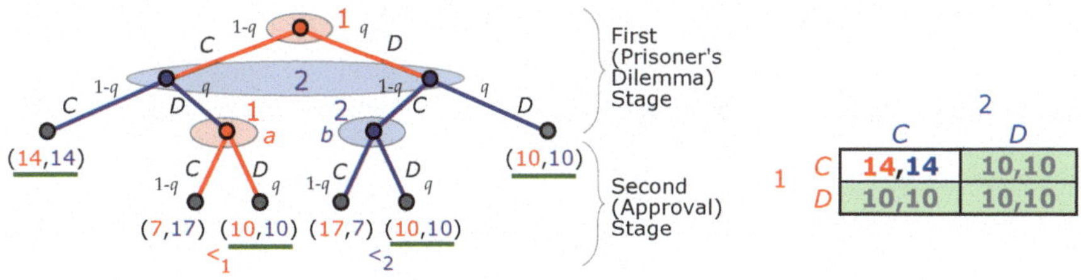

Fig. 1 The *SAM* and its reduced normal form game when $n = 2$

of choices, such as *CDC*, shows a strategy path. Huang et al. (2017) showed the following properties of the *SAM*.

Proposition 1 (*i*) *The simplified approval mechanism implements cooperation in SPEWDS, and* (*ii*) *the simplified approval mechanism cannot implement cooperation in subgame perfect equilibrium (SPE).*

"Cooperation" in Proposition 1 indicates that all players choose *C* in the reduced normal form game. As the reduced normal form game in Fig. 1 shows, (*D*,*D*) is also a *subgame perfect equilibrium* (*SPE*) outcome, and hence, the *SAM* cannot implement cooperation in *SPE*.

Thus far, we have supposed that every player chooses the alternative under *SPEWDS*, but we consider the cases where some player might deviate from it. For simplicity, we assume that every player deviates with a probability of q, with $0 < q < 1$ at each node. As shown in Fig. 1, the success probability achieving (14,14) that follows path *CC* is $(1 - q)^2$. Let $C_{SAM}(n,q)$ be the *success probability function* of the *SAM*, where n is the number of players. Then, $C_{SAM}(n,q) = (1 - q)^n$. Figure 2 shows the case of $n = 2$. The horizontal axis displays the probability of deviation and the vertical axis the probability of all players cooperating. Since $\partial C_{SAM}(2,0)/\partial q = -2$ and $\partial C_{SAMST}(2,0,1)/\partial q = 0$, the success probability of the *SAM* decreases as q rises around zero, whereas the success probability of the *SAMST* stays at a probability of one as q increases around zero. Next, fix any q. Since $C_{SAMST}(2,q,\cdot)$ is always higher than $C_{SAM}(2,q)$ because of $2q(1 - q)^2$ except for $q = 0$ or 1, the success probability of the *SAMST* is always better than that of the *SAM* excluding the end points. That is, the *SAMST* is relatively robust enough to handle deviation by players.

Huang et al. (2017) conducted experiments of the *SAM* with each group consisting of three subjects. In total, 63 subjects played the *SAM* for 15 periods. The groups were formed randomly in each period. The cooperation rates for the

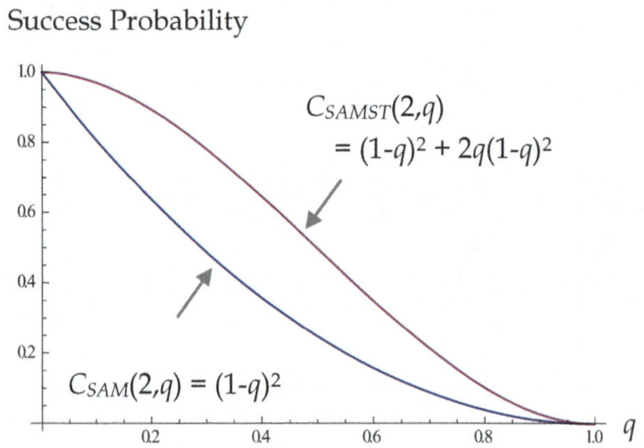

Fig. 2 Success probability functions when $n = 2$

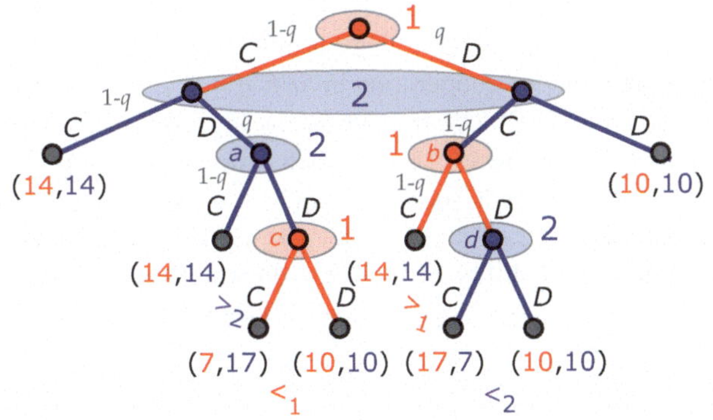

		2: C	2: D
1	C	14,14	14,14
	D	14,14	10,10

Fig. 3 The *SAMST* and its reduced normal form game when $n = 2$

first four to seven periods were between 64.9% and 77.7%, and they rose above 90% thereafter.[3] In order to improve the low cooperation rates in the early rounds in the experiment of Huang et al. (2017), we introduce the one more stage called the second thought stage in the following manner. Every player chooses either C or D in the first stage simultaneously. If all players choose either C or D, the game ends. If the number of D players is at least one and at most $n - 1$, then D players have the chance to change from D to C sequentially, knowing all the choices made in the first stage. The order of the choices of D players is determined exogenously, for example, based on the numbering assigned to players. By observing the choices of all D players in the second thought stage, C players can change their choices from C to D simultaneously except for the case when all D players change their choices in the second thought stage. This stage is called the third stage although the second thought stage might have several stages. When all D players change their choices, the game ends, and the outcome is that all players choose C. We call this the *simplified approval mechanism with second thoughts* (*SAMST*).

Figure 3 shows an example with $n = 2$, $\alpha = 0.7$, and $w = 10$. Consider subgame a. By observing player 1's choice C, player 2 (who has the chance to change his or her choice) must consider player 1's choice in subgame c. Since $10 > 7$, namely, C is dominated by D, player 1 will choose D in subgame c. By understanding this fact, player 2 in subgame a thus chooses C since $14 > 10$. Therefore, the outcome in subgame a is (14,14), which differs from the outcome of the *SAM*. By applying the same argument, we have (14,14) in subgame b. That is, the outcomes except for that at (D,D) are (14,14), although (14,14) is achieved at (C,C) only in the *SAM*.

When the number of players is two, each player chooses C in the second thought stage, and thus the payoff outcome at (C,D) and (D,C) in the reduced normal form

[3]Huang et al. (2017) used the ex post cooperation rate. For example, even though a player chose C in the first stage, this was not counted in the cooperation rate if that player changed his or her decision from C to D in the second stage.

game is ($2\alpha w$,$2\alpha w$). Since the payoff outcomes at (*C*,*C*) and (*D*,*D*) are ($2\alpha w$,$2\alpha w$) and (w,w), respectively, and $2\alpha w > w$, the *SPE* strategy profiles are (*C*,*C*), (*C*,*D*), and (*D*,*C*). That is, the *SAMST* implements cooperation in *SPE* if $n = 2$.

Consider that players deviate from *SPEWDS*. In contrast, as Fig. 3 shows, three paths, namely, *CC*, *CDC*, and *DCC*, achieve (14,14) when we use the *SAMST*. The probability of the paths *CDC* and *DCC* is $(1 - q)q(1 - q)$ and $q(1 - q)(1 - q)$, respectively. Hence, $C_{SAMST}(2,q) = (1 - q)^2 + 2q(1 - q)^2$. Since $\partial C_{SAMST}(2,0)/\partial q = 0$, the success probability of the *SAM* does not decreases as q rises around zero. As Fig. 2 shows, $C_{SAMST}(2,q) > C_{SAM}(2,q)$ for all $q \in (0, 1)$.

3 The Simplified Approval Mechanism with Second Thoughts for $n = 3$

This section illustrates the three player case that basically contains problems that should be handled for the general case. Figure 4 illustrates the *SAMST* with $n = 3$, $\alpha = 0.7$ (0.4 or 0.5), and $w = 10$. The bold face numbers show the payoffs with $\alpha = 0.7$, the numbers in braces show the payoffs with $\alpha = 0.4$, and the numbers in parentheses in the braces show them with $\alpha = 0.5$. Since the entire tree is relatively large, we only show the subgames with *CCC*, *CCD*, *CDD*, and *DDD*, which are sufficient to understand the entire tree. Consider first the case of $\alpha = 0.7$. Look at subgame *a* where players 1 and 2 chose *C*, but player 3 chose *D*. Player 3, who faces the second thought stage, must consider what would happen in subgame *c*. Players who chose *C* in the first stage face a *PD* game in subgame *c*, and hence, both choose *D*. In this sense, players who chose *C* in the first stage can *burden* players who chose *D* in the second thought stage, although this hurts every player. By understanding this fact, player 3 compares 21 with *C* and 10 with *D*. Since *C* dominates *D*, player 3 chooses *C* in subgame *a*. That is, player 3, who chose *C* at node *a*, can obtain the *bonus* from players 1 and 2, who chose *C* in the first stage. Therefore, the outcome of subgame *CCD* is (21,21,21).

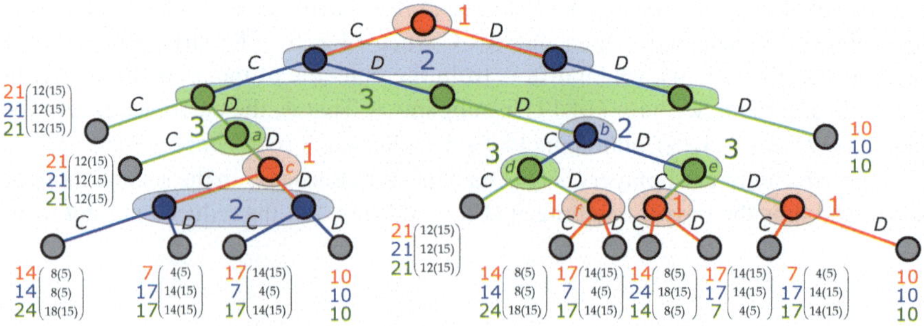

Fig. 4 The *SAMST* when $n = 3$

Consider the next subgame *b* where player 1 chose *C* but players 2 and 3 chose *D*. Players 2 and 3 face the second thought stage sequentially. Pay attention to the last nodes or the third stage where player 1 faces the choice between *C* and *D*. Although the number of players is one, players who arrive at the nodes face *PD* games, and hence, they always choose *D* at each node.

Moreover, consider subgame *e* where player 2 did not change his or her choice in subgame *b*. Since player 1 chooses *D* in the following subgames, player 3 chooses *D* in subgame *e*, and the payoff is 10. Consider subgame *d*. In contrast, player 3, who can take advantage of the bonus effect in subgame *d*, chooses *C* since player 1 in the following subgame will choose *D* if player 3 chooses *D*. That is, 21 > 17. Knowing this process, player 2 chooses *C* since 21 > 10. Therefore, all payoff outcomes other than *(D,D,D)* are (21,21,21), and hence, the final outcome is *(C,C,C)* under *SPEWDS*, as the reduced normal form game in Fig. 5a shows. In contrast, as Fig. 4b shows, (21,21,21) appears only in subgame *CCC* under the *SAM*.

The sequentiality of *D* players is important to implement cooperation.[4] If nodes *d* and *e* were in the same information set, the payoff of player 3 from choosing *C* in the information set would be (21,7), and the payoff from choosing *D* would be (17,10); hence, both would survive by using the elimination of weakly dominated strategies.

Let us next look at the case of $\alpha = 0.4$. Consider node *f* where player 1 chooses *D*. Then, player 3 at node *d* chooses *D* since 14 > 12. Knowing this fact, player 2 chooses *D* since 10 > 4. That is, the payoff in subgame *b* becomes (10,10,10). From the viewpoint of player 3, since α is too small, the player cannot take advantage of the bonus effect at node *d*.

As the reduced normal form game in Fig. 5c shows, the payoff outcome with subgames where two players choose *C* and one player chooses *D* is (12,12,12), and the payoff outcome with subgames where one player chooses *C* and two players choose *D* is (10,10,10), but the final outcome is still *(C,C,C)* under *SPEWDS*. In contrast, as Fig. 5d shows, (12,12,12) appears only in subgame *CCC* under the *SAM*.

Consider the case when $\alpha = 0.5$. The payoff outcomes at paths *CDDCDD* and *CDDCC* in Fig. 3 are (15,5,15) and (15,15,15), respectively. If this were the case, player 3 at node *d* would be indifferent between *C* and *D*. That is, both *C* and *D* survive by using the elimination of weakly dominated strategies at node *d*. This influences the decision of player 2 at node *b*. Figure 6 shows the reduced normal form game at node *b* excluding player 1.

Note that the payoff at *(D,C)* in Fig. 6 should be (10,10) when player 2 chooses *D*. That is, player 3's choice does not matter, and hence, both *C* and *D* survive by using the elimination of weakly dominated strategies for player 2. To sum up, the payoff outcome in subgame *b* or *CDD* is (15,15,15), (15,5,15), or (10,10,10). At (15,5,15), player 1 changes from *C* to *D* at the third stage knowing that player 3 kept the choice at *D*. That is, only player who chose *C* is player 2. Similarly, the payoff outcome in subgame *DCD* or *DDC* is (15,15,15), (5,15,15), or (10,10,10). Since the

[4]We thank Xiaochuan Huang for indicating this fact.

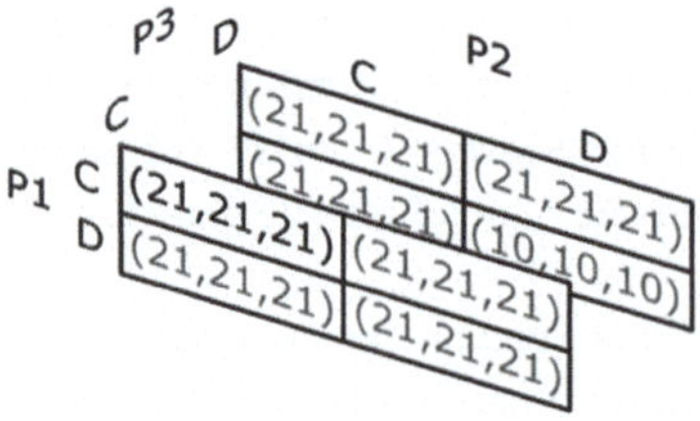

(a) α = 0.7 with the *SAMST*.

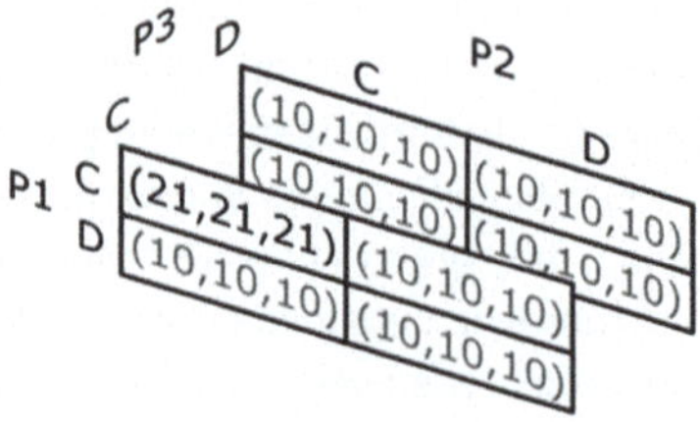

(b) α = 0.7 with the *SAM*.

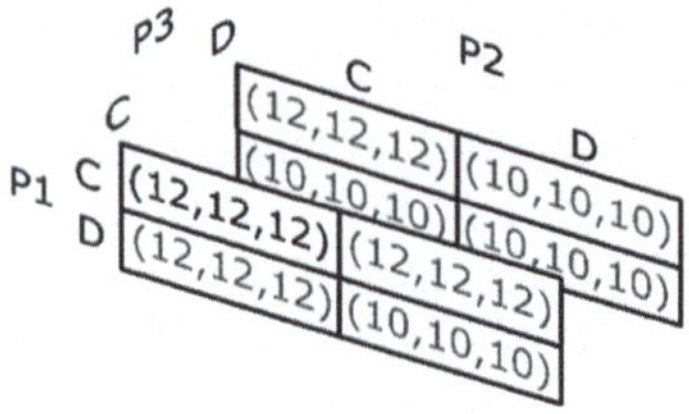

(c) α = 0.4 with the *SAMST*.

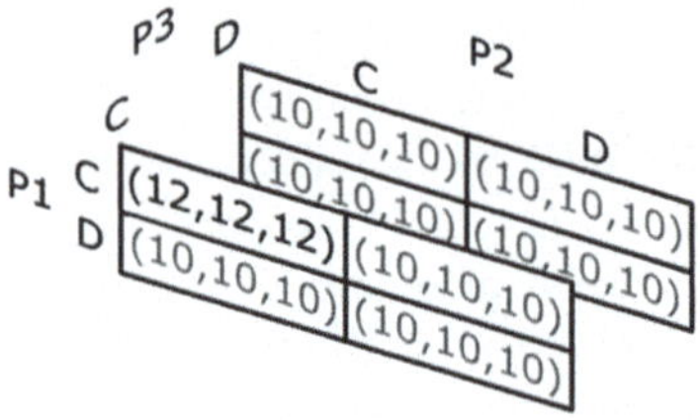

(d) α = 0.4 with the *SAM*.

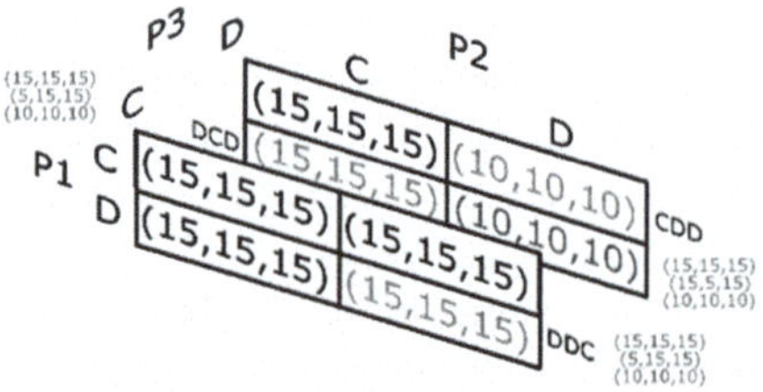

(e) α = 0.5 with the *SAMST*.

Fig. 5 The reduced normal form games of the *SAMST* and *SAM* when $n = 3$. (**a**) $\alpha = 0.7$ with the *SAMST*. (**b**) $\alpha = 0.7$ with the *SAM*. (**c**) $\alpha = 0.4$ with the *SAMST*. (**d**) $\alpha = 0.4$ with the *SAM*. (**e**) $\alpha = 0.5$ with the *SAMST*

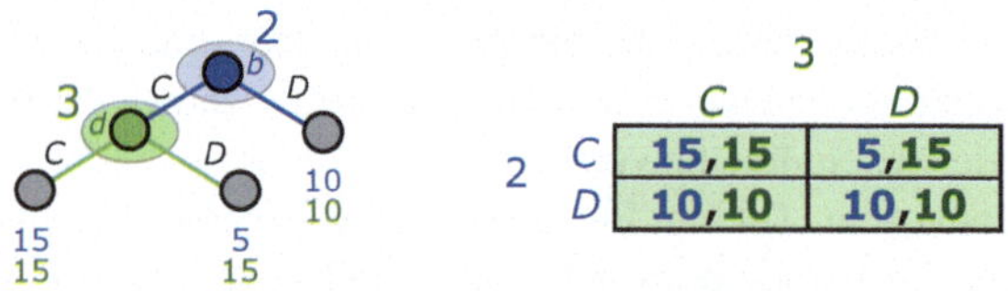

Fig. 6 Subgame *b* and its reduced normal form game at node *b*

first *D* player in the second thought stage at *DCD* and *DDC* is player 1, the payoff outcome should be (5,15,15) when player 1 changes from *D* to *C* at the second thought stage and the rest choose *D*. That is, there are $3^3 = 27$ reduced normal form games when $\alpha = 0.5$. Look at Fig. 5e. Since there are three possible payoff outcomes, for example, at *CDD*, we write these payoff outcomes under *CDD*. Since one of them should be chosen at a reduced normal form game, we chose (10,10,10) in Fig. 5e. Similarly, there are three possible payoff outcomes at the left-hand side of *DCD* and at the right-hand side of *DDC*.

Consider player 1. There is a case where both *C* and *D* survive by using the elimination of weakly dominated strategies for player 1: (10,10,10) at *CDD* and (15,15,15) at *DCD* and *DDC*. Then, both players 2 and 3 choose *C*, and hence, the *SPEWDS* strategy profile of the reduced normal form game is (*C*,*C*,*C*) or (*D*,*C*,*C*). Although player 1 chooses *D* at (*D*,*C*,*C*), the player will change from *D* to *C* in the second thought stage; hence, the payoff outcome is (15,15,15). The real problem is that player 1 cannot tell which reduced normal form game player 1 faces, and hence, both *C* and *D* survive when player 1 must make a decision in the first stage.

Note also that the payoff at *C* and the payoff at *D* of player 1 are the same for all possible choices of players 2 and 3 in Fig. 5e. In other words, player 1 cannot distinguish between *C* and *D*. We say that strategies *A* and *B* of a player are *indistinguishable* if the payoffs at *A* and *B* are the same for all possible choices of the other players.

Consider all possible reduced normal form games at each node. Strategy *A* *weakly rules* strategy *B* if *A* weakly dominates *B* in some reduced normal form game and strategies *A* and *B* are indistinguishable in the remainder of the reduced normal form games. Then, we can define a refinement of *SPEWDS* by using the *weak rule* instead of the weak dominance in the definition of *SPEWDS*, which we denote it backward elimination of weakly ruled strategies (*SPEWRS*).

Look at (c) in Fig. 5 and consider (*D*,*D*,*D*). This is also an *SPE* strategy profile, and hence, the *SAMST* cannot implement cooperation in *SPE* if $n = 3$. If $n > 3$, it is easy to find similar examples by choosing α to satisfy $1/(n-1) > \alpha$.

4 The Simplified Approval Mechanism with Second Thoughts

The following proposition shows that the *SAMST* implements cooperation in *SPEWDS* or *SPEWRS*.

Proposition 2 *(i) If* $\alpha \notin \{1/(n-1), 1/(n-2), \ldots, 1/2\}$*, the SAMST implements cooperation in SPEWDS, and (ii) the SAMST implements cooperation in SPEWRS.*

Proof See Appendix 1.

Consider the meaning of inequality $\alpha > 1/(m+1)$, i.e., $m > (1/\alpha) - 1$. If $\alpha = 0.7$, $(1/\alpha) - 1 = 3/7$. That is, the minimum number $\underline{m}$ of the *C* players in the first stage where the "bonus" effect is activated is at least one. If $\alpha = 0.4$, $\underline{m}(\alpha) = 2$, which shows that the cooperative outcome in subgame *b* in Fig. 3 cannot be realized since only one *C* player is in the first stage. In contrast, $\bar{l}(\alpha) = n - \underline{m}(\alpha)$ is the maximum number of D players in the first stage, thus leading to the cooperative outcome. The proof of Proposition 2 shows the following corollary.

Corollary 1 *If there is an indistinguishable player in a reduced normal form game at the beginning node, no other players are indistinguishable.*

Suppose that $(1/\alpha) - 1$ is an integer. Then, there are $\bar{l}+1$ possible payoff outcome profiles in a subgame in the second thought stage. The total number of subgames in the second thought stage where the number of D players is $\bar{l}$ is ${}_nC_{\underline{m}}$, and hence, the total number of all possible reduced normal form games at the beginning node is $(\bar{l}+1)^{{}_nC_{\underline{m}}}$, where ${}_nC_k$ is the number of k combinations from n players. Among these subgames, each player faces just one reduced form game in which C and D are indifferent. As Fig. 5e shows, the total number of all possible reduced normal form games at the node is $(2+1)^3 = 27$ when $n = 3$ and $\underline{m} = 1$. If $n = 5$ and $\underline{m} = 2$, it is 4^{10}. If this were the case, α must be $1/3$, and the chance of a player being indistinguishable would be $1/4^{10}$.

Consider the case of $n = 3$. Although C_{SAMST} with $n = 2$ does not depend on $\bar{l}$, C_{SAMST} with at least three players depends on $\bar{l}$, which is determined by α. That is, C_{SAMST} is a function of n, q, and $\bar{l}$, and thus, we write it as $C_{SAMST}(n, q, \bar{l})$. Consider the case of $\alpha = 0.4$. Then, $\bar{l}(0.4) = 1$, and there are two types of success paths. The first one is the success paths up to $\bar{l}$. These are *CCC*, *CCDC*, *CDCC*, and *DCCC*, and their probabilities are $(1-q)^3$, $(1-q)^2q(1-q)$, $(1-q)q(1-q)^2$, and $q(1-q)^3$, respectively. The second one is the success paths beyond $\bar{l}$. Look at node b in Fig. 4. Although player 2 should choose D when $\alpha = 0.4$, the player might choose C because of deviation. If player 3 also chooses C after player 2's choice induced by deviation, the path *CDDCC* is also a success path that has a probability of $(1-q)q^4$. Since there are two other paths of this kind, $C_{SAMST}(3,q,1) = (1-q)^3(1+3q) + 3(1-q)q^4$. In contrast, if $\alpha = 0.7$, the probability of *CDDCC* is $(1-q)^3q^2$. That is, since $\bar{l}(0.7) = 2$, deviation in the second thought stage must lead players to choose D. Therefore, $C_{SAMST}(3,q,2) = (1-q)^3\,(1+3q+3q^2)$. Figure 7 shows this case. In order to avoid the indeterminacy case, let us assume $\alpha \notin \{1/(n-1), 1/(n-2), \ldots, 1/2\}$. Then, by summarizing the above argument, we have $C_{SAM}(n,q) = (1-q)^n$, and $C_{SAMST}(n, q, \bar{l}) = (1-q)^n \sum_{k=0}^{\bar{l}} {}_nC_k q^k + \sum_{k=\bar{l}+1}^{n-1} {}_nC_k (1-q)^{n-k} q^{2k}$.

Proposition 3 (*i*) $\partial C_{SAM}(n,0)/\partial q = -n$ and $\partial C_{SAMST}(n, 0, \bar{l})/\partial q = 0$ for all $\bar{l}$; and (*ii*) for any $1 \le l \le n-1$, $C_{SAMST}(n,q,l) > C_{SAM}(n,q)$ on $(0,1)$.

Proof See Appendix 2.

The fact that $C_{SAMST}(n, q, \bar{l}) > C_{SAM}(n,q)$ on in $(0,1)$ shows that the *SAMST* is always better than the *SAM* with respect to the success probability of cooperation. Since $\bar{l}\,(\alpha)$ is a non-decreasing function, roughly speaking, the success probability increases as α rises.

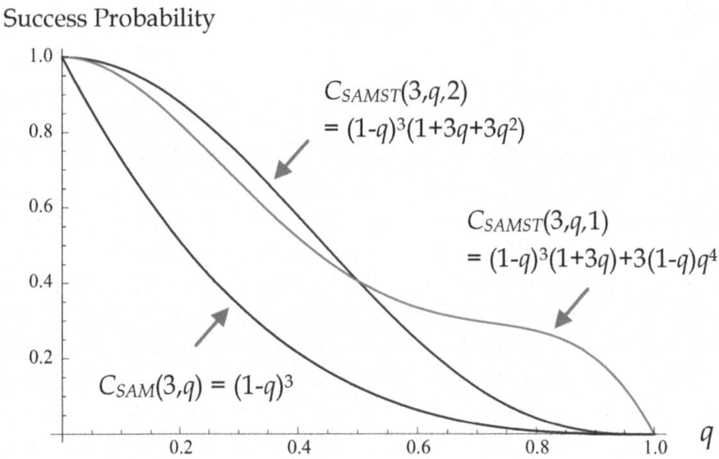

Fig. 7 Success probability functions when $n = 3$

5 Concluding Remarks

Second thoughts are a powerful tool in implementation theory. They change the payoff structure of the game in favor of cooperation. Furthermore, the mechanism with second thoughts is robust even when players deviate from *EWDS*.

Our approach is different from the trend in implementation theory where finding some conditions on social goals such as social choice correspondences is a major research goal. Instead, we fix the social goal as Pareto efficiency in social dilemma, then construct a mechanism incorporating social dilemma. That is, introducing second thoughts in mechanisms implementing a social goal in some equilibrium concept is still an open question.

The validity of second thoughts should be confirmed in experiments. Although it is still an early stage, we started confirming that second thoughts make subjects cooperative in even early rounds in experiments.

Acknowledgments The author thanks Yoshitaka Oakano for his helpful comments and suggestions. This research was supported by Scientific Research A (24243028 and 17H00980) and Challenging Exploratory Research (16K13354) of the Japan Society for the Promotion of Science; the Research Institute for Humanity and Nature (RIHN Project Number 14200122); and "Experimental Social Sciences: Toward Experimentally-based New Social Sciences for the 21st Century," a project funded by a Grant-in-Aid for Scientific Research on Priority Areas from the Ministry of Education, Science and Culture of Japan.

Appendix 1: The Proof of Proposition 2

(i) Let m and l be the numbers of C and D players in the first stage, respectively. If $m = n$, the outcome is $(\alpha nw, \alpha nw, \ldots, \alpha nw)$. If $l = n$, the outcome is $(w, w, \ldots, w)$.

Suppose $1 \leq l < n$. Consider the choice of players who chose C in the first stage after observing the choices of D players in the second thought stage. Let $0 \leq l' < l$ be the number of D players who change their choices from D to C in the second thought stage. Since $\alpha(m' + l' + 1)w < \alpha(m' + l')w + w$ for all $1 \leq m' \leq m - 1$, where m' is the number of C players in the first stage who remains to choose C in the subgame after the second thought stage, D is better than C for any C player in the third stage after observing the choices in the second thought stage. That is, all players who chose C in the first stage choose D after the second thought stage.

Consider any strategy path on which at least one D player chose D again in the second thought stage. If this were the case, every C player after the second thought stage would choose D. Keeping this fact in mind, let us choose the youngest D player (e.g., by names or numbers assigned to players) who chose D in the second thought stage. Then, the subgame after the choice of the youngest D player is a sequential social dilemma game, and hence, every D player after the choice chooses D.

We now identify the payoff outcome of every subgame constructed by the end nodes in the first stage. Choose any subgame except for the cases where all players chose C or all players chose D in the first stage. Suppose that every D player except for the last one changed his or her choice from D to C in the second thought stage. Consider next the choice of the last D player. If the player chooses C, the payoff is αnw, whereas if the player chooses D, the payoff is $\alpha(l - 1)w + w$. Since $\alpha nw - \alpha(l - 1)w - w = \alpha\{n - (l - 1)\}w - w$ and $n - (l - 1) = m + 1$:

If $\alpha > 1/(m + 1)$, then the last D player chooses C.
If $\alpha = 1/(m + 1)$, then the last D player is indifferent between C and D.
If $\alpha < 1/(m + 1)$, then the last D player chooses D.

Suppose $\alpha > 1/(m + 1)$. If the penultimate player chooses D, then the payoff is $\alpha(l - 2)w + w$ since the last D player chooses D in the second thought stage and every C player in the first stage chooses D in the third stage. If the player chooses C, then it is αnw since the last player chooses C. Since $\alpha nw - \{\alpha(l - 2)w + w\} = \{\alpha(n - l + 2) - 1\}w = \{\alpha(m + 2) - 1)\}w > 0$, the player chooses C. Since $\alpha nw - \{\alpha(l - 2)w + w\} > 0$, $\alpha nw - \{\alpha(l - k)w + w\} > 0$ for all $2 \leq k \leq l$. That is, the k-th player to last chooses C, and hence, all D players choose C in the second thought stage, and the payoff outcome is $(\alpha nw, \ldots, \alpha nw)$.

Suppose $\alpha < 1/(m + 1)$. Then, the last D player chooses D, and hence, the payoff of the penultimate player is $\alpha(l - 1)w$ if the player chooses C. If the player chooses D, then the payoff is $\alpha(l - 2)w + w$. Since $\alpha(l - 2)w + w - \alpha(l - 1)w = (1 - \alpha)w > 0$, the player chooses D. Since $\alpha(l - k)w + w - \alpha(l - k + 1)w = (1 - \alpha)w > 0$ for all $2 \leq k \leq l$, the k-th player to last chooses D, and hence, no D players in the first stage change their decisions in the second thought stage, and the payoff outcome is $(w, \ldots, w)$.

Take any α satisfying $1/n < \alpha < 1$ and $\alpha \notin \{1/(n-1), 1/(n-2), \ldots, 1/2\}$. Consider the case of $\alpha > 1/2$. Then, $\alpha > 1/2 \geq 1/(m+1)$ for all $m \geq 1$, and hence, the payoff outcome of every subgame other than $(D,D,\ldots,D)$ is $(\alpha nw,\ldots,\alpha nw)$: without loss of generality, consider player 1. The payoff in subgame $(C,D,D,\ldots,D)$ is αnw and that in subgame $(D,D,\ldots,D)$ is w. Since $\alpha nw > w$, C is better than D. Since $\alpha > 1/(m+1)$ for all $m \geq 1$, the outcome of the two subgames $(C,\cdot)$ and $(D,\cdot)$ is $(C,C,\ldots,C)$ where "$\cdot$" shows that at least one player's choice is C. That is, player 1 is indifferent between the outcomes of subgames $(C,\cdot)$ and $(D,\cdot)$. Therefore, C weakly dominates D for all players, and hence, $(C,C,\ldots,C)$ is the *SPEWDS* outcome.

Consider next the case of $1/2 \geq 1/(k+1) > \alpha > 1/(k+2) \geq 1/n$. Consider player 1. Let "$\cdot$" indicate that the number of C is k. Then, the payoff in subgame $(C,\cdot)$ is αnw since $\alpha > 1/\{(k+1)+1\}$, and that in subgame $(D,\cdot)$ is w since $1/(k+1) > \alpha$. That is, C is better than D. Since the outcome of the two subgames $(C,\cdot)$ and $(D,\cdot)$ is the same where "$\cdot$" indicates that the number of C is not k, C weakly dominates D for all players, and hence, $(C,C,\ldots,C)$ is the *SPEWDS* outcome.

Thus, if $\alpha \notin \{1/(n-1), 1/(n-2), \ldots, 1/2\}$, the *SAMST* implements cooperation in *SPEWDS*.

(*ii*) Suppose $\alpha = 1/(m+1)$. Then, the last D player is indifferent between C and D since $\alpha nw = \alpha(l-1)w + w$. Suppose that the penultimate player chooses C. Then, the payoff of the penultimate player is αnw if the last D player chooses C and is $\alpha(l-1)w$ if the last D player chooses D. If the penultimate player chooses D, then the payoff is $\alpha(l-2)w + w$. Since $\alpha nw - \{\alpha(l-2)w + w\} = \alpha w > 0$, $\alpha nw > \alpha(l-2)w + w > \alpha(l-1)w$. That is, both C and D survive by using the elimination of weakly dominated strategies. Since $\alpha nw > \alpha(l-k-1)w + w > \alpha(l-k)w$ for all $k = 1, \ldots, l-1$, both C and D survive by using the elimination of weakly dominated strategies for all D players.

Let $\underline{m}(\alpha) = \lceil (1/\alpha) - 1 \rceil$ where $\lceil a \rceil$ is the smallest integer not less than a. Since $1/n < \alpha < 1$, $1 \leq \underline{m}(\alpha) \leq n-2$. Suppose that $(1/\alpha) - 1$ is an integer. Then, $\underline{m} = (1/\alpha) - 1$. The following case shows that there exists a player who is indifferent between C and D when the number of C players is $\underline{m}$ or $\underline{m} - 1$. Consider two cases:

Case 1: Suppose that the number of C players is $\underline{m}$. Choose any player who is not a member of the C players. If the player chooses C, the payoff outcome is αnw. If the player chooses D, the maximum possible payoff is that all D players other than the player change from D to C and the player is the last D player since all C players change from C to D after the second thought stage. Then, the payoff is $\alpha(\bar{l}-1)w + w$, and hence, $\alpha nw - \{\alpha(\bar{l}-1)w + w\} = \{\alpha n - \alpha(\bar{l}-1) - 1\}w = \{\alpha(\underline{m}+1) - 1\}w = 0$, where $\bar{l} = n - \underline{m}$ and $\bar{l} \geq 2$ since $n \geq \underline{m} + 2$. That is, the payoff of C is the same as the payoff of D for the player.

Case 2: Suppose that the number of C players is $\underline{m} - 1$. Choose any player who is not a member of the C players. If the player chooses C in the first stage, we show that the payoff outcome should be at least w. Since $\bar{l} \geq 2$, there must be at least one D player. If all D players change from D to C in the second thought stage, the

C player obtains αnw. If at least one D player chooses D in the second thought stage, the C player obtains at least w by changing from C to D after the second thought stage. If the player chooses D, the payoff is w. That is, the payoff of C can be the same as the payoff of D for the player.

Thus, there is a possibility that C and D are indistinguishable for some players. Let player 1 be such a player and suppose that the first $\underline{m}$ players choose C. Then, since C and D are indistinguishable,

$$\text{the payoff outcome of subgame } (\underbrace{C, C, \ldots, C}_{\underline{m}}; \underbrace{D, \ldots, D}_{\bar{l}})$$
$$= \text{the payoff outcome of subgame } (D, \underbrace{C, \ldots, C}_{\underline{m}-1}; \underbrace{D, \ldots, D}_{\bar{l}}).$$

Since the payoff outcome of the latter should be $(w,w,\ldots,w)$, each of the last $\bar{l}$ players in the former can obtain αnw by changing from D to C. That is, C weakly dominates D for the last $\bar{l}$ players.

In contrast, compare the payoff outcome of subgame $(D, \underbrace{C, \ldots, C}_{\underline{m}}; \underbrace{D, \ldots, D}_{\bar{l}-1})$ with the payoff outcome of subgame $(\underbrace{C, C, \ldots, C}_{\underline{m}+1}; \underbrace{D, \ldots, D}_{\bar{l}-1})$. The latter payoff outcome should be $(\alpha nw,\ldots,\alpha nw)$, and hence, the payoff of player 1 should be αnw. Since player 1 in the former should obtain αnw, which is more than w, at least one player changes from D to C in the second thought stage, and hence, all C players who change from C to D after the second thought stage should obtain strictly more than w. Then, each of the same $\underline{m}$ players obtains w by changing from C to D. That is, C weakly dominates D for the $\underline{m}$ players. Thus, C and D are indistinguishable for player 1, and C weakly dominates D for the rest.

Suppose that C and D are indistinguishable for player 1. Then, there exists another reduced normal form game in the first stage where C weakly dominates D for player 1. Since C and D are indistinguishable for player 1, the payoff of player 1 in subgame $(C,C,\ldots,C;D,\ldots,D)$ is either αnw or w. Since the payoff outcome in this subgame can be either $(\alpha nw,\ldots,\alpha nw)$ or $(w,\ldots,w)$, there is another reduced normal form game where C weakly dominates D for player 1.

Since the choice of a player who faces indistinguishability is arbitrary, C weakly rules D for all players. That is, the *SAMST* implements cooperation in *SPEWRS*. ■

Appendix 2: The Proof of Proposition 3[5]

(i) Let $f(q) = (1-q)^n$, $g(q) = \sum_{k=0}^{\bar{l}} {}_nC_k q^k$ and $h(q) = \sum_{k=\bar{l}+1}^{n-1} {}_nC_k (1-q)^{n-k} q^{2k}$. Then, $f(0) = 1$. Since $f'(q) = -n(1-q)^{n-1}$, $f'(0) = -n$. Since $\bar{l} \in \{1, \ldots, n-1\}$ and $g(q) = {}_nC_0 q^0 + {}_nC_1 q^1 + \sum_{k=2}^{\bar{l}} {}_nC_k q^k = 1 + nq + \sum_{k=2}^{\bar{l}} {}_nC_k q^k$, $g(0) = 1$ and $g'(0) = n$. Since $h(q) = q^2 r(q)$, where $r(q) = \sum_{k=\bar{l}+1}^{n-1} {}_nC_k (1-q)^{n-k} q^{2(k-1)}$, $h'(0) = 0$. Since $C_{SAMST}(n, q, \bar{l}) = f(q) g(q) + h(q)$,

$$\frac{\partial C_{SAMST}(n, 0, \bar{l})}{\partial q} = f'(0)g(0) + f(0)g'(0) + h'(0) = -n + n + 0 = 0.$$

(ii) By definition, since $C_{SAMST}(n, q, \bar{l})$ has a positive part in addition to $C_{SAM}(n,q)$ on $(0,1)$, we have the result. ∎

References

Andreoni, J., & Varian, H. R. (1999). Preplay contracting in the prisoners' dilemma. *Proceedings of the National Academy of Sciences, 96*(19), 10933–10938.

Fehr, E., Powell, M., & Wilkening, T. (2015). *Behavioral limitations of subgame-perfect implementation.*

Huang, X., Masuda, T., & Saijo, T. (2017). *Cooperation among behaviorally heterogeneous players in social dilemma with stay or leave decisions* (University of Arizona Department of Economics working paper series 2017-16).

Hurwicz, L. (1979). Outcome functions yielding Walrasian and Lindahl allocations at Nash equilibrium points. *The Review of Economic Studies, 46*(2), 217–225.

Hurwicz, L., & Schmeidler, D. (1978). Construction of outcome functions guaranteeing existence and Pareto optimality of Nash equilibria. *Econometrica: Journal of the Econometric Society, 46*, 1447–1474.

Kalai, E. (1981). Preplay negotiations and the prisoner's dilemma. *Mathematical Social Sciences, 1*(4), 375–379.

Maskin, E. (1999). Nash equilibrium and welfare optimality. *The Review of Economic Studies, 66*(1), 23–38.

Masuda, T., Okano, Y., & Saijo, T. (2014). The minimum approval mechanism implements the efficient public good allocation theoretically and experimentally. *Games and Economic Behavior, 83*, 73–85.

Moore, J., & Repullo, R. (1988). Subgame perfect implementation. *Econometrica: Journal of the Econometric Society, 56(5)*, 1191–1220.

Saijo, T., & Shen, J. (2018). Mate choice mechanism for solving a quasi-dilemma. *Journal of Behavioral and Experimental Economics, 72*, 1–8.

Saijo, T., Masuda, T., & Yamakawa, T. (2018). Approval mechanism to solve prisoner's dilemma: Comparison with Varian's compensation mechanism. *Social Choice and Welfare, 51(1)*, 65–77.

Varian, H. R. (1994). A solution to the problem of externalities when agents are well-informed. *American Economic Review, 84*(5), 1278–1293.

[5]The author would like to thank Yoshitaka Okano for supporting the proof.

Part III
Markets

Allocation Mechanisms, Incentives, and Endemic Institutional Externalities

Peter J. Hammond

1 Introduction

1.1 Hurwicz on Mechanism Design

Much of Leo Hurwicz's long and distinguished career was devoted toward discovering how market and other economic institutions could be designed in order to improve the effect of individual agents' economic decisions on the well-being of society.

Leo's early work on this topic appeared as Hurwicz (1960, 1972), much of which was synthesized in Hurwicz (1986)—see also Arrow and Hurwicz (1977). Late enough in his life for him to have been invited to deliver the Richard T. Ely Lecture to the American Economic Association, Hurwicz (1973) did a great deal to promote the systematic exploration of incentive compatible allocation mechanisms for resource allocation. This was very useful to me when working on Hammond (1979), especially the typical incentive incompatibility of lump-sum redistribution of the kind needed to support typical first-best Pareto efficient allocations. This and the earlier articles by Hurwicz were a source of inspiration for many of the other contributions to the *Review of Economic Studies* "Symposium on Incentive Compatibility" that I edited, including inter alia Hurwicz (1979) and Dasgupta et al. (1979).[1]

[1]A confession may be in order. As the deadline for sending the papers for the symposium to the production editor loomed, there was only a still incomplete version of Leo's contribution sitting

P. J. Hammond (✉)
Department of Economics, and CAGE (Competitive Advantage in the Global Economy), University of Warwick, Coventry, UK
e-mail: p.j.hammond@warwick.ac.uk

W. Trockel (ed.), *Social Design*, Studies in Economic Design,
https://doi.org/10.1007/978-3-319-93809-7_10

1.2 Hurwicz on Institutions and Externalities

Several years later, Leo Hurwicz (1995, 1999) wrote specifically about externalities, including the Coase theorem. In Hurwicz (1996), he wrote about "institutions as families of game forms," and in Hurwicz (1998) on "the design of mechanisms and institutions," which appeared in a volume with the title "designing institutions for environmental and resource management." In his Nobel Memorial Prize lecture, Hurwicz (2008), he revisited this idea of the link between institutions and game forms.

I take this background as inspiration for using this opportunity to write about externalities and mechanism design, though from a perspective that is no doubt very different.

1.3 Outline

The purpose of this paper is to relate different concepts of externality to the economic institutions which determine, or at least influence, what outcome to the participating agents emerges.

As argued in Sect. 2, classical externalities come about as a departure from the standard "neoclassical" institutional framework, with complete and perfectly competitive markets for private goods.

Next, Sect. 3 considers pecuniary externalities. As Laffont (2008) correctly observes, unlike classical externalities, they do nothing to upset the usual efficiency properties of equilibrium allocations in competitive markets. They do, however, have a significant influence on gains from trade results.

Section 4 introduces the concept of an institutional externality. Like classical and pecuniary externalities, it captures the idea that one agent's actions can influence the possibilities open to other agents. With institutional externalities, however, the influence is more subtle. The idea is that, except when the institution can be modelled as a game form in which agents can choose dominant strategies, one agent's strategy choice can influence what other agents will want to choose. This is what we call an "institutional externality."

Hurwicz, of course, demonstrated that such strategy-proof mechanisms fail to exist in many economic environments. In this sense, institutional externalities are endemic. Nevertheless, Sect. 4 concludes with some prominent examples of economic environments in which institutional externalities can be avoided.

The final Sect. 5 attempts to put these results in a general perspective.

---

on my editorial desk. In particular, there was no introduction, though fortunately a first footnote provided most of what was needed. So, in an era when even transatlantic phone calls remained rare and expensive, Leo's paper appeared without his formal approval of this last minute change. I have heard that Leo, as one might have expected, was amused rather than offended by this course of action.

2 Classical Externalities as Constraints

2.1 Defining Classical Externalities

In what has become the standard textbook for graduate courses in microeconomic theory, Mas-Colell, Whinston and Green (1995, p. 351) write:

> Surprisingly, perhaps, a fully satisfying definition of an externality has proved somewhat elusive.

As a "serviceable departure," they offer this as a definition:

> An *externality* is present whenever the well-being of a consumer or the production possibilities of a firm are directly affected by the actions of another agent in the economy.

They also offer this additional "subtle point":

> When we say "directly," we mean to exclude any effects that are mediated by prices.

This use of the keyword "directly" contrasts markedly with the word "indirect" that is used in the definition provided at the head of Laffont's (2008) entry in the *New Palgrave Dictionary of Economics*:

> Externalities are indirect effects of consumption or production activity, that is, effects on agents other than the originator of such activity which do not work through the price system.
>
> In a private competitive economy, equilibria will not be in general Pareto optimal since they will reflect only *private* (direct) effects and not *social* (direct plus indirect) effects of economic activity.

2.2 Externalities and Constrained Efficiency

Laffont (2008) goes on to write:

> In a private competitive economy, equilibria will not be in general Pareto optimal since they will reflect only *private* (direct) effects and not *social* (direct plus indirect) effects of economic activity.

Indeed, there is the well-known relation between perfectly competitive markets for private goods, with or without lump-sum wealth redistribution, and the Pareto efficient allocation of private goods. On this topic, this is not the occasion to try to add to the survey chapter Hammond (2011). In the presence of externalities or public goods, however, given any competitive market allocation of private goods, there will usually be Pareto superior reallocations of private goods and externalities together. Thus, even perfect markets for private goods achieve at best a constrained notion of Pareto efficiency, along the lines of Hammond (1995) or Hammond and Sempere (2009).

2.3 Additional Markets for Externalities

It is commonly suggested that, even in the presence of externalities, unconstrained Pareto efficiency could be achieved by creating new markets for those externalities, with prices (positive or negative) that correspond to the appropriate Pigou subsidy or tax. This suggestion loses much of its persuasive power once one realizes that, as Starrett (1972) observes, negative externalities typically give rise to "fundamental non-convexities," which prevent existence of competitive equilibrium in a system of markets that allows externalities to be priced.

Nevertheless, the suggestion leads one to realize that the distinction between private goods and externalities, or public goods, really depends on the institutions that determine which goods are traded, and which are not. This, of course, introduces some ambiguity into the closely related definitions presented in Sect. 2.1. Looked at this way, it is institutions rather than tastes and technology that create externalities.

3 Pecuniary Externalities

3.1 Definition

In the first paragraph of his short subsection on pecuniary externalities, Laffont (2008) wrote as follows:

> During the 1930s, a confused debate occurred between economists on the relevance of pecuniary externalities, that is, on externalities which work through the price system. A quite general consensus was that pecuniary externalities are irrelevant for welfare economics: the fact that by increasing my consumption of whisky I affect your welfare through the consequent increase in price does not jeopardize the Pareto optimality of competitive equilibria.

In the penultimate sentence of the subsection, he wrote:

> When agents affect prices, they affect the welfare of the other agents by altering their feasible consumption sets or their information structures. Pecuniary externalities matter for welfare economics.

3.2 Limits to Gains from Liberalization

As an example of how pecuniary externalities can matter, it is worth considering gains from trade in international economics, notably the literature inspired by the classical results due to Samuelson (1939, 1962) and Kemp (1962). In general, moves toward freer trade are particular instances of economic liberalization or supply side policy reforms where, given a status quo allocation which would result in the absence of any liberalizing reform, there are moves away toward a more extensive market system—see, for example, Hammond and Sempere (1995).

Any such liberalizing reform typically changes the prices of goods, including the wages of workers. Such price changes will typically make some agents better off, and others worse off. To that extent, they are pecuniary externalities. The early literature often applied the Kaldor–Hicks compensation test, claiming that a reform would be beneficial provided that the gainers could compensate the losers in a way that would make all agents better off. Such compensation tests are not only ethically indefensible because they do nothing to ensure that losers actually get compensated. As Scitovsky (1941) and Gorman (1955) pointed out, they can also be logically inconsistent—see also Chipman's (2008) survey. Instead of relying on any compensation test, the real issue is whether a liberalizing reform can be made "credible" by linking it to suitably chosen policy instruments intended to limit the damage arising from negative pecuniary externalities—see Hammond (1993).

3.3 No Adverse Pecuniary Externalities

In order to ensure that there are no adverse pecuniary externalities, the classical literature on the gains from trade due to Samuelson and Kemp largely confines itself to two special cases.

In the first of these, there is a finite collection of trading nations, in each of which there is a single representative consumer. Moreover, the status quo allocation is taken to be autarkic, without any international trade. This ensures that whatever equilibrium prices result from free international trade in world markets, there can be no deterioration in any country's terms of trade. So, no nation's representative consumer can be made worse off by trade; moreover, except in the special case when the status quo allocation is already Pareto efficient, at least one nation's representative consumer will be strictly better off.

The second special case occurs when a single nation with just one representative consumer is a "small country," in the technical sense that its trade policy has no effect on prevailing world market prices. In this case, there are no pecuniary externalities at all because if the small nation liberalizes by moving to free trade at world prices, by definition this has no effect on world prices. So, except in the special case when the status quo is already a competitive equilibrium at world prices, the small nation's lone representative consumer will gain.

3.4 Mitigating Pecuniary Externalities

Though negative pecuniary externalities may be inevitable outside the two special cases just discussed, there are three particular kinds of mitigating policy that have received attention in the theoretic literature on economic liberalization. All of these mitigating policies work, moreover, without the need to assume any kind of representative consumer.

Following the work of Grandmont and McFadden (1972) in particular, the first kind of mitigating policy involves using lump-sum wealth redistribution. The idea is first to compensate each consumer for any adverse price movement, and then to share among all consumers any surplus generated by moving to a perfectly competitive allocation. Unless the status quo is already a Pareto efficient allocation, standard assumptions ensure that this surplus will be positive. So, the allocation after the reform, including this redistribution, makes every consumer strictly better off provided that they all have monotone preferences.

This kind of lump-sum redistribution, however, is typically incentive incompatible because it encourages agents to exaggerate the minimum compensation they need to ensure that they are no worse off than in the status quo, where there has been no reform. Following the ideas of Diamond and Mirrlees (1971) on optimal commodity taxation, Dixit and Norman (1986) discussed a second way to mitigate price changes. This involved fixing consumer prices at their status quo levels, while letting commodity tax rates and associated producer prices adjust to clear markets. This would then allow any surplus due to efficiency gains to be spent on a *uniform* lump-sum subsidy that is the same for all individuals. For details, see for example Hammond and Sempere (1995).

A third way of mitigating pecuniary externalities arises when the status quo has publicly known fixed quantities, as might be the case in a command economy such as China during the Maoist era. Such a status quo offers the scope for "dual-track liberalization" of the kind discussed by Lau, Qian, and Roland (1997, 2000) and by Che and Facchini (2007). The first track is the specified status quo allocation in the command economy; the second track is a competitive market economy. The two tracks are combined by first insisting that each agent receives the consumption goods and also supplies whatever is specified under the status quo. Agents, however, are also allowed to trade freely at market prices in order to determine whatever additional supply vector they want to offer in exchange for any additional consumption, etc. In effect, this dual-track policy determines a particular version of the lump-sum wealth redistribution rule considered by Grandmont and McFadden (1972), where each agent's wealth endowment equals the net value at the liberalized market prices of the fixed status quo allocation specified for them in the command economy.

4 Institutional Externalities

4.1 Strategy-Proof Allocation Mechanisms

An *economic environment* can be defined as a collection of economic agents, each of whom has a specified *individual characteristic* in the form of preferences and an endowment—possibly in the form of a production set—within a given finite-dimensional commodity space. Then, an *allocation rule* can be defined as a mapping from a given domain of possible economic environments to a codomain of allocations that are feasible in the relevant environment.

Hurwicz (1960, 1972, 1973) did much to initiate the systematic study of such allocation rules, and the information that would be needed to reach a satisfactory allocation—especially an allocation that is Pareto efficient—in each relevant environment. He considered a *principal* or *mechanism designer* who is granted the power to construct a *game form* or *allocation mechanism* in which each agent is required to send a signal from a suitably specified *signal space*, whereupon each possible profile of agents' signals is mapped into a feasible economic allocation. Notice that, when combined with agents' preferences for the economic allocation, and assuming these take the form of an expected utility function, the game form defines a game of incomplete information where each agent's payoff function is replaced by their expected utility.

A special case of particular interest is when every agent in every permissible economic environment has a dominant strategy which depends only on their own characteristic. In this case, one has a *dominant strategy* game form. The almost trivial theorem 4.4.1 of Dasgupta et al. (1979) proves that, in this case, there is an *equivalent direct mechanism* in which each agent's message is a direct signal of their individual characteristic; moreover, this direct mechanism is *strategyproof* in the sense that a dominant strategy for each agent is to announce their true characteristic.

4.2 Why Strategyproofness?

During the 1970s, Gibbard (1973) and Satterthwaite (1975) proved the general impossibility of constructing a strategy-proof social decision mechanism. Leo Hurwicz helped reinforce these negative results by considering when they held in specific economic environments, with or without public goods. Along with Groves and Ledyard (1977), Maskin (1999), and many others, he initiated the search for mechanisms whose Nash equilibria would yield Pareto efficient allocations.

Implementation in Nash equilibrium, however, can be criticized on methodological grounds. Let us exclude the very special case when a principal who is designing a mechanism lacks information which is common knowledge to all the agents who participate in the mechanism. Outside this case, it would seem that the relevant game form should involve incomplete information, thus suggesting Bayesian Nash equilibrium as a solution concept. Furthermore, it follows from Theorem 5.1 of Dasgupta et al. (1979) that, if a mechanism is not strategyproof, then the outcome it generates will be sensitive to agents' beliefs about each other—see also Ledyard (1978). Hence, except in rare cases, a mechanism that is implemented in Nash equilibrium rather than in dominant strategies generates allocations that depend not just on agents' tastes and endowments but also on their beliefs. These beliefs, moreover, concern not just other agents' tastes and endowments but also their beliefs about how these other agents will play the game form.

4.3 Strategy-Proof Exchange: When Is It Possible?

For the case of an exchange economy with two individuals, Hurwicz (1972) proved that any strategy-proof allocation rule yielding Pareto efficient outcomes must be dictatorial. Satterthwaite and Sonnenschein (1981) explored the difficulties in extending this result beyond two individuals. Serizawa (2002), along with Serizawa and Weymark (2003), showed that, even if they do not have to be dictatorial, nonetheless Pareto efficient strategy-proof rules always involve allocations that are close to being extreme—i.e., close to dictatorial. Finally, Barberà and Jackson (1995) characterized strategyproofness in general exchange economies with finite numbers of agents and goods and showed how limited they must be even if one does not insist on Pareto efficient outcomes.

Even so, there are some particular economic environments where strategy-proof exchange is possible. Apart from trivial cases, these environments have the key property that changes in agents' characteristics have no influence on the competitive equilibrium price, at least if the changes are small. It follows that institutional externalities are merely endemic, rather than universal.

4.4 Strategy-Proof Mechanism 1: An Islands Model

The first example is a static microeconomic version of the islands model, which is well known to macroeconomists following the work of Lucas (1972). There is a finite set of islands, each with a lone representative consumer. There is no possibility of trade between the islands, so each island has its own distinct commodity space of located goods specific to that island. Nor are the preferences or welfare of the representative consumer in any one island affected at all by the allocation in any other island. In this case, an obvious mechanism is to select an isolated optimal allocation separately within each island. This mechanism is clearly strategyproof because no agent's incentives are affected at all by the allocation that is chosen in any other island.

This example shows that the institutional externality that prevents strategy-proof exchange can be ascribed to the resource balance constraints that arise in a general exchange economy. In the special case of the islands model, agents are so separated that these constraints have force only within each island, so strategyproofness is possible.

4.5 Strategy-Proof Mechanism 2: Local Independence

The first case where the independence property mentioned in Sect. 4.3 holds, at least locally, is discussed by Makowski et al. (1999). They assume that at least one agent has a flat indifference surface in some neighbourhood of a Walrasian equilibrium allocation. While the economy has a Walrasian equilibrium allocation which remains in this neighbourhood, price ratios in this particular equilibrium are determined by the normal to this flat surface. Provided that this is the equilibrium chosen by the mechanism, individual agents cannot manipulate prices except by distorting their desired trades so much that they become worse off.

Section 7.5.2 of Hammond (2011) describes a second case of local independence, which holds if there is a linear technology. A particular example is when the "non-substitution theorem" holds. In its simplest form, this theorem relies on the assumptions that the economy's production possibilities are described by a finite collection of activities exhibiting constant returns to scale, a single common primary factor of production, and no joint production. These assumptions imply that equilibrium prices are independent of demand as long as demand does not change so much as to alter the pattern of goods that are inputs and goods that are outputs in any activity.

4.6 Strategy-Proof Mechanism 3: Infinitely Many Agents

The main case when strategy-proof exchange is possible, however, is when there are infinitely many agents. As acknowledged in Hammond (1979), it was Hurwicz (1972) himself who observed that the competitive mechanism is incentive compatible in a large economy. Sections 14 and 15 of Hammond (2011) are devoted to a survey of the results that hold in such environments. There is a broad class of environments in which strategy-proof exchange is possible, even in the presence of tax mechanisms such as those studied in Guesnerie (1995). Since that survey was written, the paper Hammond (2017) has appeared. It considers the complications involved in devising mechanisms that remain strategyproof even when not only are agents privately informed of their endowments but also any contracts to supply goods fail to be self-enforcing.

5 Concluding Remarks

The first part of this paper focused on both classical and pecuniary externalities, emphasizing their links to institutional features of the economic system in which they arise. Later, the paper went on to explore some implications of viewing any institution that is modelled by a game form that is not strategyproof as giving rise to institutional externalities. Specifically, as with both classical and pecuniary externalities, they arise when an agent's choice of action in the game form affects other agents' incentives.

Leo Hurwicz's early work on the difficulties of constructing strategy-proof mechanisms shows that institutional externalities, understood in this way, are endemic. The paper also explores a few very special cases where there will be no institutional externalities. Typically, these involve purely static economic environments with only private goods and either many economic agents, as Hurwicz (1972) himself had suggested; or other special environments where no individual agent has an influence over prices, such as when the non-substitution theorem holds.

Acknowledgements I had the privilege and pleasure of meeting Leo Hurwicz on several occasions, notably at summer workshops organized by the economics section of the Institute of Mathematical Studies in the Social Sciences at Stanford University. My last conversation with him, however, took place during the summer 2000 meeting of APET (The Association for Public Economic Theory) at the University of Warwick. During that meeting, Leo gave a talk on externalities which seems related to Hurwicz (1999). Thoughts provoked by his presentation may have helped me prepare an after dinner talk entitled "What *isn't* an Externality?" for a conference on "Modelling public goods and public policy: Past, present and future prospects" organized by Monique Florenzano and Sylvie Thoron at C.I.R.M. (Centre International de Rencontres Mathématiques) in Marseille-Luminy. This was held almost immediately after I took up my current position at Warwick on 1st April 2007. My thanks to the audience in Luminy for encouraging me to share more widely a significantly revised version of my remarks on that occasion, and also to Walter Trockel for providing a suitable outlet.

References

Arrow, K. J., & Hurwicz, L. (Eds.) (1977). *Studies in resource allocation processes*. Cambridge: Cambridge University Press.

Barberà, S., & Jackson, M. O. (1995). Strategy–proof exchange. *Econometrica, 63*(1), 51–88.

Che, J., & Facchini, G. (2007). Dual track reforms: With and without losers. *Journal of Public Economics, 91*, 2291–2306.

Chipman, J. S. (2008). Compensation principle. In S. N. Durlauf & L. E. Blume (Eds.), *New Palgrave dictionary of economics* (2nd edn.). Basingstoke: Palgrave Macmillan.

Dasgupta, P., Hammond, P., & Maskin, E. (1979). The implementation of social choice rules: Some general results on incentive compatibility. *Review of Economic Studies, 46*, 185–216.

Diamond, P. A., & Mirrlees, J. A. (1971). Optimal taxation and public production, I and II. *American Economic Review, 61*, 8–27 and 261–278.

Dixit, A., & Norman, V. (1986). Gains from trade without lump-sum compensation. *Journal of International Economics, 21*, 99–110.

Gibbard, A. (1973). Manipulation of voting schemes: A general result. *Econometrica, 41*(4), 587–601.

Gorman, W. M. (1955). The intransitivity of certain criteria used in welfare economics. *Oxford Economic Papers, 7*(1), 25–34.

Grandmont, J.-M., & McFadden, D. (1972). A technical note on classical gains from trade. *Journal of International Economics, 2*, 109–125.

Groves, T., & Ledyard, J. (1977). Optimal allocation of public goods: A solution to the 'Free Rider' problem. *Econometrica, 45*(4), 783–809.

Guesnerie, R. (1995). *A contribution to the pure theory of taxation*. Cambridge: Cambridge University Press.

Hammond, P. J. (1979). Straightforward individual incentive compatibility in large economies. *Review of Economic Studies, 46*, 263–282

Hammond, P. J. (1993). Credible liberalization: Beyond the three theorems of neoclassical welfare economics. In D. Bös (Ed.), *Economics in a changing world, vol. 3: Public policy and economic organization* (ch. 3, pp. 21–39) IEA Conference Volume No. 109. London: Macmillan.

Hammond, P. J. (1995). Four characterizations of constrained Pareto efficiency in continuum economies with widespread externalities. *Japanese Economic Review, 46*, 103–124.
Hammond, P. J. (2011). Competitive market mechanisms as social choice procedures. In K. J. Arrow, A. K. Sen, & K. Suzumura (Eds.), *Handbook of social choice and welfare, vol. II* (ch. 15, pp. 47–151). Amsterdam: North-Holland.
Hammond, P. J. (2017). Designing a strategy-proof spot market mechanism with many traders: Twenty-two steps to Walrasian equilibrium. *Economic Theory, 63*(1), 1–50.
Hammond, P. J., & Sempere, J. (1995). Limits to the potential gains from economic integration and other supply side policies. *Economic Journal, 105*, 1180–1204.
Hammond, P. J., & Sempere, J. (2009). Migration with local public goods and the gains from changing places. *Economic Theory, 41*, 359–377.
Hurwicz, L. (1960). Optimality and informational efficiency in resource allocation processes. In K. J. Arrow, S. Karlin, & P. Suppes (Eds.), *Mathematical methods in the social sciences, 1959: Proceedings of the first Stanford symposium* (pp. 27–46). Stanford: Stanford University Press.
Hurwicz, L. (1972). On informationally decentralized systems. In C. B. McGuire & R. Radner (Eds.), *Decision and organization: A volume in honor of Jacob Marschak* (pp. 297–336). Amsterdam: North-Holland.
Hurwicz, L. (1973). The design of mechanisms for resource allocation. *American Economic Review, Papers and Proceedings, 63*(2), 1–30.
Hurwicz, L. (1979). Outcome functions yielding Walrasian and Lindahl allocations at Nash equilibrium points. *Review of Economic Studies, 46*, 217–225.
Hurwicz, L. (1986). Incentive aspects of decentralization. In K. J. Arrow & M. D. Intriligator (Eds.) *Handbook of Mathematical Economics* (vol. 3, ch. 28, pp. 1441–1482). Amsterdam: Elsevier.
Hurwicz, L. (1995). What is the Coase theorem? *Japan and the World Economy, 7*, 49–74.
Hurwicz, L. (1996). Institutions as families of game forms. *Japanese Economic Review, 47*(2), 113–132.
Hurwicz, L. (1998). Issues in the design of mechanisms and institutions. In E. T. Loehman & D. M. Kilgour (Eds.), *Designing Institutions for Environmental and Resource Management*. Cheltenham: Edward Elgar.
Hurwicz, L. (1999). Revisiting externalities. *Journal of Public Economic Theory, 1*(2), 225–245.
Hurwicz, L. (2008). But who will guard the guardians? In K. Grandin (Ed.), *Les Prix Nobel. The Nobel Prizes 2007* (pp. 280–291). Stockholm: Nobel Foundation. .
Kemp, M. C. (1962). The gains from international trade. *Economic Journal, 72*, 803–819.
Laffont, J.-J. (2008). Externalities. In S. N. Durlauf & L. E. Blume (Ed.), *New Palgrave Dictionary of Economics* (2nd. edn.). Basingstoke: Palgrave Macmillan.
Lau, L. J., Qian, Y., & Roland, G. (1997). Pareto-improving economic reforms through dual-track liberalization. *Economics Letters, 55*, 285–292.
Lau, L. J., Qian, Y., & Roland, G. (2000). Reform without losers: An interpretation of China's dual-track approach to transition. *Journal of Political Economy, 108*, 120–143.
Ledyard, J. O. (1978). Incentive compatibility and incomplete information. *Journal of Economic Theory, 18*(1), 171–189.
Lucas, R. E. (1972). Expectations and the neutrality of money. *Journal of Economic Theory, 4*(2), 103–124.
Makowski, L., Ostroy J. M., & Segal, U. (1999). Efficient incentive compatible economies are perfectly competitive. *Journal of Economic Theory, 85*, 169–225.
Mas-Colell, A., Whinston, M. D., & Green, J. R. (1995). *Microeconomic theory*. New York: Oxford University Press.
Maskin, E. (1977, published 1999). Nash equilibrium and welfare optimality. *Review of Economic Studies, 66*(1), 23–38.
Samuelson, P. A. (1939). The gains from international trade. *Canadian Journal of Economics, 5*, 195–205.
Samuelson, P. A. (1962). The gains from international trade once again. *Economic Journal, 72*, 820–829.

Satterthwaite, M. A. (1975). Strategy-proofness and arrow's conditions: Existence and correspondence theorems for voting procedures and social welfare functions. *Journal of Economic Theory, 10*(2), 187–217.

Satterthwaite, M. A., & Sonnenschein, H. (1981). Strategy-proof allocation mechanisms at differentiable points. *Review of Economic Studies, 48*(4), 587–597.

Scitovsky, T. (1941). A note on welfare propositions in economics. *Review of Economic Studies, 9*, 77–88.

Serizawa, S. (2002). Inefficiency of strategy-proof rules for pure exchange economies. *Journal of Economic Theory, 106*(2), 219–241.

Serizawa, S., & Weymark, J. A. (2003). Efficient strategy-proof exchange and minimum consumption guarantees. *Journal of Economic Theory, 109*(2), 246–263.

Starrett, D. A. (1972). Fundamental nonconvexities in the theory of externalities. *Journal of Economic Theory, 4*(2), 180–199

The Role of (Quasi) Analyticity in Establishing Completeness of Financial Markets Equilibria

Yakar Kannai and Roberto C. Raimondo

1 Introduction

It is well-known that there is a sharp difference between an Arrow-Debreu market and a securities market. In the former, agents are allowed to trade a complete set of Arrow-Debreu contingent claims. In the case of a binary tree with T periods, one deals with 2^T Arrow-Debreu contingent claims, whereas in a securities market where one is allowed to trade in every time period, two commodities suffice. Passing to a continuous time model, one notes that there are too many (infinitely many) Arrow-Debreu contingent claims, and there is reason to conjecture that under suitable assumptions a finite number of instruments traded continuously would do. A market is said to be *dynamically complete* if agents can, by trading the given set of securities, achieve all the consumption allocations that they could achieve in an Arrow-Debreu market (see Duffie, 1986 and Duffie and Huang, 1985). A necessary condition for dynamic completeness is that the market is potentially dynamically complete. This means, if the uncertainty is driven by a multi-dimensional Brownian motion, that the number of

Y. Kannai (✉)
Department of Mathematics, Weizmann Institute of Science, Rehovot, Israel
e-mail: kannai@wisdom.weizmann.ac.il

R. C. Raimondo
Department of Mathematics and Applications, Universita' degli Studi Milano-Bicocca, Milan, Italy

Department of Economics, University of Melbourne, Parkville, Victoria, Australia
e-mail: roberto.raimondo@unimib.it; rraim@unimelb.edu.au

W. Trockel (ed.), *Social Design*, Studies in Economic Design,
https://doi.org/10.1007/978-3-319-93809-7_11

independent securities is at least one more than the dimension of this Brownian motion.

There are two important reasons for establishing completeness of financial markets. One is that with completeness one may aggregate and therefore fully justify the representative agent model so common in macroeconomics (see, e.g., Browning et al., 1999). The other major application of completeness is in derivative pricing. Only when the price processes form a dynamically complete market, options and other derivatives can be uniquely priced by arbitrage arguments and can be replicated by trading the underlying securities. If we do not have dynamic completeness, then replication is not possible, so that a unique pricing of derivatives is impossible.

In many studies of completeness the prices are introduced as given. It is desirable to let securities prices be determined endogenously as equilibrium prices, determined by market forces. But securities are not consumed. One considers a market in which there exists at least one consumption good from which agents derive utility and which can be bought by income generated by trade in securities and by dividends.

Existence of equilibrium in a continuous-time securities market in which the securities are endogenously dynamically complete has been proved for the first time under rather restrictive assumptions by Anderson and Raimondo (2008); their approach was, broadly speaking, followed by subsequent work. Given the fundamental importance of completeness for pricing and for aggregation (see, e.g., Browning et al., 1999) various extensions of this result have been proposed (Hugonnier et al., 2012; Kramkov and Predoiu, 2014; Riedel and Herzberg, 2013). However, as suggested already in the discussion of Anderson and Raimondo (2008), in all of these attempts a crucial role was played by assuming analyticity[1] of the basic economic ingredients, in particular agents utility functions, endowments, and dividends.

In all of the above-mentioned papers the role of analyticity is crucial for showing that if a market is incomplete in an open set then it must be incomplete everywhere including the terminal date. This, from the point of view of theory of complete markets and optimality, is the most important property, since optimality is lost even with a local loss of completeness. Hence the really crucial issue it is to determine exactly when local collapse brings about the global one. To achieve such a goal we show that the right notion is not analyticity but a much weaker one, known as quasi-analyticity. The relevance of this property was not recognized in the previous literature; analyticity was used instead. Let us stress that as in Anderson and Raimondo (2008) our existence result is universal rather than generic. Moreover, our result is robust w.r.t. a general class

[1]Some of the results in Kramkov and Predoiu (2014) and in the supplement of Hugonnier et al. (2012) are obtained using only time analyticity. However, as stated in Kramkov and Predoiu (2014), their assumption on the terminal condition is stronger. Such extensions are treated with our method in a forthcoming paper.

of reasonable mis-specifications, a fundamental property lacking in the analytic models.

A class of functions is said to be quasi-analytic if the local collapse of a function in the class implies global vanishing; formally, no non-zero function in the class has a zero of infinite order. It is well-known (Hörmander, 1990) that this is equivalent for the class to be determined by certain growth conditions on the derivatives. In particular, the class of analytic functions is a very special case of such classes. The existence of functions which are quasi-analytic but not analytic was established more than a century ago (see Borel, 1901; we reproduce his example in appendix 1). Properties of quasi-analytic functions have been treated recently in Bierstone and Milman (2004). It follows from Bierstone and Milman (2004) that quasi-analytic functions of several variables cannot vanish on a set of positive measure without being identically zero. Quasi-analyticity of solutions of parabolic partial differential equations such as those that appear in Finance Theory has been established recently by the authors of this paper (Kannai and Raimondo, 2013).

As is common in establishing existence of equilibrium price processes, we first construct a candidate equilibrium process. Then we prove that the candidate equilibrium price process is actually dynamically complete, and that the candidate equilibrium is in fact an equilibrium. In our case the dynamic completeness of the candidate equilibrium price process and existence of equilibrium follow from the way information is revealed by a general Ito process, and from an exogenous nondegeneracy condition on the terminal security dividends. This nondegeneracy condition is the customary one, see Anderson and Raimondo (2008) where the condition is motivated and discussed.

The model is introduced in Sect. 2. Note that we diverge from Anderson and Raimondo (2008) by allowing a much more general underlying stochastic processes and from Hugonnier et al. (2012), Kramkov and Predoiu (2014), Riedel and Herzberg (2013) by making the bare minimum assumptions, namely we assume only quasi-analyticity of the basic economic ingredients. In Sect. 3 we illustrate the effectiveness of our approach by exhibiting several economic examples. The results are discussed in Sect. 4. Properties of quasi-analytic functions of several variables and related regularity results for solutions of parabolic partial differential equations are described in Appendix 1. Proofs of the main theorems are sketched in Appendix 2. Let us point out that our results may be extended, similarly to Hugonnier et al. (2012), to the non-finite horizon framework, and that it is possible to relax the quasi-analyticity assumption w.r.t. space variables (compare Hugonnier et al., 2012 and Kramkov and Predoiu, 2014); we leave the details of these extensions as well as the detailed proofs of the results presented in the present paper to a subsequent publication. Finally we stress that in all our proofs and theorems non-standard analysis is not used, hence they should sound and look more familiar to economists.

2 The Model and Main Results

In this section we present the essentials of a standard continuous-time model with consumption and equilibrium, as described fully in Anderson and Raimondo (2008), and we highlight the main points where we differ. Here we allow for much more general dividend processes and handle more general utilities, endowments, and payoffs as well.

There is a single consumption good. Trade and consumption occur over a compact time interval $[0, T]$, endowed with a measure ν which agrees with Lebesgue measure on $[0, T)$ and such that $\nu(\{T\}) = 1$. Consumption and dividends on $[0, T)$ are flows; consumption at the terminal date T is a lump. A nondegeneracy assumption will be imposed on the lump dividend at the terminal date T.

The uncertainty in the model is described by a standard K-dimensional Brownian Motion $\{B_t\}_{t\geq 0}$ on a probability space Ω. Let X_t be the strong solution of the following SDE

$$dX_t = b(t, X_t)dt + \sigma(t, X_t)dB. \quad (1)$$

Then the primitives of the economy—dividends, endowments, and utility functions—will be described as functions of (t, X_t). We assume that the functions $b(t, x)$ and $\sigma(t, x)$ are quasi-analytic functions of $(t, x) \in (0, T) \times \mathbf{R}^K$ and continuous on the closure. (We are abusing terminology slightly; in this section, we say that a function is quasi-analytic if it belongs to a certain quasi-analytic class $\mathcal{D}(X, \{M_k\})$, where X is an open subset of R^n, see Appendix 1 for the precise definition.) It is important to notice that this assumption is considerably weaker than that of real analyticity.[2]

Eventually, prices are going to satisfy the partial differential equation.

$$\frac{\partial p_{Aj}}{\partial t} = -A(t, x, D)p_{A_j} - p_C(t, x)g_j(t, x). \quad (2)$$

As usual, we assume that the operator

$$A(t, x, D) = \sum_{j=1}^{K} b_j(t, x)\frac{\partial}{\partial x_j} + \frac{1}{2}\sum_{i,j,\ell=1}^{K} \sigma_{i\ell}(t, x)\sigma_{\ell j}(t, x)\frac{\partial^2}{\partial x_i \partial x_j} \quad (3)$$

is uniformly elliptic in $(0, T) \times \mathbf{R}^K$ and the first and the second derivatives of all coefficients with respect to all of the variables are uniformly bounded and quasi-analytic.

[2] Actually, to pass from separate analyticity in space and time variables one needs a stronger form of real analyticity, involving, e.g., suitable local extendibility such as in Hugonnier et al. (2012).

We are given a right-continuous filtration $\{\mathcal{F}_t : t \in [0, T]\}$ on $(\Omega, \mathcal{F}, \mu)$ such that $\mathcal{F}_0$ contains all null sets and B is adapted to $\{\mathcal{F}_t\}$, i.e. for all t, $B(t, \cdot)$ is measurable with respect to $\mathcal{F}_t$.

There are $K+1$ securities $A_0, A_1, \ldots, A_K$. Security j pays dividends (measured in consumption units) at a flow rate

$$D_j(t, \omega) = g_j(t, X_t)$$

at times $t \in [0, T)$, and a lump dividend $D_j(T, \omega) = G_j(X_T)$ at time T. We assume that $g : [0, T] \times \mathbf{R}^K \to \mathbf{R}_+^K$ is quasi-analytic on $(0, T) \times \mathbf{R}^K$ and that G_j is locally in L^2.

There are I agents $i = 1, \ldots, I$. Agent i has a flow rate of endowment $e_i(t, \omega) = f_i(t, X_{t,\omega})$ at times $t \in [0, T)$, and a lump endowment $e_i(T, \omega) = F_i(X_T(\omega))$, where f_i is quasi-analytic on $(0, T) \times \mathbf{R}^K$ and F_i is continuous almost everywhere on $\mathbf{R}^K$. Agent i is initially endowed with deterministic security holdings $e_{iA} = (e_{iA_0}, \ldots, e_{iA_J}) \in \mathbf{R}^{J+1}$ satisfying

$$\sum_{i=1}^{I} e_{iA_j} = \eta_j$$

where the security $j = 0, \ldots, K$ is assumed to be in net supply $\eta_j \in [0, 1]$. We impose the same condition on the initial holdings as in Anderson and Raimondo (2008). Let $e(t, \omega) = \sum_{i=1}^{I} e_i(t, \omega)$ denote the aggregate endowment.

We assume the endowments and dividends satisfy the following mild growth conditions: for every $\epsilon > 0$

$$\int_{\mathbf{R}^K} |u_0(x)| \exp\left(-\epsilon \sum_{i=1}^{K} |x_i|^2\right) dx < \infty, \tag{4}$$

and

$$\int_{\mathbf{R}^K} |v(t, x)| \exp\left(-\epsilon \sum_{i=1}^{K} |x_i|^2\right) dx < \infty \tag{5}$$

uniformly in $[0, T]$, where $u_0(x)$ is either of the functions $F_i(T, x)$ or $G_j(T, x)$, and $v(t, x)$ is either of the functions $f_i(T, x)$ or $g_j(T, x)$, $\frac{\partial f_i(t,x)}{\partial x}$, $\frac{\partial g_j(t,x)}{\partial x}$. Actually, we need some technical growth conditions on high time derivatives of the function v as well (see Theorem 7). Observe that these growth conditions are weaker than the ones implicitly assumed in Anderson and Raimondo (2008). Note that we do not require G_j to be even continuous. Standard option payoffs are not differentiable at the strike price, and other derivatives need not be continuous.

The utility functions are von-Neumann Morgenstern utility functions, expectations of functions of the consumption and the process X which are quasi-analytic on $(0, T) \times \mathbf{R}^K$. More formally, given a measurable consumption function $c_i : [0, T] \times \Omega \to \mathbf{R}_{++}$, the utility function of the agent is

$$U_i(c) = E_\mu \left[\int_0^T h_i(c_i(t, \cdot), X(t, \cdot))dt + H_i(c_i(T, \cdot), X(T, \cdot)) \right] \tag{6}$$

where the functions $h_i : \mathbf{R}_{++} \times \left([0, T) \times \mathbf{R}^K\right) \to \mathbf{R} \cup \{-\infty\}$ and $H_i : \mathbf{R}_{++} \times \left(\{T\} \times \mathbf{R}^K\right) \to \mathbf{R} \cup \{-\infty\}$ are quasi-analytic on $\mathbf{R}_{++} \times \left((0, T) \times \mathbf{R}^K\right)$ and C^2 on $\mathbf{R}_{++} \times \left(\{T\} \times \mathbf{R}^K\right)$, respectively, and satisfy the standard Inada conditions as well as standard monotonicity and concavity assumptions as elaborated in Anderson and Raimondo (2008). As usual we assume that the space of consumption processes is in $L^2\left([0, T] \times \mathbf{R}^K, dv \otimes d\mu\right)$. We also impose joint growth assumptions on the utility functions and the social consumption in order to avoid imposing that the consumption is bounded away from zero. Namely, we assume that there exists $r \in \mathbf{R}^K$ such that the functions $\frac{\partial H_i}{\partial c}_{c(T,x)/I}$ and $\frac{\partial h_i}{\partial c}_{c(t,x)/I}$ are bounded by $r + e^{r|x|}$ where

$$c(t, x) = \begin{cases} \sum_{i=1}^I f_i(t, x) + \sum_{j=0}^J \eta_j g_j(t, x) & \text{if } t < T \\ \sum_{i=1}^I F_i(t, x) + \sum_{j=0}^J \eta_j G_j(t, x) & \text{if } t = T \end{cases}.$$

and we assume $\frac{\partial H_i}{\partial c}_{e_i(T,x)}, \frac{\partial h_i}{\partial c}_{e_i(t,x)} \in L^2\left([0, T] \times \mathbf{R}^K, dv \otimes d\mu\right)$. As customarily in continuous-time models we assume that the zeroth security is a money-market account, in other words, it is instantaneously risk-free.

We make the following non-degeneracy assumption: there is an open set $V \subset \mathbf{R}^K$ such that $G_0(T, x) > 0$ for all $x \in V$ and for $j = 1, \ldots, J$ and $i = 1, \ldots, I$

$$G_j, F_i \in C^1(V) \text{ and } \forall x \in V \text{ rank} \begin{pmatrix} \left.\frac{\partial(G_1/G_0)}{\partial X}\right|_{(T,x)} \\ \vdots \\ \left.\frac{\partial(G_J/G_0)}{\partial X}\right|_{(T,x)} \end{pmatrix} = K \tag{7}$$

Note that if the zeroth security is a bond, the rank condition is equivalent to assuming that the $K \times K$ matrix

$$\begin{pmatrix} \left.\frac{\partial G_1}{\partial X}\right|_{(T,x)} \\ \vdots \\ \left.\frac{\partial G_J}{\partial X}\right|_{(T,x)} \end{pmatrix}$$

is nonsingular. This simply says that there is some possible terminal value of the process whose differential is $dX_t = b(t, X_t)dt + \sigma(t, X_t)dB$ so that the dividends of securities $A_1, \ldots, A_J$ are locally linearly independent.

The definitions of budget constraints, trading strategies and equilibrium are the standard ones for models where the securities are priced *cum dividend*. In order to define the budget set of an agent, we need to have a way of calculating the capital gain the agent receives from a given trading strategy. In other words, we need to impose conditions on prices and strategies that ensure that the stochastic integral of a trading strategy with respect to a price process is defined. The essential requirements are that the trading strategy at time t does not depend on information which has not been revealed by time t, and the trading strategy times the variation in the price yields a finite integral. Specifically, a consumption price process is an Itô process $p_C(t, \omega)$. A securities price process is an Itô process $p_A = (p_{A_0}, \ldots, p_{A_J}) : \Omega \times [0, T] \to \mathbf{R}^{J+1}$ such that the associated cumulative gains process

$$G_j(t, \omega) = p_{A_j}(t, \omega) + \int_0^t p_C(s, \omega) A_j(s, \omega)\, ds$$

is a martingale. Given a securities price process p_A, an admissible trading strategy for agent i is a process z_i which is Itô integrable with respect to G and such that $\int z_i\, dG$ is a martingale. Given a securities price process p_A and a consumption price process p_C, the budget set for agent i is the set of all consumption plans c_i such that there exists an admissible trading strategy so that c_i and t_i satisfy the budget constraint

$$\begin{aligned}
&p_A(t, \omega) \cdot z_i(t, \omega) \\
&= p_A(0, \omega) \cdot e_{iA}(\omega) + \int_0^t z_i\, dG + \int_0^t p_C(s, \omega)(e_i(s, \omega) - c_i(s, \omega))ds \\
&\quad \text{for almost all } \omega \text{ and all } t \in [0, T) \\
0 &= p_A(0, \omega) \cdot e_{iA}(0, \omega) + \int_0^T z_i\, dG + \int_0^T p_C(s, \omega)(e_i(s, \omega) - c_i(s, \omega))ds \\
&\quad + p_C(T, \omega)(e_i(T, \omega) - c_i(T, \omega))) \\
&\quad \text{for almost all } \omega
\end{aligned}$$

Given a price process p, the demand of the agent is a consumption plan and an admissible trading strategy which satisfy the budget constraint and such that the consumption plan maximizes utility over the budget set.

Of course, we also use the following standard

Definition 1 A Radner equilibrium (Radner, 1972) is a set of price processes $(p_C, p_{A_0}, p_{A_1}, \ldots, p_{A_K})$, a consumption allocation $(c_i)_{i=1}^I$, and a set of strategies $\left((z_{iA_0}, \ldots, z_{iA_K})\right)_{i=1}^I$ such that

(a) The plan c_i maximizes U_i over the budget set and is financed by $\left(z_{iA_0}, \ldots, z_{iA_K}\right)$,
(b) All markets clear.

The following is a considerable extension of the main results of the previous literature, establishing existence of an effectively dynamically complete equilibrium pricing process. (This means that derivatives can be uniquely priced by arbitrage arguments and can be replicated by trading the underlying securities.) Namely, previous work applies to the special cases where $dX_t = b(t, X_t)dt + \sigma(t, X_t)dB$ with $b(t, X_t) = 0$ and $\sigma(t, X_t)$ is the identity matrix in $\mathbb{R}^K$ (Anderson and Raimondo, 2008) or when $b(t, x)$ and $\sigma(t, x)$ are analytic functions (Hugonnier et al., 2012; Riedel and Herzberg, 2013); in Kramkov and Predoiu (2014) only time-analyticity is assumed, however the rank condition (7) has to hold almost everywhere in $\mathbf{R}^K$.

Our main result is the following

Theorem 2 *The continuous-time finance model just described has an equilibrium, which is Pareto optimal. The equilibrium pricing process is effectively dynamically complete, and the admissible replicating strategies are unique. Moreover the prices for assets and goods are quasi-analytic.*

Note that (besides following the Negishi method Negishi, 1960 , standard for infinitely many commodities, see, e.g., Dana and Jeanblanc, 2002) the two main ingredients of the proof of Anderson and Raimondo (2008) and subsequent work are the implicit function theorem for real analytic functions and the fact that a nonzero real analytic function of several variables cannot vanish on a set of positive measure. It turns out that both ingredients continue to be valid for quasi-analytic classes of functions (see Appendix 1).

We emphasize that in our model, as in Anderson and Raimondo (2008), the prices are given by

$$p_A(t, X) = E^{t,X}\left(p_C(T, X_T)A(T, X_T) + \int_t^T p_C(s, X(s))A(s, X(s))\,ds\right), \quad (8)$$

where $p_C(s, X(s))$ is the relative price of consumption at equilibrium. As in Anderson and Raimondo (2008) this price is determined by the implicit function theorem. Applying now the implicit function theorem for quasi-analytic classes (Bierstone and Milman, 2004; see Theorem 5 in Appendix 1) we find that the function $p_C(t, x)$ is (jointly) quasi-analytic in (t, x). The asset prices determined by (8) satisfy the partial differential equation (2)

$$\frac{\partial p_{Aj}}{\partial t} = -A(t, x, D)p_{A_j} - p_C(t, x)g_j(t, x) \quad (9)$$

with the boundary condition

$$p_{Aj}(T, x) = G_j(T, X(T))p_C(T, x) \quad (10)$$

where

$$A(t,x,D) = \sum_{j=1}^{K} b_j(t,x)\frac{\partial}{\partial x_j} + \frac{1}{2}\sum_{i,j,\ell=1}^{K} \sigma_{i\ell}(t,x)\sigma_{\ell j}(t,x)\frac{\partial^2}{\partial x_i \partial x_j}. \tag{11}$$

The transitions probabilities are just the fundamental solutions of (11). We use these representations to show that the prices are quasi-analytic. This is the main technical point and, as specified in the introduction, it is addressed in Appendices 1 and 2. We stress that quasi-analyticity of prices in both space and time is established from the assumptions on the basic underlying economic ingredients.

3 Examples

In our model and examples securities are described by their dividend processes, rather than by their price processes. Similar models and examples already appear in the literature but they all hold under special assumptions on the dividends, in particular either analyticity is assumed or, worse, completeness is postulated. In the following all functions (drift, instantaneous volatility, utilities, dividends, endowments) are assumed just to be quasi-analytic (more general and much more natural than analytic).

In all examples below there are I agents and each agent i is endowed with a flow rate of endowment e_i for $t \in [0,T)$, and a lump endowment $e_i(T)$ at time T. There are $K+1$ securities, which pay dividends at times $t \in [0,T)$ and which pay lump dividends at time T. The zeroth security is a zero coupon bond which pays a lump dividend of one unit of consumption at time T. The zeroth security is in zero net supply, while the remaining securities are in net supply one. At time zero, agent i has an initial holding of $\delta_i \in \mathbf{R}^{K+1}$ units of the securities so that the $\delta_i s$ with $\delta = (\delta_1, \ldots, \delta_I)$ are in

$$H_\delta = \left\{ (\delta_1, \ldots, \delta_I) \in \mathbb{R}^{(K+1)I} \;\middle|\; \sum_{i=1}^{I} \delta_i = (0,1,\ldots,1) \right\}.$$

Our examples are presented in parametric form in order to account for possible mis-specification when the theory is used in applications. We get robustness even if analyticity of the functional form is itself mis-specified. For example, even if the assumed functional form is analytic, it may well cease to be so if mis-specified.

Example 3 This example is very close to the ones used already in this context (see Merton, 1973 and Merton, 1990). Each agent i is endowed with a flow rate of endowment e_i for $t \in [0,T)$

$$e_i(t,\omega) = f_i(X(t,\omega)) + \epsilon_{1,i}\widetilde{f}_i(X(t,\omega))$$

where the first term denotes a fixed process which is perturbed by the second term, and a lump endowment $e_i(T)$ at time T. Agent i's utility for a stream of consumption c_i is

$$U_i(c_i) = E\left(\int_0^T c_i(s)^{\alpha_i}\, ds + c_i(T)^{\alpha_i}\right)$$

where c^{α_i} is a state-independent CRRA utility function with coefficient of relative risk aversion α_i (one could use any CRRA). There are $K + 1$ securities, which pay dividends at times $t \in [0, T)$ and which pay lump dividends at time T. The dividends of the risky securities are given as follows if $t \in [0, T)$

$$D_j(t, \omega) = g_j(X(t, \omega)) + \epsilon_{2,j}\widetilde{g_j}(X(t, \omega))$$

where once again the first term denotes a fixed process which is perturbed by the second term. If $t = T$, then the dividend is given by $e^{\sigma B(T)}$, where $\sigma = \left[\sigma_{j\ell}\right]_{i,j=1\ldots k}$ is constant $K \times K$ matrix such that

$$\det\left[\sigma_{j\ell}\right]_{i,j=1\ldots k} \neq 0$$

i.e., it is nonsingular so that the dividends are terminal values of a K-dimensional geometric Brownian Motion. Of course, the main question is for what set of parameters does this securities market have an equilibrium. If an equilibrium exists, is it dynamically complete? At this level of generality there is no known result that could give us an answer even for a single value of the parameters, unless all of the $\alpha_i's$ are equal, but this means that we have just one agent! Our main result, Theorem 2, implies that if σ is nonsingular, then for *every* value of the other parameters, i.e. every $(\epsilon_1, \epsilon_2, \alpha, \delta_1, \ldots, \delta_I, T)$ in the space

$$\mathbf{R}^I_{++} \times \mathbf{R}^K_{++} \times (0, 1)^I \times H_\delta \times \mathbf{R}_{++}$$

an equilibrium exists and is dynamically complete.

In the next example we do not make any special assumptions on utilities, instead we let the agents be as heterogeneous as possible. The importance of heterogeneity is by now widely accepted in the literature. It seems that this very simple illustration of our result is widely applicable.

Example 4 Each agent i is endowed with a flow rate of endowment e_i for $t \in [0, T)$

$$e_i(t, \omega) = f_i(X(t, \omega))$$

and a lump endowment e_i at time T. Agent i's utility for a stream of consumption c_i is

$$E\left(\int_0^T u_i(c_i(s), \omega)\, ds + u_i(c_i(T), \omega)\right)$$

where each u_i is an arbitrary quasi-analytic utility function. (**We emphasize that there is no relation whatsoever between the various u_i's.**) There are $K + 1$ securities, which pay dividends at times $t \in [0, T)$ and which pay lump dividends at time T. The dividends of the risky securities are given by

$$D_j(t, \omega) = g_j(X(t, \omega))$$

if $t \in [0, T)$, and at $t = T$ by $e^{\sigma B(T)}$, where $\sigma = \left[\sigma_{j\ell}(T, \omega)\right]_{i,j=1\ldots k}$ is a random $K \times K$ matrix such that

$$\det\left[\sigma_{j\ell}(T, \omega)\right]_{i,j=1\ldots k} \neq 0$$

i.e., it is nonsingular almost surely. Of course, given our (arbitrary) level of heterogeneity, the main question is for what set of primitives does this securities market have a dynamically complete equilibrium. There is no known result that could give us an answer. Once again, our main result implies that if σ is nonsingular, then for *every* choice of value of the other parameters, i.e. every profile of utilities $(u_i, \ldots, u_I)$ there exists a dynamically complete equilibrium.

These examples are only indicative. One could consider more general perturbations, in particular it is possible to perturb the utility functions. Moreover, with extra effort we could consider much more general examples by noting that the results of Kannai and Raimondo (2013) are stable under sufficiently small perturbations of a certain kind.

4 Existence, Completeness, and Optimality of Financial Equilibrium

Our existence proof deals with the case where the market is potentially dynamically complete. The hardest part of this proof is to guarantee the dynamic completeness of the model at equilibrium. In order to explain why our approach is the proper one we need to recast the Anderson and Raimondo's work in a different way. In fact, we can say that Anderson and Raimondo studied a model where the local collapse of completeness would force a global collapse. This, from the point of view of theory of complete markets and optimality, is the most important case, since optimality is lost even with a local loss of completeness. Hence the really crucial issue is to determine exactly when local collapse brings the global one. Of course, this comes at a cost since we are not looking for a sufficient condition (such as analyticity) but for the necessary and sufficient condition (quasi-analyticity).

In order to achieve the correct satisfactory level of generality we had to create the proper mathematical setup, departing from the previous analytic setup. In the first step we apply a new result in the theory of partial differential equations, developed by the authors Kannai and Raimondo (2013), that allows us to prove that prices have a much greater regularity than previously expected. In Appendix 1 we present the relevant results, and discuss other tools we use in order to prove our theorem; tools that are technical in nature. We stress the fact that this approach leads to a much more general and natural formulation of the problem. Moreover, we are able to recover all previous results without making any extraneous assumptions.

In Appendix 2 we present the existence proof. This makes essential use of the tools presented in Appendix 1 and of course of the nondegeneracy condition at terminal time.

The result we prove about existence of a contigent Arrow-Debreu equilibrium allows us to derive, in a standard way, the existence of the Radner equilibrium.

Observe that analyticity of a function means that locally the function is really a restriction of a holomorphic function of complex variables. This is clearly too strong assumption with no clear economic content, unsuitable as a requirement for the basic economic data. On the other hand, quasi-analyticity has nothing to do with complex extendability and is the precise framework for passing from local to a global collapse.

In light of this, statements such as *the requirement that the candidate prices be real analytic in time cannot be relaxed* are somewhat misleading.

A crucial technical point in (most of) previous work was to establish joint space-time analyticity. This was achieved essentially by working with complex variables. In our method no ad hoc considerations are involved.

Acknowledgements Raimondo's work was supported by grant DP0558187 from the Australian Research Council.

Parts of the research reported on in this paper were performed while Y. K. was visiting the University of Melbourne.

Appendix 1: Quasi-Analytic Functions of Several Variables

In this Appendix, we summarize the results on quasi-analytic functions of several variables used in our proofs.

Let the sequence of positive numbers $\{M_k\}_{k=0}^{\infty}$ satisfy the following conditions (see Tanabe, 1979): there exist positive numbers d_0, d_1, and d_2 such that

1. $\{M_k\}_{k=0}^{\infty}$ is logarithmically convex (Q1),

2. $M_{k+1} \leq d_0^k M_k$ for all $k \geq 0$ (Q2),

3. $\binom{k}{j} M_{k-j} M_j \leq d_1 M_k$ for $0 \leq j \leq k$ (Q3),

4. $M_k \leq M_{k+1}$ for all $k \geq 0$ (Q4),

5. $M_{k+j} \leq d_2^{k+j} M_k M_j$ for all $k, j \geq 0$ (Q5).

Let Ω be an open subset of $\mathbb{R}^n$. We denote by $\mathcal{D}(\Omega, \{M_k\})$ the set of all infinitely differentiable functions u defined on Ω such that for each compact set $\mathfrak{K} \subset \Omega$ there exist positive constants C_0 and C so that for every multi-index α we have the inequality

$$\max_{x \in \mathfrak{K}} \left| D^{\alpha} u(x) \right| \leq C_0 C^{|\alpha|} M_{|\alpha|}. \tag{12}$$

The class $\mathcal{D}(\Omega, \{k!\})$ coincides with the class of real analytic functions on Ω. A class $\mathcal{D}$ of functions defined on Ω is said to be *quasi-analytic* if the only function u in the class $\mathcal{D}$ such that $D^{\alpha} u(x_0) = 0$ for all α for a fixed $x_0 \in \Omega$ is the identically zero function, $u \equiv 0$. It is well-known (Carleman-Denjoy Theorem) (see Hörmander, 1990 and Tanabe, 1979) that the class $\mathcal{D}(\Omega, \{M_k\})$ is quasi-analytic if and only if $\sum_{k=0}^{\infty} (M_k)^{-\frac{1}{k}} = \infty$. The sequence $M_k = (k \log k)^k$ determines a non-real analytic, but a quasi-analytic, class and satisfies $Q1 - Q5$. The following class of functions which are nowhere real analytic in the real line but are nevertheless determined uniquely by the values of all the derivatives at *one point* was given by Borel (1901):

$$g(x) = \sum_{q=1}^{\infty} \sum_{p=-\infty}^{\infty} \sum_{p'=-\infty}^{\infty} \frac{\varphi(p, p', q')}{x + i\sqrt{2} - \frac{p+ip'}{q}}$$

with

$$\left| \varphi(p, p', q') \right| < e^{-e^{p^8 + p'^8 + q^8}}$$

and the coefficients $\varphi(p, p', q')$'s are nowhere zero (for real functions consider the class $f(x) = |g(x)|^2$).

Note the well-known facts that if $f \in \mathcal{D}(\Omega, \{M_k\})$ then $D^{\alpha} f \in \mathcal{D}(\Omega, \{M_k\})$ for any α, and if $f, g \in \mathcal{D}(\Omega, \{M_k\})$ then the product $f \cdot g \in \mathcal{D}(\Omega, \{M_k\})$. We are going to make a constant use of these facts.

Quasi-analytic functions of several variables have been studied extensively in Bierstone and Milman (2004). They prove an implicit function theorem for quasi-analytic functions.

Theorem 5 (The Quasi-Analytic Implicit Function Theorem) *Suppose that U is open in $\mathbf{R}^n \times \mathbf{R}^p$ (with product coordinates $(x, y) = (x_1, \ldots, x_n, y_1, \ldots, y_p)$). Suppose that $f_1, \ldots, f_p$ are quasi-analytic, $(a, b) \in U$, $f(a, b) = 0$ and*

$$(\partial f/\partial y)(a, b)$$

is invertible, where $f = (f1, \ldots, fp)$. Then there is a product neighborhood $V \times W$ of (a, b) in U and a quasi–analytic mapping $g : V \to W$ such that $g(a) = b$ and

$$f(x, g(x)) = 0 \,\forall x \in V.$$

The main subject of Bierstone and Milman (2004) is proving the possibility of resolution of singularities for quasi-analytic functions. As observed in Bierstone and Milman (2004, Corollary 5.13), if U is an open set in $\mathbf{R}^n$ and $f : U \to \mathbf{R}$ is quasi-analytic, then for every $x_0 \in U$, there exists a neighborhood V_{x_0} of x_0 such that either $F(x) = 0$ for all $x \in V_{x_0}$ or $\{x \in \mathcal{V}_{x_0} : F(x) = 0\}$ is a finite union of quasi-analytic varieties of dimension $< n$. From this follows the following

Corollary 6 *Let $\mathcal{O} \subset \mathbf{R}^n$ be open and convex, $f : \mathcal{O} \to \mathbf{R}$ is quasi-analytic. If $\{x \in \mathcal{O} : f(x) = 0\}$ has positive Lebesgue measure, then f is identically zero on $\mathcal{O}$.*

Proof If $f(x) = 0$ for all $x \in \mathcal{V}_{x_0}$ and $y \in \mathcal{O}$, there is a ray that passes through $\mathcal{V}_{x_0}$ then the function f has to vanish identically on the connected set $\mathcal{O}$. On the other hand, if $\{x \in \mathcal{V}_{x_0} : f(x) = 0\}$ is a finite union of quasi-analytic varieties of dimension $< n$, $\{x \in \mathcal{V}_{x_0} : f(x) = 0\}$ has Lebesgue measure zero. There is a countable collection $\{x_n : n \in \mathbf{N}\}$ such that $\cup_{n \in \mathbf{N}} \mathcal{V}_{x_n} \supset U$, so $\{x \in U : f(x) = 0\}$ has Lebesgue measure zero. ■

Quasi-analyticity of solutions of parabolic partial differential equations such as those that are satisfied by securities prices has been obtained recently in Kannai and Raimondo (2013) in a form sufficient for establishing dynamic completeness. The relevant definitions and assumptions are stated explicitly in the cited paper.

Theorem 7 *Let $f(t, x) \in \mathcal{D}(\mathbb{R}^n \times (0, T), \{M_k\})$, $u_0(x) \in L^1_{\text{loc}}(\mathbb{R}^n)$ be such that for every $\delta > 0$*

$$\int |u_0(x)| \exp\left(-\delta \sum_{i=1}^{n} |x_i|^{\frac{m}{m-1}}\right) dx < \infty, \tag{13}$$

and there exist constants C_0, C such that

$$\int \left| \left(\frac{\partial}{\partial t}\right)^k f(t, x) \right| \exp\left(-\delta \sum_{i=1}^{n} |x_i|^{\frac{m}{m-1}}\right) dx < C_0 C^k M_k \tag{14}$$

uniformly in $[0, T]$, *for every non-negative integer* k. *Let* A *satisfy the assumptions of Lemma 3 of Kannai and Raimondo (2013). Then the solution* $u(t, x)$ *of the differential equation*

$$\frac{\partial u}{\partial t} + Au = f \tag{15}$$

with the initial condition

$$u(x, 0) = u_0 \tag{16}$$

given by the formula

$$u(x,t) = \int_{\mathbb{R}^n} U(x,t,y,0)u_0(y)dy + \int_0^t \int_{\mathbb{R}^n} U(x,t,y,s)f(s,y)dyds \tag{17}$$

is quasi-analytic in $\mathbb{R}^n \times (0, T)$.

Observe that in our case $m = 2$.

Appendix 2: Sketch of Proof of Main Theorem

Proof Here we only sketch the proof, as it follows the one in Anderson and Raimondo (2008), emphasizing only the points where we differ. In order to prove existence we observe that (see Dana, 1993) the consumption is given by the solution of the social planner's problem

$$\max_{\sum_{i=1,\dots,I} c_i \le e} \sum_{i=1}^{I} \lambda_i U_i(c_i),$$

where the parameters λ_i are Negishi weights. Since we have von Neumann-Morgenstern utility functions the solution is equivalent to the solutions of the problem state-by-state. Therefore the solution is given by

$$u(\lambda_1, \dots, \lambda_I, t, x) = \max_{\sum_{i=1,\dots,I} x_i \le x} \sum_{i=1}^{I} \lambda_i h_i(t, x_i)$$

and this is equivalent to the solution of the following (Lagrange multipliers) system

$$\begin{cases} \lambda_1 \frac{\partial h_1}{\partial c} & = \widetilde{\mu} \\ \lambda_2 \frac{\partial h_2}{\partial c} & = \widetilde{\mu} \\ \vdots & \\ \lambda_I \frac{\partial h_I}{\partial c} & = \widetilde{\mu} \\ \sum_{i=1,\dots,I} x_i & = x \end{cases}$$

The Implicit Function Theorem of Bierstone and Milman (2004, see theorem 5 in Appendix 1) implies that there exist quasi-analytic functions $x_1, \ldots, x_I, \widetilde{\mu}$ which are solutions of the system and this implies, by a well-known result (Dana, 1993), that

$$(c_1, \ldots, c_I, p_C) = (x_1(t, e), \ldots, c_I(t, e), \widetilde{\mu}(t, e))$$

is a contingent Arrow Debreu equilibrium with quasi-analytic data, in particular the relative price of consumption at equilibrium $p_C(t, x)$ is quasi-analytic. It follows that the securities prices $p_A(t, X) = \left(p_{A_0}(t, X), \ldots, p_{A_K}(t, X)\right)$ are given by the expectations (8) and satisfy the partial differential equation (2) with the final condition

$$p_{Aj}(T, x) = D_j(T, X(T))p_C(T, x) \tag{18}$$

and Theorem 7 of Appendix 1 applies. Hence the security prices are quasi-analytic.

Finally quasi-analyticity and non-degeneracy of the dispersion of the final lump dividend (7), together with corollary (6), imply the non-degeneracy of the price dispersion matrix. Hence (see Karatzas and Shreve, 1998) the equilibrium is dynamically complete. ■

References

Anderson, R. M., & Raimondo, R. C. (2008). Equilibrium in continuous-time financial markets: Endogenously dynamically complete markets. *Econometrica, 76*(4), 841–907.

Bierstone, E., & Milman, P. D. (2004). Resolution of singularities in Denjoy-Carleman Classes. *Selecta Mathematica 10*(1), 1–28.

Borel, E. (1901). Sur les séries depolynomes et de fractions rationnelles. *Acta Mathematica, 24*, 309–382.

Browning, M., Hansen, L. P., & Heckman, J. J. (1999). Micro data and general equilibrium models. In J. B. Taylor & M. Woodford (Eds.), *Handbook of macroeconomics* (vol. 1, Part A, pp. 543–633). Amsterdam: Elsevier.

Dana, R. A. (1993). Existence and uniqueness of equilibria when preferences are additively separable. *Econometrica, 61*, 953–957.

Dana, R. A., & Jeanblanc, M. (2002). *Financial markets in continuous time*. Springer: Berlin.

Duffie, D. (1986). Stochastic equilibria: Existence, spanning number and the 'No Expected Financial Gain From Trade' hypothesis. *Econometrica, 54*, 1161–1183

Duffie, D., & Huang, C.-F. (1985). Implementing arrow–debreu equilibria by continuous trading of few long–lived securities. *Econometrica, 53*, 1337–1356

Hörmander, L. (1990). *The analysis of linear partial differential operators I* (2nd edn.). Berlin: Springer.

Hugonnier, J., Malamud, S., & Trubowitz, E. (2012). Endogenous completeness of diffusion driven equilibrium markets. *Econometrica, 80*, 1249–1270. Supplement: In econometrica supplementary material.

Kannai, Y., & Raimondo, R. C. (2013). Quasi-analytic solutions of linear parabolic equations. *Journal d' Analyse Mathematique, 119*, 115–145

Karatzas, I., & Shreve, S. (1998). *Methods of mathematical finance*. New York: Springer.

Kramkov, D., & Predoiu, S. (2014). Integral representation of martingales motivated by the problem of endogenous completeness in financial models. *Stochastic Processes and Their Applications, 124*, 81–100.

Merton, R. C. (1973). An intertemporal capital asset pricing model. *Econometrica, 41*, 867–887.

Merton, R. C. (1990). A complete-markets general equlibrium theory of finance in continuous time. In R. C. Merton (Ed.), *Continuous-time finance* (chap. 16). Cambridge: Basil Blackwell. 524–576.

Negishi, T. (1960). Welfare economics and existence of an equilibrium for a competitive economy. *Metroeconomica, 12*, 92–97.

Radner, R. (1972). Existence of equilibrium of planes, prices and price expectations in a sequence of markets. *Econometrica, 40*, 289–303.

Riedel, F., & Herzberg, F. (2013). Existence of financial equilibria in continuous time with potentially complete markets. *Journal of Mathematical Economics, 49*, 398–404.

Tanabe, H. (1979). *Equations of evolution*. London: Pitman.

Are We There Yet? Mechanism Design Beyond Equilibrium

Matthew Van Essen and Mark Walker

In mechanism design, Leo Hurwicz created a new field of economics. He used the theory to bring informational and incentive issues to the fore, and to address two fundamental questions in economics: in classical allocation problems, what can economic institutions achieve, and what can't they achieve? (Hurwicz, 1979; Hurwicz and Walker, 1990). The tool he used, and the one everyone has used ever since, is game theoretic equilibrium.

The two classical allocation problems that Leo addressed were price formation in the pure exchange problem, and the "free rider problem" with public goods. In both problems it's noteworthy that we often implement outcomes in real time, as our institutions produce them, rather than waiting to attain an equilibrium and asking along the way "Are we there yet?"

Realistically, we are probably never really at an equilibrium. Equilibrium predictions are useful when we think we will at least be "close to" an equilibrium, reasonably quickly, or when our interest is primarily in a system's long-run state.

But if outcomes are going to be implemented in or out of equilibrium, then clearly we need to know something about disequilibrium outcomes. Knowing only about equilibrium outcomes is not good enough.

We view this as a variation on Wilson's argument for "robust" mechanism design (Wilson, 1987). Wilson's emphasis was on the theory's assumption of common knowledge of the participants' preferences and information. Milgrom subsequently went further, maintaining that "the behavior of [a mechanism's participants] cannot

M. Van Essen
Department of Economics, Finance, and Legal Studies, University of Alabama, Tuscaloosa, AL, USA
e-mail: mjvanessen@cba.ua.edu

M. Walker (✉)
Department of Economics, University of Arizona, Tucson, AZ, USA
e-mail: mwalker@arizona.edu

W. Trockel (ed.), *Social Design*, Studies in Economic Design,
https://doi.org/10.1007/978-3-319-93809-7_12

be regarded as perfectly predictable" (Milgrom, 2004). Milgrom's larger point was that when mechanism design is required to actually perform well on the ground, "mechanisms that are optimized to perform well when the assumptions are exactly true may still fail miserably in the much more frequent cases when the assumptions are untrue."

We take this view a step further, treating the idea that the players in a game will play equilibrium strategies as an additional assumption. In some cases the equilibrium assumption is fruitful: the equilibrium prediction is close enough to the choices the players actually make, and this tells us with some degree of accuracy what the welfare implications will be and what will happen if we change the rules of the game or if some features of the environment change. But this requires either that we determine, for any given mechanism, whether actual participants will play "close enough" to equilibrium, or else that we design the mechanism in the first place to be "robust" to non-equilibrium play.

This is the approach we have taken in some recent research we describe here on public-good mechanisms. In a paper with Lazzati, we conducted an experiment with three well-known Lindahl mechanisms, mechanisms whose equilibria produce Lindahl allocations and prices. Over the course of many plays, by many subjects, equilibrium play was never observed in the experiment. Worse, the play that *was* observed often produced infeasible outcomes and outcomes with welfare properties that were much worse than the welfare properties of equilibrium outcomes. There are reasons to believe that the performance of these three mechanisms would be at least roughly representative of other public-goods mechanisms that economists have proposed. In order to provide the necessary incentives, the mechanisms use unintuitive outcome functions that do not seem to lead participants to the mechanisms' equilibria. Moreover, the features that provide these incentives also have the potential to create infeasible and otherwise undesirable outcomes when not in equilibrium.

These experimental results led us to devise a mechanism that would be intuitive, and if not necessarily optimal, at least satisfactory, whether in or out of equilibrium. The approach we adopted was to emulate simple mechanisms for attaining a Walrasian equilibrium, such as the mechanism introduced by Dubey (1982). The motivation here was that when there are only two persons, the problem of selecting an amount of a public good, together with the allocation of its cost, is exactly equivalent to the standard Edgeworth box two-person two-good exchange economy. (To see this geometrically, compare the Edgeworth box and the Kolm triangle.) Mechanisms such as Dubey's are transparent, relying on price and quantity proposals, so participants might be expected to play an equilibrium, or at least close to an equilibrium; the mechanisms also have desirable equilibria; and their outcomes are well-defined when not in equilibrium. Following this idea, we first defined a price-quantity mechanism for the two-person Edgeworth box problem, then reinterpreted it for the two-person public-good problem, and then generalized the public-good version of the mechanism to an arbitrary number of participants.

Does the new price-quantity mechanism perform any better than existing public-good mechanisms? Or, since it has many equilibria, most of which are not Pareto optimal, does it actually perform worse? We devised and conducted an experiment to answer that question.

We begin with a brief description of our experiment with three Lindahl mechanisms. We follow that with a description of the price-quantity mechanism. And we follow that in turn with a description of our experiment using the new price-quantity mechanism, and a comparison of its performance to the performance of the three Lindahl mechanisms.

1 An Experiment: Three Lindahl Mechanisms

In Van Essen, Lazzati, and Walker (2012; henceforth VLW) we conducted an experiment to evaluate the performance of three mechanisms designed to achieve Lindahl allocations at their equilibria, the mechanisms introduced by Walker (1981), Kim (1993), and Chen (2002). The mechanisms were applied to the following simple public-good allocation problem: three participants must choose the quantity q of a public good and also how to allocate among themselves the cost of providing the q units.

Each of the three mechanisms requires each participant to choose an action, or message, m_i, and produces an outcome $(q, t_1, t_2, t_3) \in \mathbb{R}_+ \times \mathbb{R}^3$, where t_i denotes the tax to be paid by participant i; if $t_i < 0$, then i is *paid* $|t_i|$ dollars. The particular mechanism is defined by the domain from which participants may choose their messages m_i and by the outcome function φ that maps message profiles (m_1, m_2, m_3) into outcomes (q, t_1, t_2, t_3).

The subjects in the experiment were divided into groups of three and each three-person group used one of the three mechanisms repeatedly, forty times, to determine an outcome (q, t_1, t_2, t_3) in each of the 40 periods. Each group used the same mechanism for all 40 periods; each subject played the same role, $i = 1, 2$, or 3, at each of the 40 periods; and the subjects were paid for all 40 outcomes at the end of the experimental session.

The public good cost the group twelve experimental dollars (E\$) per unit. Each group member $i = 1, 2, 3$ received a benefit of $v_i(q) = a_i q - q^2$ E\$ when q units were provided, where $a_1 = 22, a_2 = 16, a_3 = 28$. The Pareto allocations are the ones that maximize the economic surplus $S(q) = \sum_1^3 v_i(q) - 12q$, *viz.* $\widehat{q} = 9$ and $S(\widehat{q}) = 243$. The Lindahl outcome is unique and independent of the mechanism: the Lindahl quantity is $q = 9$, the unique Pareto public good level, and the Lindahl taxes are $t_1 = 36, t_2 = -18, t_3 = 90$.

A total of 81 subjects participated in the experiment: nine three-subject groups for each mechanism. This provided, for each mechanism, 360 "plays" (9 groups times 40 periods) and 1080 individual decisions and outcomes (3 times 360). Altogether, for the three mechanisms, there were 1080 plays and 3240 individual decisions and outcomes.

Table 1 Means and standard deviations of the economic surplus $S(q)$ attained by each mechanism (360 plays in each mechanism)

	Mean	Std dev
Kim	164.4	34.0
Chen	162.7	69.0
Walker	79.4	71.8

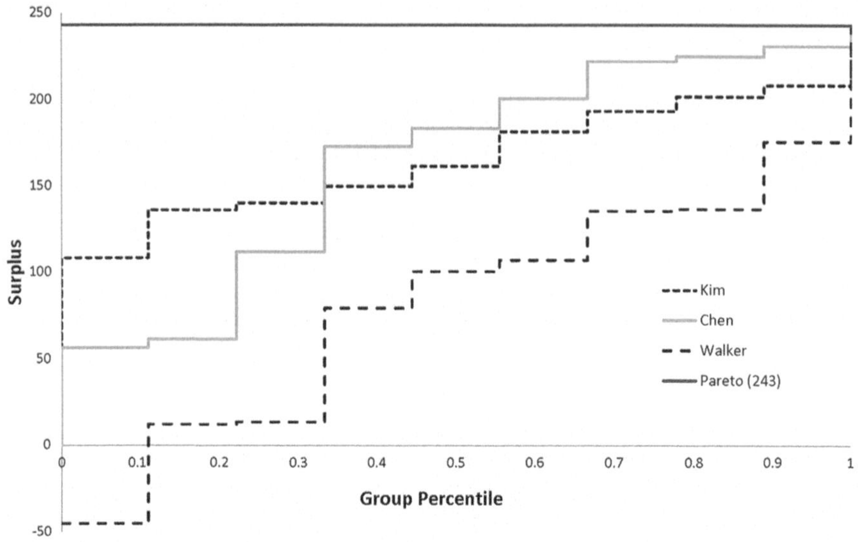

Fig. 1 Distributions of economic surplus in three Lindahl mechanisms

We provide a brief summary of the results:

Equilibrium, Lindahl, and Pareto In 1080 plays, Nash equilibrium and the Lindahl outcome were never observed; all outcomes were disequilibrium outcomes. The frequencies of the Pareto outcome $q = 9$ were 25 plays in the Chen mechanism (6.9% of the 360 plays); 19 plays in the Kim mechanism (5.3%); and 13 plays in the Walker mechanism (3.6%).

Economic Surplus Figure 1 and Table 1 describe the distributions of the economic surplus $S(q)$ earned by the groups: the Kim and Chen mechanisms produced very similar means, much larger than the Walker mechanism's mean; and the Kim mechanism produced much less variability across groups than the other two mechanisms.

Budget Imbalances and Infeasible Outcomes While the economic surplus the Chen and Kim mechanisms produced was, on average, only about 30% below the optimal level of E$ 243, the mechanisms experienced a much more serious failure: the budget was balanced in only five of the 360 plays in the Chen mechanism, and only twelve times out of 360 plays in the Kim mechanism. About half the time (54% and 51%, respectively) there was a budget surplus: the participants were required to pay more in taxes than the cost of the public good. Since these excess taxes cannot be rebated to the participants without altering the mechanism, we must count them as an additional cost of the mechanism, thereby reducing the economic surplus the mechanism generates for its participants. Conversely, in the case of a budget deficit (the remaining 44% and 46% of plays, respectively), implementation of the mechanism's outcome requires an infusion of resources from external sources—again, an additional cost of implementing the mechanism.

Taking these additional costs into account, the economic surplus produced by the Kim mechanism was reduced, on average, by 38%, from E$ 164 to E$ 101. The magnitudes of the budget imbalances were far more serious in the Chen mechanism: in more than 90% of the 360 plays the budget imbalance was greater than E$ 100; it was more than E$ 1,000 in one-third of the plays; and it was occasionally more than E$ 10,000. Deducting the budget imbalances from the (direct) economic surplus $S(q)$ reduced the Chen mechanism's average (net) economic surplus to a *negative* E$ 1,051 (163–1214).

The budget is identically balanced in the Walker mechanism, but that mechanism's average economic surplus was still well below the Kim mechanism's average net economic surplus of E$ 101.

Individual Rationality We say that an outcome is *acceptable* to a participant if it leaves him at least as well off as he would be at the status quo outcome in which $q = 0$ and there are no taxes or transfers. When there is a natural status quo outcome like this, or where there is a need to guarantee participation in a mechanism, outcomes are often called *individually rational* if they are acceptable to every participant. Lindahl allocations are always individually rational, so each of the mechanisms in our experiment always produces an individually rational outcome at a Nash equilibrium.

But we've seen that these mechanisms never produced a Nash equilibrium. And in each of the mechanisms, out-of-equilibrium profiles (m_1, m_2, m_3) of messages may yield outcomes that are not individually rational. This occurred with considerable frequency in our experiment: 39% of the 1080 individual outcomes in the Chen mechanism were unacceptable to a participant, 11% were unacceptable in the Kim mechanism, and 29% were unacceptable in the Walker mechanism.

Summarizing These three mechanisms all yield the Lindahl outcome at their equilibria, but in our experiment none of the mechanisms was ever in equilibrium. Disequilibrium is certainly not bad per se: if outcomes are not far from equilibrium, and if welfare is not far from what it would be in a good equilibrium, that would generally be considered a success. But the three Lindahl mechanisms we included in our experiment mostly failed, rather badly, to satisfy several criteria that we would generally regard as essential when designing an allocation mechanism. As we describe in the following section, we used these failures as a guide to design a mechanism that would be more robust to out-of-equilibrium behavior—a mechanism that would be relatively successful even if typically out of equilibrium.

2 The PQ Mechanism

In Van Essen and Walker (2017) we defined a mechanism, which we call the price-quantity or PQ mechanism, in which participants make quantity-and-price proposals (q_i, π_i). These proposals are the arguments of an outcome function φ that determines the level q at which a public good will be provided, as well as the amount t_i each participant i will pay to finance the public good. The price proposal π_i and the tax t_i may be any real numbers; if π_i is negative, $|\pi_i|$ is a proposed per-unit subsidy to be paid to i; if t_i is negative, $|t_i|$ is a proposed total payment to i.

We assume that each participant's maximum ability to pay (for example, his income or wealth) is observable and we denote it by $\mathring{y}_i$. The mechanism restricts participant i to proposals that satisfy $\pi_i q_i \leqq \mathring{y}_i$. For any $\mathring{y} \in \mathbb{R}_+$ we denote the set of all such proposals by $\psi(\mathring{y})$:

$$\psi(\mathring{y}) = \{(q, \pi) \in \mathbb{R}^2 \mid q \geqq 0 \text{ and } \pi q \leqq \mathring{y}\},$$

and the set of all profiles of admissible proposals as Ψ:

$$\Psi = \times_{i=1}^{n} \psi(\mathring{y}_i).$$

We assume that the cost of providing q units of the public good[1] is $C(q) = cq$. The PQ mechanism's outcome function $\varphi : \Psi \to \mathbb{R}^{N+1}$ is defined as follows:

$$q = \begin{cases} \min\{q_1, \ldots, q_N\}, & \text{if } \sum_i \pi_i \geqq c \\ 0, & \text{otherwise;} \end{cases}$$

$$t_i = p_i q, \quad \text{where} \quad p_i = \frac{1}{N}c + \pi_i - \frac{1}{N}\sum_{j=i}^{n} \pi_j \quad (i = 1, \ldots, N).$$

Thus, if the participants' price proposals π_i cover the cost of production (*i.e.*, $\sum_i \pi_i \geqq c$), then the mechanism produces the smallest quantity anyone has proposed. If the price proposals don't cover the cost, the mechanism produces zero. Consequently, if $q > 0$, then $p_i \leqq \pi_i$, and if $q = 0$, then $t_i = 0$. Therefore each participant never pays more than the amount $\pi_i q_i$ he has proposed.

2.1 *The PQ Mechanism's Properties*

We denote profiles $\big((q_1, \pi_1), \ldots, (q_N, \pi_N)\big)$ of proposals by ξ. For every profile ξ of proposals, whether it's an equilibrium or not, the mechanism's outcome $(q, \mathbf{t}) = (q, t_1, \ldots, t_N) = \varphi(\xi)$ has the following properties:

[1]For a more general cost function $C(q)$, c is replaced by $C(q)/\min\{q_1, \ldots, q_N\}$ in the equations defining the outcome function. Some of the properties described below do not hold for a nonlinear cost function.

(P1) The budget is balanced—*i.e.*, $\sum_{i=1}^{N} t_i = C(q)$—because $\sum_{i=1}^{N} p_i \equiv c$.
(P2) No participant pays more than his proposed price π_i per unit of the public good.
(P3) As a consequence of (P1) and (P2) and the fact that $\pi_i q_i \leqq \mathring{y}_i$, the outcome is both individually feasible and collectively feasible—*i.e.*, $t_i \leqq \mathring{y}_i$ for each $i = 1, \ldots, N$, and $C(q) \leqq \sum_1^N \mathring{y}_i$.

Assume that each participant's preference over outcomes is represented by a utility function $u_i(q, t_i)$ which is strictly quasiconcave, strictly increasing in q, and strictly decreasing in t_i. (Note that this is equivalent to saying the participant has a strictly quasiconcave, strictly increasing utility function over pairs $(q, y) \in \mathbb{R}^2_+$, where y_i is his after-tax dollar holdings, $\mathring{y}_i - t_i$.)

As noted above, we say that an outcome $(q, \mathbf{t}) = (q, t_1, \ldots, t_N)$ is **acceptable to** i if $u_i(q, t_i) \geqq u_i(0, 0)$—*i.e.*, if participant i is at least as well off at the outcome $(q, \mathbf{t})$ as he is at the status quo outcome—and a *proposal* $\xi_i = (q, \pi_i)$ is **acceptable to** i if $u_i(q, \pi_i q) \geqq u_i(0, 0)$. For each i and each $\xi_i = (q, \pi_i) \in \psi(\mathring{y}_i)$, let $\varphi_i(\xi_i)$ denote the set of all outcomes that can occur if i chooses the proposal ξ_i:

$$\varphi_i(\xi_i) := \{(q, \mathbf{t}) \in \mathbb{R}^{N+1} \mid (q, \mathbf{t}) = \varphi(\widetilde{\boldsymbol{\xi}}) \text{ for some } \widetilde{\boldsymbol{\xi}} \in \Psi \text{s.t.} \widetilde{\xi}_i = \xi_i \}.$$

We say that a proposal $\xi_i \in \psi(\mathring{y}_i)$ is **uniformly acceptable** to i if every outcome in $\varphi_i(\xi_i)$ is acceptable to i.

(P4) If preferences are quasiconcave, then under the outcome function φ every proposal $\xi_i = (q_i, \pi_i)$ that satisfies $u_i(q_i, \pi_i q_i) \geqq u_i(0, 0)$ is uniformly acceptable to player i. In other words, any proposal that's acceptable to a participant is uniformly acceptable to him. If he makes only proposals that are acceptable to him, then the outcome under φ (whether in equilibrium or not) will always be acceptable to him.

The properties (P1)–(P4) hold for all profiles of proposals and therefore for all outcomes of the mechanism, not merely for the equilibrium outcomes. This is in contrast to the three Lindahl mechanisms in the VLW experiment: although (P1) and (P3) hold for equilibrium outcomes in those mechanisms, (P1)–(P4) fail to hold in general. And indeed, in the VLW experiment the mechanisms' outcomes often violated these properties, generally by large amounts.

The PQ mechanism's equilibria also have several properties worth noting. Recall that an outcome is **individually rational** if it is acceptable to every participant $i = 1, \ldots, n$.

(P5) It follows from (P4) that a Nash equilibrium of the PQ mechanism is individually rational.
(P6) The Lindahl outcome is an equilibrium outcome.

The PQ mechanism has many Nash equilibria, in fact a continuum of them. (These are described in some detail in Van Essen and Walker (2017)). In particular, there are equilibria in which the public good level is zero. In order to gain some insight into the outcomes that participants in the mechanism will actually attain, we conducted an experiment.

3 An Experiment: The PQ Mechanism

In Van Essen and Walker (2018) we report on an experiment we conducted to compare the performance of the PQ mechanism with the results in the VLW experiment. In order to generate results that can be directly compared to the results in the VLW experiment, we used the same public-good problem: three participants, with the same valuation functions $v_i(x)$ for each $i = 1, 2, 3$ and the same cost function, $C(x) = 12x$. Therefore the unique Pareto level of the public good is the same, $\widehat{x} = 9$; the maximum possible economic surplus is $S(\widehat{x}) =$ E\$ 243; and the Lindahl taxes are $t_1 = 36$, $t_2 = -18$, $t_3 = 90$. Each participant's surplus $v_i(\widehat{x}) - t_i$ at the Lindahl outcome is E\$ 81. The experiment had 81 subjects, divided into 27 three-person groups.

We describe the results of this experiment along several dimensions, in each case comparing the results to those described above in the VLW experiment.

3.1 Equilibrium

The PQ mechanism's participants played equilibrium profiles in 282 of the 1080 plays (26%), and in 44% of the 270 later-period plays, from period 31 to period 40. Recall that the participants in the Lindahl mechanisms never played an equilibrium, out of 360 plays in each mechanism. The numbers are perhaps misleading, however, because the PQ mechanism has many equilibria while each of the other three mechanisms has only one equilibrium. The high frequency of equilibrium play in the PQ mechanism might be mostly due to nothing more than the presence of so many equilibria.

Nine of the twenty-seven groups attained one of the equilibria and continued to play that equilibrium in nearly every subsequent period. Each of these instances of equilibrium play produced public good levels of either 6 or 7 units, with E\$216 or E\$231 of economic surplus, somewhat less than the Pareto level of E\$243. Clearly, none of these observed equilibria was the PQ mechanism's Lindahl equilibrium, since the public good levels they achieved were smaller than the Pareto public good level of 9 units.

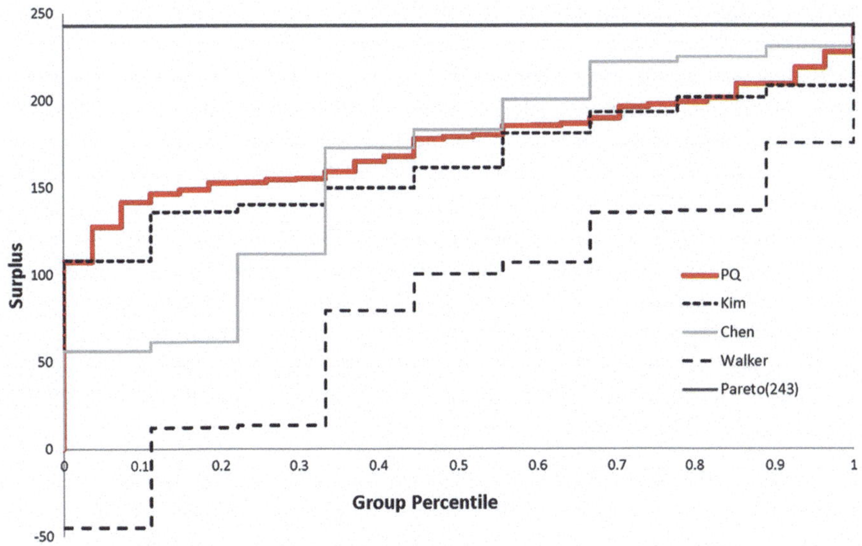

Fig. 2 Observed distributions of economic surplus

3.2 Economic Surplus

Figure 2 depicts the average surplus attained, over all 40 periods, by each of the 27 groups in our experiment, as well as by each of the 9 groups in each of the three Lindahl mechanisms in the VLW experiment. Each of the graphs orders the 27 or 9 observed levels of surplus from smallest to largest, left to right—so the graphs are the empirical cdf's of the observed levels of surplus, with the cumulative frequencies (or percentiles) on the horizontal axis.

It's clear from the graphs in Fig. 1 that the groups who did the worst in the Chen mechanism did much worse than the corresponding worst-performing groups in the PQ mechanism, and the groups who did the best in the Chen mechanism did slightly better than the corresponding best-performing groups in the PQ mechanism. The welfare distributions generated by the Kim and the PQ mechanisms were quite close, and the distribution of welfare produced by the Walker mechanism was significantly dominated by the PQ distribution. The mean surplus achieved by the PQ mechanism was E$ 175 and the standard deviation of the distribution is 28.8. The mean is not statistically different from the means for the Chen and Kim mechanisms that appear in Table 1. The standard deviation is lower than the standard deviation of 69 for the Chen mechanism at the 5% significance level, and is not statistically different than the Kim distribution's standard deviation of 34.

Measured by the direct economic surplus the mechanisms produced, the PQ mechanism seems to have performed at least as well as the Chen and Kim mechanisms, and clearly better than the Walker mechanism.

Budget Balance and Feasibility Recall that the issues of budget imbalance and infeasibility of outcomes were a serious problem in the Chen and Kim mechanisms: the budget was almost never balanced in either mechanism, and when we took these costs into account in measuring the economic surplus the mechanisms produced, the reduction in surplus was significant for the Kim mechanism (reducing the surplus from E$ 164 to E$ 101, and overwhelming in the Chen mechanism (reducing the surplus from E$ 163 to negative E$ 1,051).

The property (P1) of the PQ mechanism—that the budget is always balanced, whether in or out of equilibrium—therefore appears to be an important advantage. The economic surplus the PQ mechanism produced was as large as, and no more variable than, the surplus produced by the Chen and Kim mechanisms directly. And when we take account of the additional costs imposed on mechanisms by budget imbalances, the net surplus of the other two mechanisms falls well below the E$ 175 surplus produced by the PQ mechanism.

Individual Rationality Recall that in the Chen and Kim mechanisms many of the outcomes were not individually rational. Because the PQ mechanism has the uniform acceptability property (P4) and each participant's valuation function is concave, a participant in the PQ mechanism, by always choosing a proposal that's acceptable to him, can ensure that the outcome will always make him at least as well off as the status quo. Only 32 of the 3240 proposals made by the 81 subjects in our experiment were not acceptable (less than one percent), and only one of the 1080 outcomes failed to be individually rational, by failing to be acceptable to only one participant (less than one-tenth of one percent).

Summarizing Ignoring budget imbalances and infeasibility of outcomes in the Lindahl mechanisms, and in spite of the multiple non-optimal equilibria of the PQ mechanism, the PQ-mechanism performed at least as well as the Lindahl mechanisms we had examined in our earlier experiment. If we then take account of unbalanced budgets—and include their costs as reductions in welfare—the PQ mechanism clearly outperformed the Lindahl mechanisms.

4 Concluding Remarks

In the theory of mechanism design, equilibrium analysis has paid enormous dividends, illuminating myriad issues, from the possibility of providing economic agents with differing incentives, to the important roles of information and beliefs—all of which were anticipated by Hurwicz in the earliest stages of his development of the theory.

In *Putting Auction Theory to Work*, Milgrom almost exclusively puts mechanism design's *equilibrium theory* to work. But at the outset he points out that "the equilibrium analysis of game theory is an abstraction based on a sensible idea" which "relies on stark and exaggerated assumptions to reach theoretical conclusions that can sometimes be fragile." He lists assumptions about players perfectly maximizing, about players' information, and about their beliefs about other players' maximization, information, and beliefs, and points out that "these assumptions are extreme."

To Milgrom's list of assumptions we would add the "assumption" of equilibrium. Without denying the power and influence of the equilibrium assumption (like all of us, the authors have made careers from a reliance on it), we suggest that it would be fruitful to incorporate disequilibrium analysis into the theory as well. We mean not merely that we should ask whether disequilibria will converge over time, or how long convergence will take—*i.e.*, "Are we there yet?" Rather, we should recognize that we're never actually going to get there—it's the journey that matters, not the destination. As we've suggested above, we regard this idea as an extension of the "Wilson doctrine" that mechanisms should be "robust." The notion of **universal acceptability** that we introduced here, and which we applied to *all* behavior, disequilibrium as well as equilibrium, in the PQ mechanism for a public good, is a first attempt at this approach.

References

Chen, Y. (2002). A family of supermodular Nash mechanisms implementing Lindahl allocations. *Economic Theory, 19*, 773–790.

Dubey, P. (1982). Price quantity strategic market games. *Econometrica, 50*, 111–126.

Hurwicz, L. (1979). Outcome functions yielding Walrasian and Lindahl allocations at Nash equilibrium points. *The Review of Economic Studies, 46*, 217–224.

Hurwicz, L., & Walker, M. (1990). On the generic nonoptimality of dominant-strategy allocation mechanisms: A general theorem that includes pure exchange economies. *Econometrica, 58*, 683–704.

Kim, T. (1993). A stable Nash mechanism implementing Lindahl allocations for quasi-linear environments. *Journal of Mathematical Economics, 22*, 359–371.

Milgrom, P. (2004). *Putting auction theory to work*. Cambridge: Cambridge University Press.

Van Essen, M., Lazzati, N., & Walker, M. (2012). Out-of-equilibrium performance of three lindahl mechanisms: Experimental evidence. *Games and Economic Behavior, 74*, 366–381.

Van Essen, M., & Walker, M. (2017). A simple market-like allocation mechanism for public goods. *Games and Economic Behavior, 107*, 6–19.

Van Essen, M., & Walker, M. (2018). An experimental evaluation of a price-quantity mechanism for public goods. University of Arizona working paper.

Walker, M. (1981). A simple incentive compatible scheme for attaining Lindahl allocations. *Econometrica, 49*, 65–71.

Wilson, R. (1987). Game theoretic anayses of trading processes. In T. F. Bewley (Ed.), *Advances in Economic Theory: Fifth World Congress* (pp. 33–70). Cambridge: Cambridge University Press.

Part IV
Rules

Formation of Committees Through Random Voting Rules

Souvik Roy, Soumyarup Sadhukhan, and Arunava Sen

1 Introduction

A classic paper in the theory of mechanism design is Hurwicz (1972). It considered an exchange economy with at least two agents and demonstrated the impossibility of constructing an allocation rule that satisfied strategy-proofness, efficiency, and individual rationality. The paper inspired an enormous and rapidly expanding literature that analyzes socially desirable goals that can be achieved in the presence of private information and strategic agents, in a wide variety of models. The present paper contributes to that literature by investigating the structure of rules that permit randomization in the well-known model of committee formation.

The committee formation model is due to Barberà et al. (1991). The problem is one of choosing a committee from a set of available candidates based on the preferences of agents who have the responsibility of selecting the committee. The preferences of each agent are assumed to be separable, i.e. if the agent "likes" a candidate, she strictly prefers a committee where this candidate is included to one where she is excluded, the status of all other candidates remaining unchanged. A committee formation rule or a social choice function is a map that associates every collection of (separable) agent preferences with a committee. Agent preferences are private information—a fact that necessitates the elicitation of these preferences via voting. A social choice function is strategy-proof if truth-telling is an optimal strategy for each agent irrespective of her beliefs about how other agents may vote. The main result of Barberà et al. (1991) is that strategy-proof social choice functions (that additionally satisfy a weak efficiency property called unanimity)

S. Roy · S. Sadhukhan
Economic Research Unit, Indian Statistical Institute, Kolkata, India

A. Sen (✉)
Economics and Planning Unit, Indian Statistical Institute, Delhi, India

W. Trockel (ed.), *Social Design*, Studies in Economic Design,
https://doi.org/10.1007/978-3-319-93809-7_13

must be decomposable. In other words, the decision on each candidate's inclusion must be taken independently of the decisions on others and must be based only on preferences that agents have over the candidate (called marginal preferences). The decomposability condition on social choice function rules out many plausible rules. For instance, if there are two candidates, we could start with candidate 1 and consider candidate 2 only if 1 is not selected. Breton and Sen (1999) show that the decomposability property of strategy-proof social choice functions is very general—it holds for all multi-dimensional models with separable preferences.

In our paper we consider the same model as in Barberà et al. (1991) but analyze committee formation rules that permit randomization. A random social choice functions is a map that associates a collection of (separable) agent preferences with a probability distribution over committees. Randomization is a natural way to resolve conflicts of interest amongst agents especially in models where compensation via monetary transfers is not feasible. The analysis of randomized mechanisms in voting models was initiated in Gibbard (1977). Once randomization is allowed, the evaluation of truth-telling versus misrepresentation involves the comparison of lotteries. This evaluation typically involves domain restrictions on preferences over lotteries (i.e., all preferences over lotteries are not allowed) as a result of which the class of strategy-proof social choice functions expands (see Chatterji et al., 2014).[1]

According to our characterization result, a random social choice function is strategy proof and satisfies unanimity[2] if and only if it satisfies the properties of monotonicity and marginal decomposability. Monotonicity is a familiar property in mechanism design theory. In our model, it requires the probability of the inclusion of a candidate in every possible committee to be non-decreasing as more agents approve the candidate. Furthermore, if no agent approves a candidate, the candidate is never selected; on the other hand, if all agents approve a candidate, she is always selected.

Consider an arbitrary subset of candidates and two preference profiles where all agents agree in their opinions over this subset of candidates (they may differ in their opinion of other candidates). Marginal decomposability is satisfied if the marginal probability distribution over the subset of candidates is the same in the two profiles. Suppose there are three agents and five candidates. Consider the set of the first three candidates and two preference profiles where all agents agree in their opinions over the first three candidates. Pick any subset of the first three candidates, say candidates one and three. If marginal decomposability is satisfied, the probability of candidates one and three being selected in the committee at the two profiles must be the same. Note that marginal decomposability only guarantees that marginal probabilities will be uniquely determined by marginal preferences, but does not say anything about the joint probability distribution. Thus decomposability in the sense of Breton and

[1]There are several ways in which this can be done. Here we follow the standard stochastic dominance approach developed in Gibbard (1977).

[2]A random social choice function satisfies unanimity if it picks a committee that is first-ranked by all agents, with probability one.

Sen (1999) is not guaranteed. However, marginal decomposability is equivalent to decomposability when we restrict attention to deterministic social choice functions thus getting back the decomposability result of Breton and Sen (1999) in our model.

Finally we consider the special problem of forming a committee with a number of members. A random social choice function is onto if every committee of the required size is selected with probability one at some preference profile. We show that every onto and strategy-proof RSCF in this case is a random dictatorship in an appropriate sense. This result follows from an application of applying the main result of Gibbard (1977).

2 The Model

Let $M = \{1, \ldots, m\}$ be a finite set of m components. For each component k, $A^k = \{0, 1\}$ is the set of alternatives available in component k. For any $K \subseteq M$, $A^K = \prod_{k \in K} A^k$ denotes the set of alternatives available in components in K. The set of (multi-dimensional) alternatives is given by A^M. For ease of presentation, we write A instead of A^M. Note that the number of alternatives in A is 2^m. Throughout this paper, we do not use braces for singleton sets.

In the model M denotes the set of possible candidates from which a committee has to be formed. Thus each component refers to a possible candidate for a committee, where the numbers 0 and 1 for a component refer to the social states where the corresponding member is excluded and included in the committee, respectively. Similarly, every alternative $a = (a^1, \ldots, a^m) \in A$ refers to a committee in which the member k is present if and only if $a^k = 1$.

Let $N = \{1, \ldots, n\}$ be a set of finite set of n agents. Each agent i has a strict preference ordering P_i over the elements of A. We assume that all P_i's are separable, i.e. for all $a^{-k}, b^{-k} \in A^{M-k}$ and all $x^k, y^k \in A^k$, $(x^k, a^{-k})P_i(y^k, a^{-k})$ holds if and only if $(x^k, b^{-k})P_i(y^k, b^{-k})$. We denote by P_i^k the marginal preference induced by P_i over component k. The existence of marginal preference orderings is guaranteed by separability. We let $\tau(P_i)$ and $\tau(P_i^k)$ denote the top-ranked alternative in P_i and the top-ranked alternative in the kth component according to the marginal ordering P_k^i. In general, $r_t(P_i)$ the t-th ranked alternative in P_i where $t \in \{1, 2, \ldots, 2^m\}$. The upper contour set of an alternative a at preference P_i denoted by $U(a, P_i)$ is defined as follows: $U(a, P_i) = \{b \mid bP_ia\} \cup a$. Let $\mathcal{D}$ denote the set of all separable preferences over A. An element P_N of $\mathcal{D}^n$ is called a (preference) profile.

A random social choice function (RSCF) φ is a mapping $\varphi : \mathcal{D}^n \to \Delta A$ where ΔA denotes the set of probability distributions over A. We define some important properties of an RSCF most of which are familiar from the literature.

Definition 2.1 An RSCF $\varphi : \mathcal{D}^n \to \Delta A$ is unanimous if for all P_N and all $a \in A$,

$$[\tau(P_i) = a \text{ for all } i \in N] \implies [\varphi_a(P_N) = 1].$$

If all agents have a common top-ranked committee at a profile, a unanimous RCSF picks that committee at that profile. It is clearly a weak form of efficiency.

Definition 2.2 An RSCF $\varphi : \mathcal{D}^n \to \Delta A$ is strategy-proof if for all $i \in N$, all $P_i, P_i' \in \mathcal{D}$, and all $P_{-i} \in \mathcal{D}^{n-1}$, $\varphi(P_i, P_{-i})$ first order stochastically dominates $\varphi(P_i', P_{-i})$ according to P_i, that is,

$$\sum_{t=1}^{j} \varphi_{r_t(P_i)}(P_i, P_{-i}) \geq \sum_{t=1}^{j} \varphi_{r_t(P_i)}(P_i', P_{-i}) \text{ for all } j = 1, \ldots, 2^m.$$

Our notion of strategy-proofness for RSCFs is the standard one of first-order stochastic dominance introduced in Gibbard (1977). No agent can strictly increase the aggregate probability over any upper contour set according to her true preferences. If it were possible to do, there would exist a utility representation of her true preferences with the property that the expected utility from misrepresentation strictly exceeds that from truth-telling.

3 Formation of Arbitrary Committees

In this section, we consider the problem of forming a committee by random voting rules. We assume that there are no restrictions on the committee that is to be formed.[3] A few additional concepts are required for the analysis.

Let $\mathcal{N}$ denote the set of all subsets (power set) of N. For any $K \subseteq M$, S^K denotes a collection $(S^k)_{k\in K}$, where $S^k \subseteq N$ for all $k \in K$. Also $\mathcal{N}^K$ denotes the set of all such collections. Note that the cardinality of $\mathcal{N}^K$ is $(2^n)^{|K|}$. We illustrate these notions by means of an example.

Example 3.1 Suppose $N = \{1, 2, 3, 4\}$, $M = \{1, 2, 3\}$ and $K = \{2, 3\}$. An example of $S^{\{2,3\}}$ is (S^2, S^3) where $S^2 = \{1, 2, 4\}$ and $S^3 = \{2, 3\}$. Also, $\mathcal{N}^{\{2,3\}}$ is the collection of all (S^2, S^3) where S^2 and S^3 are arbitrary subsets of $\{1, 2, 3, 4\}$.

Consider an arbitrary $K \subseteq M$ and profile $P_N \in \mathcal{D}^n$. Then $S^K(P_N)$ denotes an element $(S^k)_{k\in K}$ of $\mathcal{N}^K$ such that for all $k \in K$, we have $i \in S^k$ if and only if $\tau(P_i^k) = 1$. In other words S^k consists of the agents who have 1 as the top-ranked element in component k at the profile P_N. Hence S^k consists of exactly those agents who approve candidate k for the committee at the profile P_N.

Example 3.2 Suppose $N = \{1, 2, 3, 4\}$ and $M = \{1, 2, 3\}$. Consider the profile P_N where the top-ranked alternatives of the agents are as follows: $((1, 0, 1), (0, 0, 1), (1, 1, 0))$. Let $K = \{1, 3\}$ or $\{1, 2, 3\}$ Then, $S^{\{1,3\}}(P_N) = (\{1, 3\}, \{1, 2\})$ and $S^{\{1,2,3\}}(P_N) = (\{1, 3\}, \{3\}, \{1, 2\})$.

[3]We will consider one such problem in the next section.

For $K \subseteq M$, $a^K \in A^K$ and $P_N \in \mathcal{D}^n$, we define $\varphi_{a^K}(P_N) = \sum_{\{b \in A | b^K = a^K\}} \varphi_b(P_N)$. Thus $\varphi_{a^K}(P_N)$ is the total probability of realizing outcomes whose kth component agrees with the kth component of a^K for all $k \in K$, in the probability distribution $\varphi(P_N)$.

3.1 Characterization

In this section, we identify properties that characterize unanimous and strategy-proof RSCFs in our model. The first property is marginal decomposability. Roughly speaking, it says that the marginal probability distribution generated by the RSCF over an arbitrary set of components depends only on the preferences of the agents over those components. In particular, it does not change if agents change their preferences over the other components.

Definition 3.1 An RSCF $\varphi : \mathcal{D}^n \to \Delta A$ is marginally decomposable if for all $K \subseteq M$, $P_N, \bar{P}_N \in \mathcal{D}^n$ with $S^K(P_N) = S^K(\bar{P}_N)$, and all $a^K \in A^K$, we have

$$\varphi_{a^K}(P_N) = \varphi_{a^K}(\bar{P}_N).$$

Marginal decomposability is weaker than decomposability as defined in Breton and Sen (1999). As mentioned earlier, marginal decomposability requires the marginal probability distribution over a set of components at a profile to be completely determined by the marginal preference profile over those components. Importantly, it does not say anything about the joint probability distribution. Clearly, a marginally decomposable RSCF is decomposable if the joint probability distribution is given by the product of marginal probability distributions, i.e. if the joint probability distribution is independent over components. In our model, unanimity and strategy-proofness imply marginal decomposability; however, they do not imply independence over components.

We illustrate the notion of marginal decomposability by means of the following example.

Example 3.3 Let $N = \{1, 2\}$ and $M = \{1, 2\}$. Consider the RSCF $\varphi : \mathcal{D}^n \to \Delta A$ given in Table 1. Here, rows are indexed by the $S^1(P_N)$ and columns are by $S^2(P_N)$. The matrix, say X, corresponding to row $\hat{S}^1$ and column $\hat{S}^2$ gives the

Table 1 Outcomes of φ

$1\backslash 2$	$\emptyset$	$\{1\}$	$\{2\}$	$\{1,2\}$
$\emptyset$	$\begin{pmatrix} 1 & 0 \\ 0 & 0 \end{pmatrix}$	$\begin{pmatrix} 0.3 & 0.7 \\ 0 & 0 \end{pmatrix}$	$\begin{pmatrix} 0.5 & 0.5 \\ 0 & 0 \end{pmatrix}$	$\begin{pmatrix} 0 & 1 \\ 0 & 0 \end{pmatrix}$
$\{1\}$	$\begin{pmatrix} 0.4 & 0 \\ 0.6 & 0 \end{pmatrix}$	$\begin{pmatrix} 0.2 & 0.2 \\ 0.1 & 0.5 \end{pmatrix}$	$\begin{pmatrix} 0.3 & 0.1 \\ 0.2 & 0.4 \end{pmatrix}$	$\begin{pmatrix} 0 & 0.4 \\ 0 & 0.6 \end{pmatrix}$
$\{2\}$	$\begin{pmatrix} 0.7 & 0 \\ 0.3 & 0 \end{pmatrix}$	$\begin{pmatrix} 0.15 & 0.55 \\ 0.15 & 0.15 \end{pmatrix}$	$\begin{pmatrix} 0.25 & 0.45 \\ 0.25 & 0.05 \end{pmatrix}$	$\begin{pmatrix} 0 & 0.7 \\ 0 & 0.3 \end{pmatrix}$
$\{1,2\}$	$\begin{pmatrix} 0 & 0 \\ 1 & 0 \end{pmatrix}$	$\begin{pmatrix} 0 & 0 \\ 0.3 & 0.7 \end{pmatrix}$	$\begin{pmatrix} 0 & 0 \\ 0.5 & 0.5 \end{pmatrix}$	$\begin{pmatrix} 0 & 0 \\ 0 & 1 \end{pmatrix}$

Table 2 Outcomes of φ_1^1

1	φ_1^1
Ø	0
{1}	0.6
{2}	0.3
{1, 2}	1

Table 3 Outcomes of φ_1^2

2	Ø	{1}	{2}	{1, 2}
φ_1^2	0	0.7	0.5	1

value of $\varphi(P_N)$, where $S^1(P_N) = \hat{S}^1$, $S^2(P_N) = \hat{S}^2$, and $\varphi_{(0,0)}(P_N) = X_{11}$, $\varphi_{(0,1)}(P_N) = X_{12}$, $\varphi_{(1,0)}(P_N) = X_{21}$, and $\varphi_{(1,1)}(P_N) = X_{22}$. For instance, $\varphi_{(0,1)}((0,1),(1,0)) = 0.55$, where $((0,1),(1,0))$ denotes the profile P_N with $r_1(P_1) = (0,1)$ and $r_1(P_2) = (1,0)$.

We argue that φ satisfies marginal decomposability. Consider, for instance, the row corresponding to the set {2}. Note that for each matrix X in this row, $X_{21} + X_{22} = 0.3$, that is, the marginal probability that candidate 1 is elected is 0.3, as required by marginal decomposability. It can be readily verified that φ satisfies this constant marginal property for other rows and columns. Consequently the RSCF is marginally decomposable.

We now argue that the φ is not decomposable in the sense of Breton and Sen (1999). For $k \in \{1,2\}$, let φ^k be the marginal RSCF on the k-th component that is induced by φ by means of marginal decomposability. In Tables 2 and 3, we present φ^1 and φ^2, respectively.

Consider a profile P_N with $r_1(P_1) = (0,1)$ and $r_1(P_2) = (1,0)$, that is, $S^1(P_N) = \{2\}$ and $S^2(P_N) = \{1\}$. If φ were decomposable, then $\varphi_{(1,0)}(P_N)$ must be $0.3 \times 0.3 = 0.09$. However, as given in Table 1, $\varphi_{(1,0)}(P_N) = 0.15$, which means φ is not decomposable.

Next, we define a monotonicity property for an RSCF. This is a standard property in the literature on strategy-proof social choice functions which says that the likelihood of an outcome increases as agents become more "favorable" to that outcome.

Definition 3.2 An RSCF $\varphi : \mathcal{D}^n \to \Delta A$ satisfies the monotonicity property if for all $k \in M$, all $a^{-k} \in A^{M-k}$ and all $P_N, \bar{P}_N \in \mathcal{D}^n$ such that $S^l(P_N) = S^l(\bar{P}_N)$ for all $l \in M \setminus k$ and $S^k(P_N) \subseteq S^k(\bar{P}_N)$, we have

(i) $\varphi_{(1,a^{-k})}(P_N) \leq \varphi_{(1,a^{-k})}(\bar{P}_N)$, and
(ii) if $S^k(P_N) = \emptyset$ and $S^k(\bar{P}_N) = N$, then $\varphi_{(1,a^{-k})}(P_N) = 0$ and $\varphi_{(1,a^{-k})}(\bar{P}_N) = 1$.

Suppose that some agents change preferences in favor of some candidate while maintaining their position on all other candidates. According to (i) of the monotonicity property,the probability of *each* committee including that candidate must increase. According to (ii) a candidate not approved by any agent is not selected with certainty and a candidate approved by all agents is selected with probability one. The monotonicity property is illustrated below.

Example 3.4 Consider the RSCF φ given in Table 1. We argue that it satisfies monotonicity properties. To see this, take, for instance, the profiles indexed by $(\{1\}, \{2\})$ and $(\{1, 2\}, \{2\})$. Note that agent 2 has joined agent 1 in approving candidate 1 from the former profile to the latter, while keeping his/her stand unchanged for candidate 2. By monotonicity, the probability of each committee that includes candidate 1 must increase (weakly). This is indeed the case here since $\varphi_{(1,0)}(\{1\}, \{2\}) = 0.2 < \varphi_{(1,0)}(\{1, 2\}, \{2\}) = 0.5$ and $\varphi_{(1,1)}(\{1\}, \{2\}) = 0.4 < \varphi_{(1,1)}(\{1, 2\}, \{2\}) = 0.5$. It can be directly verified that φ satisfies this condition for other relevant cases. Hence it is monotonic.

Now, we present our characterization result for unanimous and strategy-proof RSCFs.

Theorem 3.1 *An RSCF $\varphi : \mathcal{D}^n \to \Delta A$ is unanimous and strategy-proof if and only if it is monotone and marginally decomposable.*

Proof (If part) Let $\varphi : \mathcal{D}^n \to \Delta A$ be monotone and marginally decomposable. We show φ is unanimous and strategy-proof. Unanimity follows from (ii) in Definition 3.2. We proceed to show that φ is strategy-proof.

Take $b \in A$ and let P_i and $\bar{P}_i$ be two arbitrary preferences of some agent i. It is enough to show that

$$\varphi_{U(b,P_i)}(P_N) \geq \varphi_{U(b,P_i)}(\bar{P}_i, P_{-i}). \tag{1}$$

We assume without loss of generality that there exists $\hat{m} < m$ such that $r_1(P_i^k) = 1$ and $r_1(\bar{P}_i^k) = 0$ for all $k \in \{1, \dots, \hat{m}\}$ and $r_1(P_i^k) = r_1(\bar{P}_i^k)$ for all $k \in \{\hat{m} + 1, \dots, m\}$. For $t = 0, 1, \dots, \hat{m}$, let $P_i(t) \in \mathcal{D}$ be such that $r_1(P_i^l(t)) = 1$ if $l \leq t$, $r_1(P_i^l(t)) = 0$ if $t < l \leq \hat{m}$, and $r_1(P_i^l(t)) = r_1(P_i) = r_1(\bar{P}_i)$ if $\hat{m} < l$. Note that $P_i(\hat{m}) = P_i$ and $P_i(0) = \bar{P}_i$.

Claim 3.1 $\varphi_{U(b,P_i)}(P_i(k), P_{-i}) \geq \varphi_{U(b,P_i)}(P_i(k-1), P_{-i})$ *for all* $k = 1, \dots, \hat{m}$.

For all $a^{-k} \in A^{-k}$, marginal decomposability implies

$$\varphi_{a^{-k}}(P_i(k), P_{-i}) = \varphi_{a^{-k}}(P_i(k-1), P_{-i}), \tag{2}$$

while monotonicity implies

$$\varphi_{(1,a^{-k})}(P_i(k), P_{-i}) \geq \varphi_{(1,a^{-k})}(P_i(k-1), P_{-i}). \tag{3}$$

Pick $k \in \{1, \dots, \hat{m}\}$. Since $r_1(P_i^l) = 1$ for all $l \in \{1, \dots, \hat{m}\}$, it must be true that $(1, a^{-k}) P_i (0, a^{-k})$ for all $a^{-k} \in A^{-k}$. This means $(0, a^{-k}) \in U(b, P_i)$ implies $(1, a^{-k}) \in U(b, P_i)$. In view of this, we can write $U(b, P_i) = B \cup C$, where B consists of a collection of pairs of alternatives of the form $(1, a^{-k})$, $(0, a^{-k})$ for some $a^{-k} \in A^{-k}$ and C consists of alternatives of the form $(1, a^{-k})$ for some $a^{-k} \in$

A^{-k} such that $(0, a^{-k})$ is not in $U(b, P_i)$. More formally, $B = \{(0, a^{-k}), (1, a^{-k}) \mid (0, a^{-k}) \in U(b, P_i)\}$ and $C = \{(1, a^{-k}) \in U(b, P_i) \mid (0, a^{-k}) \notin U(b, P_i)\}$.

By (2),

$$\varphi_B(P_i(k), P_{-i}) = \varphi_B(P_i(k-1), P_{-i}).$$

Further, by (3),

$$\varphi_C(P_i(k), P_{-i}) \geq \varphi_C(P_i(k-1), P_{-i}).$$

Combining, we have

$$\varphi_{U(b,P_i)}(P_i(k), P_{-i}) \geq \varphi_{U(b,P_i)}(P_i(k-1), P_{-i}).$$

This completes the proof of Claim 3.1.

By applying Claim 3.1 sequentially for $k = \hat{m}, \hat{m}-1, \ldots, 1$, we get

$$\varphi_{U(a,P_i)}(P_i(\hat{m}), P_{-i}) \geq \varphi_{U(b,P_i)}(P_i(\hat{m}-1), P_{-i}) \geq \ldots \geq \varphi_{U(b,P_i)}(P_i(0), P_{-i}),$$

which shows (1).

(Only-if part) Let $\varphi : \mathcal{D}^n \to \Delta A$ be a unanimous and strategy-proof RSCF. It follows from Chatterji and Zeng (2018) Theorem 1 and Proposition 3 that φ is *tops-only*, that is, $\varphi(P_N) = \varphi(\bar{P}_N)$ for all $P_N, \bar{P}_N \in \mathcal{D}^n$ with $r_1(P_i) = r_1(\bar{P}_i)$ for all $i \in N$.

The following claim establishes a crucial property of φ.

Claim 3.2 *Let $k \in \{1, \ldots, m\}$ and let $P_N, \bar{P}_N \in \mathcal{D}^n$ be such that $S^l(P_N) = S^l(\bar{P}_N)$ for all $l \in M \setminus k$ and $S^k(P_N) \subseteq S^k(\bar{P}_N)$. Then, for all $a^{-k} \in A^{M-k}$, we have*

(i) $\varphi_{a^{-k}}(P_N) = \varphi_{a^{-k}}(\bar{P}_N)$, and
(ii) $\varphi_{(1,a^{-k})}(\bar{P}_N) \geq \varphi_{(1,a^{-k})}(P_N)$.

Proof Let $k \in \{1, \ldots, m\}$. Take $P_N, \bar{P}_N \in \mathcal{D}^n$ such that $S^l(P_N) = S^l(\bar{P}_N)$ for all $l \in M \setminus k$ and $S^k(P_N) \subseteq S^k(\bar{P}_N)$. It is enough to prove the claim for the case where $S^k(\bar{P}_N) = S^k(P_N) \cup i$ for some $i \in N$. Since φ is tops-only, we can further assume that

(i) $P_{-i} = \bar{P}_{-i}$, and
(ii) for all $b^{-k} \in A^{M-k}$,

  (a) $(1, b^{-k})$ and $(0, b^{-k})$ are consecutively ranked in both $P_i, \bar{P}_i$, and
  (b) $(0, b^{-k}) P_i (1, b^{-k})$ and $(1, b^{-k}) \bar{P}_i (0, b^{-k})$.[4]

[4]To see that it is possible to construct such a preference ordering, consider a lexicographic (and hence separable) preference over A where k is the lexicographic worst component (details may be found in Chatterji et al., 2012).

It is easy to verify that P_i and $\bar{P}_i$ satisfy separability. Take $a^{-k} \in A^{-k}$. By our assumption on P_i and $\bar{P}_i$,

$$U((0, a^{-k}), P_i) \setminus (0, a^{-k}) = U((1, a^{-k}), \bar{P}_i) \setminus (1, a^{-k}).$$

By applying strategy-proofness at (P_i, P_{-i}) via $\bar{P}_i$ and at $(\bar{P}_i, P_{-i})$ via P_i, this means

$$\varphi_{U((0,a^{-k}),P_i)\setminus(0,a^{-k})}(P_i, P_{-i}) = \varphi_{U((1,a^{-k}),\bar{P}_i)\setminus(1,a^{-k})}(\bar{P}_i, P_{-i}). \tag{4}$$

Using a similar argument, we have

$$\varphi_{U((1,a^{-k}),P_i)}(P_i, P_{-i}) = \varphi_{U((0,a^{-k}),\bar{P}_i)}(\bar{P}_i, P_{-i}). \tag{5}$$

Subtracting (4) from (5), we get

$$\varphi_{a^{-k}}(P_N) = \varphi_{a^{-k}}(\bar{P}_i, P_{-i}),$$

which proves (i) of Claim 3.2.

Since $\varphi_{(0,a^{-k})}(P_N) + \varphi_{(1,a^{-k})}(P_N) = \varphi_{(0,a^{-k})}(\bar{P}_i, P_{-i}) + \varphi_{(1,a^{-k})}(\bar{P}_i, P_{-i})$ and $(1, a^{-k})\bar{P}_i(0, a^{-k})$, it follows by an application of strategy-proofness that $\varphi_{(1,b^{-k})}(\bar{P}_N) \geq \varphi_{(1,b^{-k})}(P_N)$, which proves (ii) of Claim 3.2. ■

We return to the proof that φ satisfies monotonicity and marginally decomposability. Condition (i) in the definition of monotonicity (Definition 3.2) follows from Claim 3.2. In what follows, we prove condition (ii) in Definition 3.2.

It suffices to show $\sum_{a^{-1}\in A^{-1}} \varphi_{(0,a^{-1})}(P_N) = 0$ for all $P_N \in \mathcal{D}^n$ with $S^k(P_N) = \emptyset$. Take P_N such that $S^k(P_N) = \emptyset$. Without loss of generality, assume $k = 1$. Let $\bar{P}_N$ be the profile such that $S^2(\bar{P}_N) = \emptyset$ and $S^l(\bar{P}_N) = S^l(P_N)$ for all $l \neq 2$. By Claim 3.2, $\varphi_{a^{-2}}(P_N) = \varphi_{a^{-2}}(\bar{P}_N)$ for all $a^{-2} \in A^{-2}$. Note that

$$\sum_{a^{-1}\in A^{-1}} \varphi_{(0,a^{-1})}(P_N) = \sum_{a^{-\{1,2\}}\in A^{-\{1,2\}}} \varphi_{(0,0,a^{-\{1,2\}})}(P_N) + \varphi_{(0,1,a^{-\{1,2\}})}(P_N). \tag{6}$$

Take $a^{-2} = (0, a^{-\{1,2\}}) \in A^{-2}$. By applying Claim 3.2, we have

$$\varphi_{(0,a^{-2})}(P_N) + \varphi_{(1,a^{-2})}(P_N) = \varphi_{(0,a^{-2})}(\bar{P}_N) + \varphi_{(1,a^{-2})}(\bar{P}_N), \tag{7}$$

Combining (6) and (7), we have $\sum_{a^{-1}\in A^{-1}} \varphi_{(0,a^{-1})}(P_N) = \sum_{a^{-1}\in A^{-1}} \varphi_{(0,a^{-1})}(\bar{P}_N)$. Continuing in this manner, it follows that

$$\sum_{a^{-1}\in A^{-1}} \varphi_{(0,a^{-1})}(P_N) = \sum_{a^{-1}\in A^{-1}} \varphi_{(0,a^{-1})}(\hat{P}_N), \tag{8}$$

where $S^l(\hat{P}_N) = \emptyset$ for all $l \in \{1, \ldots, m\}$. By unanimity, $\varphi_{(0,a^{-1})}(\hat{P}_N) = 0$ for all $a^{-1} \in A^{-1}$. This, together with (8) implies $\sum_{a^{-1} \in A^{-1}} \varphi_{(0,a^{-1})}(P_N) = 0$, which shows (ii) in Definition 3.2.

Finally we show that φ is marginally decomposable. Let $K \subseteq M$ and let P_N and $\bar{P}_N$ be such that $S^K(P_N) = S^K(\bar{P}_N)$. Assume without loss of generality that $K = \{k+1, \ldots, m\}$ for some $k < m$. Take $a^K \in A^K$. Consider a sequence of profiles $\{P_N^l\}_{l=0}^k$ such that $P_N^0 = P_N$, $P_N^k = \bar{P}_N$, and for all $1 \le l \le k$, $S^{\{1,\ldots,l\}}(P_N^l) = S^{\{1,\ldots,l\}}(\bar{P}_N)$ and $S^{\{l+1,\ldots,m\}}(P_N^l) = S^{\{l+1,\ldots,m\}}(P_N)$. By (i) of Claim 3.2, for all $1 \le l \le k$, $\varphi_{b^{-l}}(P_N^{l-1}) = \varphi_{b^{-l}}(P_N^l)$ for all $b^{-l} \in A^{-l}$. Since $l \notin K = \{k, \ldots, m\}$, an argument similar to the one used in the derivation of (6), implies $\varphi_{a^K}(P_N^{l-1}) = \varphi_{a^K}(P_N^l)$. Therefore, $\varphi_{a^K}(P_N) = \varphi_{a^K}(\bar{P}_N)$, completing the proof of the only-if part. ■

Theorem 3.1 suggests a procedure for constructing *all* unanimous and strategy-proof RSCF on $\mathcal{D}^n$. We can start with marginal probability distributions over all subsets of components that satisfy monotonicity. We can then arbitrarily specify the appropriate joint probabilities of each alternative that generate the chosen marginal distributions.

A question that has received attention in the literature is whether a domain of preferences satisfies the deterministic extreme point property, i.e. whether every unanimous and strategy-proof RSCF on $\mathcal{D}^n$ can be written as a convex combination of deterministic social choice functions (DSCFs) satisfying those properties (see, for instance, Peters et al., 2014; Chatterji et al., 2012; Picot and Sen, 2012; Pycia and Unver, 2015). Unfortunately, we are unable to provide such a characterization in our model, though we suspect it does hold. We note that there are significant technical difficulties involved in proving such a decomposability result. The set of deterministic rules is extremely large (the number of DSCFs when there are 3 agents and m components is 17^m). Furthermore the DSCFs have a complicated structure based on minimal winning coalitions. In any case, we believe that our direct characterization is both simple and intuitive.

Consider the application of our result in the simplest case where there is exactly one component (or candidate). Marginal decomposability is vacuously true in this case. Therefore, an RSCF is unanimous and strategy-proof if and only if it is monotonic, i.e. (1) whenever nobody approves the candidate, he/she is never selected (i.e., selected with zero probability), (2) whenever everybody approves the candidate, he/she is always selected (i.e., selected with probability 1), and (3) whenever the set of agents who approve the candidate increases, the probability that the candidate is selected also increases. As we have remarked earlier, this description of an RSCF is perhaps simpler to understand than expressing it as a convex combination of deterministic rules.

4 Formation of Committees of Fixed Size

In this section, we consider the problem of forming a committee with a predetermined number of members. The size of a committee is defined as the number of members in it. Formally, the size of an alternative $a \in A$ is $|a| = |\{k \mid a^k = 1\}|$. For $l < m$, $A(l)$ is the set of all committees with size l, i.e. $A(l) = \{a \in A \mid |a| = l\}$. In this section, we consider RSCFs $\varphi : \mathcal{D}^n \to \Delta A(l)$ for some $l < m$. By definition, these RSCFs give positive probabilities only to the elements of $A(l)$.

Clearly unanimity is incompatible with this range restriction. We therefore need to replace unanimity by the onto property.

Definition 4.1 An RSCF $\varphi : \mathcal{D}^n \to \Delta A(l)$ is onto if for all $a \in A(l)$, there is $P_N \in \mathcal{D}^n$ such that $\varphi_a(P_N) = 1$.

Our next theorem characterizes the set of onto strategy-proof RSCFs for selecting a committee with a predetermined size. It says that every such rule is random dictatorial restricted to $A(l)$.

Definition 4.2 A DSCF $f : \mathcal{D}^n \to A(l)$ is $A(l)$-restricted dictatorial if there exists $i \in N$ such that $f(P_N)$ chooses the most preferred alternative of agent i from the set $A(l)$. An RSCF is called random $A(l)$-restricted dictatorial if it is a convex combination of $A(l)$-restricted dictatorial DSCFs.

Theorem 4.1 *Let $l < m$. Then, an RSCF $\varphi : \mathcal{D}^n \to \Delta A(l)$ is onto and strategy-proof if and only if it is random $A(l)$-restricted dictatorial.*

Proof First we prove a claim.

Claim 4.1 *Let $P_N, \bar{P}_N$ be such that $P_i|_{A(l)} = \bar{P}_i|_{A(l)}$ for all $i \in N$. Then $\varphi(P_N) = \varphi(\bar{P}_N)$.*

Proof We show that $\varphi(P_N) = \varphi(\bar{P}_i, P_{-i})$ where $P_i|_{A(l)} = \bar{P}_i|_{A(l)}$. Suppose not. Let $b \in A(l)$ be such that $\varphi_b(P_N) \neq \varphi_b(\bar{P}_i, P_{-i})$ and $\varphi_a(P_N) = \varphi_a(\bar{P}_i, P_{-i})$ for all $a \in A(l)$ with $a P_i b$. In other words, b is the maximal element of $A(l)$ according to P_i that violates the assertion of the claim. Without loss of generality, assume that $\varphi_b(P_N) < \varphi_b(\bar{P}_i, P_{-i})$. However, since $\varphi_a(P_N) = \varphi_a(\bar{P}_i, P_{-i})$ for all $a \notin A(l)$ with $a P_i b$, we have $\varphi_{U(b,P_i)}(P_N) < \varphi_{U(b,P_i)}(\bar{P}_i, P_{-i})$. This means agent i manipulates at P_N via $\bar{P}_i$, which is a contradiction. This completes the proof of the claim. ■

Consider an RSCF $\varphi : \mathcal{D}^n \to \Delta A(l)$. For $P \in \mathcal{D}$, define $P|_{A(l)} \in \mathbb{L}(A(l))$ as follows: for all $a, b \in A(l)$, $a P|_{A(l)} b$ if and only if $a P b$. Let $\mathcal{D}|_{A(l)} = \{P|_{A(l)} \mid P \in \mathcal{D}\}$. Construct the RSCF $\hat{\varphi} : (\mathcal{D}|_{A(l)})^n \to \Delta A(l)$ as follows: for all $\hat{P}_N \in \mathcal{D}|_{A(l)}$, $\hat{\varphi}(\hat{P}_N) = \varphi(P_N)$ where $P_N \in \mathcal{D}^n$ is such that $P_i|_{A(l)} = \hat{P}_i$ for all $i \in N$. This is well-defined by Claim 4.1. Because φ is strategy-proof, $\hat{\varphi}$ is also strategy-proof. Moreover, since φ is onto with range $A(l)$, strategy-proofness of φ implies $\hat{\varphi}$ is unanimous. In what follows, we show $\mathcal{D}|_{A(l)}$ is an unrestricted domain.

Claim 4.2 *The domain $\mathcal{D}|_{A(l)}$ is unrestricted.*

Proof Take $P \in \mathcal{D}$ such that $r_1(P^l) = 1$ for all $l \in M$. Consider arbitrary $a, b \in A(l)$ such that $a \neq b$. For $x \in \{a, b\}$, let $I(x) = \{k \in M \mid x^k = 1\}$. By definition, $|I(x)| = l$ for all $x \in \{a, b\}$. Moreover, since a and b are distinct, it must be that $I(a)$ and $I(b)$ are also distinct. This, together with the fact that $|I(a)| = |I(b)| = l$, implies there must be $k, \hat{k} \in M$ such that $k \in I(a) \setminus I(b)$ and $\hat{k} \in I(b) \setminus I(a)$. This means $a^k = r_1(P^k)$ but $a^{\hat{k}} = r_1(P^{\hat{k}})$ and $b^k = r_1(P^k)$ but $b^{\hat{k}} = r_1(P^{\hat{k}})$. Therefore, responsive does not put any restriction on the relative ordering of a and b at P, and consequently, every preference in $\mathcal{D}|_{A(l)}$ can be achieved by considering a suitable preference with the alternative $(1, \ldots, 1)$ as the top-ranked element. This completes the proof of the claim. ■

Since $\mathcal{D}|_{A(l)}$ is unrestricted and $\hat{\varphi}$ is unanimous and strategy-proof, it follows from Gibbard (1977) that $\hat{\varphi}$ is random dictatorial. By the construction of $\hat{\varphi}$, this means φ is random dictatorial restricted to $A(l)$. This completes the proof of Theorem 4.1. ■

It is known that strategy-proof and onto DSCFs on $A(l)$-restricted domains are dictatorial (for a general version of this result, see Barberà et al., 2005 and Aswal et al., 2003). Unfortunately, there is no escape from this negative result if we consider random rather than deterministic rules.

5 Conclusion

In this paper, we have provided a characterization of random unanimous and strategy-proof rules in the well-known committee formation model in terms of two properties: marginal decomposability and monotonicity. We also show that if committees of a predetermined size have to be chosen, an onto and strategy-proof rule must be an appropriate random dictatorship.

References

Aswal, N., Chatterji, S., & Sen, A. (2003). Dictatorial domains. *Economic Theory, 22*, 45–62.

Barberà, S., Massó, J., & Neme, A. (2005). Voting by committees under constraints. *Journal of Economic Theory, 122*, 185–205.

Barberà, S., Sonnenschein, H., & Zhou, L. (1991). Voting by committees. *Econometrica, 59*, 595–609.

Breton, M. L., & Sen, A. (1999). Separable preferences, strategyproofness, and decomposability. *Econometrica, 67*, 605–628.

Chatterji, S., Roy, S., & Sen, A. (2012). The structure of strategy-proof random social choice functions over product domains and lexicographically separable preferences. *Journal of Mathematical Economics, 48*, 353–366.

Chatterji, S., Sen, A., & Zeng, H. (2014). Random dictatorship domains. *Games and Economic Behavior, 86*, 212–236.

Chatterji, S., & Zeng, H. (2018). On random social choice functions with the tops-only property. *Games and Economic Behavior, 109*, 413–435.

Gibbard, A. (1977). Manipulation of schemes that mix voting with chance. *Econometrica, 45*, 665–681.

Hurwicz, L. (1972). On informationally decentralized systems. In C. B. McGuire & R. Radner (Eds.), *Decision and Organization*. North Holland: Amsterdam.

Peters, H., Roy, S., Sen, A., & Storcken, T. (2014). Probabilistic strategy-proof rules over single-peaked domains. *Journal of Mathematical Economics, 52*, 123–127

Picot, J., & Sen, A. (2012). An extreme point characterization of random strategy-proof social choice functions: The two alternatives case. *Economics Letters, 115*, 49–52.

Pycia, M., & Unver, U. (2015). Decomposing random mechanisms. *Journal of Mathematical Economics, 61*, 21–33.

Equal Area Rule to Adjudicate Conflicting Claims

William Thomson

1 Introduction

When a firm goes bankrupt, how should its liquidation value be divided among its creditors? A "rule" is a mapping that specifies, for each situation of this kind, which we call a "claims problem," a division of this value. Alternatively, the problem may be that of specifying the contributions that a group of taxpayers should make to the cost of a public project as a function of their incomes. The formal literature on the subject, whose goal is to identify the most desirable rules, originates in O'Neill (1982).[1]

In the search for rules to solve any type of resource allocation problems, it is a common strategy to invoke concepts from the theory of cooperative games, bargaining games or coalitional-form games. The allocation problems under consideration are mapped into games, a solution defined on the class of games to which these games belong is applied, and the allocations whose images are the resulting payoff vectors are selected for the allocation problems.

For claims problems, this strategy has been followed by Dagan and Volij (1993), who proposed a simple way of mapping claims problems into bargaining games (Nash, 1950), and then focused on commonly used solutions to the bargaining problem, the Nash solution and its weighted versions, and the Kalai-Smorodinsky solution. Other solutions have been defined for bargaining games that are based on measuring in some fashion the sacrifice imposed on each player at a proposed payoff

In homage to Leo Hurwicz, who laid the foundations of the theory of economic design. I thank Patrick Harless for his extensive comments.

[1] For surveys, see Thomson (2003, 2015, 2018).

W. Thomson (✉)
University of Rochester, Department of Economics, Rochester, NY, USA
e-mail: william.thomson@rochester.edu

W. Trockel (ed.), *Social Design*, Studies in Economic Design,
https://doi.org/10.1007/978-3-319-93809-7_14

vector, and in selecting a vector at which sacrifices are equal across players. The "equal area" solution is a two-player solution of this type. Given a game, a player's sacrifice at a payoff vector is simply measured by the area of the set of feasible vectors at which his payoff is larger (Anbarci, 1993; Anbarci and Bigelow, 1994; Calvo and Peters, 2000; Thomson, 1996).[2] As an argument why they are not getting enough at a proposed compromise, people often point to the alternatives at which they could get more, how numerous these alternatives are, how far the compromise would place them from their most preferred alternative, as compared to how others would be treated according to such criteria.

The equal area solution is not as central in the theory of bargaining, but it enjoys a number of appealing properties. In particular, being quite sensitive to the shape of the feasible set, it does not suffer from the occasional paradoxical behaviors of other rules. This sensitivity is a disadvantage in other respects: applying the equal area solution requires the knowledge of the entire feasible set. By the same token, it prevents the solution from satisfying certain invariance properties that one may be interested in. Thus, the rule provides another illustration of the familiar tradeoff in the design of allocation rules between sensitivity and simplicity.

Here, following Ortells and Santos (2011), we apply the equal area solution to solve two-claimant claims problems, obtaining a rule we call the equal area rule. The complexity issue just discussed does not arise in the context of claims problems because the boundary of the feasible set is linear and in fact, an explicit algebraic formula can be given for the equal area rule.[3] Dagan and Volij's choice of the Nash and Kalai-Smorodinsky solutions led them to well-known rules for claims problems, but the equal area rule is new. We begin by studying its properties.

We find that it satisfies all of the basic properties that have been formulated in the literature on claims problems, including all monotonicity properties. The properties that it does not satisfy are mainly invariance properties, which should not be surprising, in the light of our earlier comments on its sensitivity to the shape of the feasible set. One property of that type that it does satisfy however is invariance with respect to truncation of claims at the endowment.[4]

We then turn to problems with more than two claimants. There is more than one way of generalizing the two-claimant equal area bargaining solution to arbitrarily many players, and we briefly discuss the reasons why. These difficulties apply here as well. In the face of this multiplicity, we invoke an important property of allocation rules, called consistency, which has successfully guided the search for extensions of two-agent rules in a great variety of contexts. For claims problems, its expression is particularly simple: a rule is consistent if for each problem, the awards vector

[2]A family of rules are introduced by Young (1987) under the name of "equal sacrifice" rules." Our solution is not a member of this family.

[3]An application of the idea to classical fair allocation problems is proposed and studied by Velez and Thomson (2012).

[4]Incidentally, this property is necessary and sufficient condition for a rule to be obtainable as the composition of two mappings: one is O'Neill's mapping from claims problems to transferable utility coalitional games; the other is a solution for this class of games (Curiel et al., 1987).

it selects is such that for each subgroup of claimants, it selects the restriction of that vector to this population for the problem of allocating among them the amount that remains available after the other claimants have collected their awards and left. Unfortunately, as we show, the two-claimant equal area rule has no consistent extension. In the light of this negative result, we turn to the weaker notion of average consistency (Dagan and Volij, 1997), which still captures much of what consistency itself conveys. This notion allows an extension, and this extension is unique. We discuss some of its properties.

2 The Model and the Equal Area Rule

A group of **agents**, N, have **claims**, $(c_i)_{i\in N}$, on an infinitely divisible resource. These claims add up to more than what is available, the **endowment**, E. Thus, a **claims problem** is a pair $(c, E) \in \mathbb{R}^N_+ \times \mathbb{R}_+$ such that $\sum_{i\in N} c_i \geq E$.[5] Let $\mathcal{C}^N$ denote the class of all claims problems.

An **awards vector** for (c, E) is a vector $x \in \mathbb{R}^N_+$ satisfying the **non-negativity** and **claims boundedness** inequalities $0 \leqq x \leqq c$ and the **balance** equality $\sum x_i = E$. We refer to the line of equation $\sum x_i = E$ as a **budget line**. A **rule** is a mapping that associates with each problem in $\mathcal{C}^N$ an awards vector for it. The **path of awards of a rule** S for a claims vector $c \in \mathbb{R}^N_+$ is the locus of the awards vector S selects for (c, E) as E ranges from 0 to $\sum c_i$. We denote it $p^S(c)$.

For our purposes, it will suffice to define a **bargaining game** with player set N (Nash, 1950) as a convex, compact, and comprehensive[6] subset of $\mathbb{R}^N_+$ that contains at least one point whose coordinates are all positive.[7] A **bargaining solution** associates with each such game a point of it. Let $\mathcal{B}^N$ be the class of all bargaining games.

The bargaining solution that is our point of departure is defined for two players. Let $N \equiv \{1, 2\}$. Given $S \in \mathcal{B}^N$, the **equal area solution**, A, selects the undominated point of S with the property that the area $\alpha_1(S, x)$ of the set of points of S of abscissa greater than x_1 is equal to the area $\alpha_2(S, x)$ of the set of points of S of ordinate greater than x_2 (Fig. 1).

Given a claims problem $(c, E) \in \mathcal{C}^N$, its associated bargaining game $B(c, E)$ consists of the points of $\mathbb{R}^N_+$ that are dominated by c and lie below the budget line. The equal area bargaining solution leads directly to the following rule for claims problems (Ortells and Santos, 2011):

[5] We denote by $\mathbb{R}^N_+$ the cartesian product of $|N|$ copies of $\mathbb{R}_+$ indexed by the members of N. The superscript N may also indicate some object pertaining to the set N. Which interpretation is the right one should be clear from the context. We allow the equality $\sum_{i\in N} c_i = E$ for convenience.

[6] A subset S of $\mathbb{R}^N_+$ is comprehensive if for each $x \in S$ and each $0 \leq y \leq x$, $y \in S$.

[7] The usual specification of a bargaining game includes a disagreement point, and our formulation amounts to assuming that it is the origin. This assumption is justified if the theory is required to be independent of the choice of origin for the utility functions that are used to represent the opportunities available to the agents.

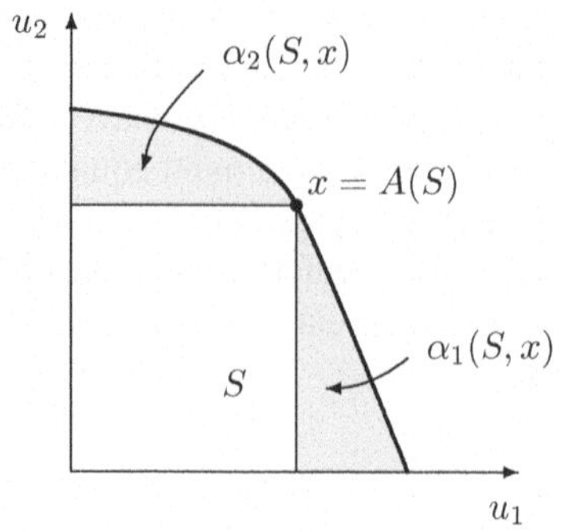

Fig. 1 For two players, the equal area solution. The equal area solution selects the undominated point x of S at which the two curvi-linear triangles defined by the boundary of S and lines parallel to the axes through x have equal areas: $\alpha_1(S, x) = \alpha_2(S, x)$

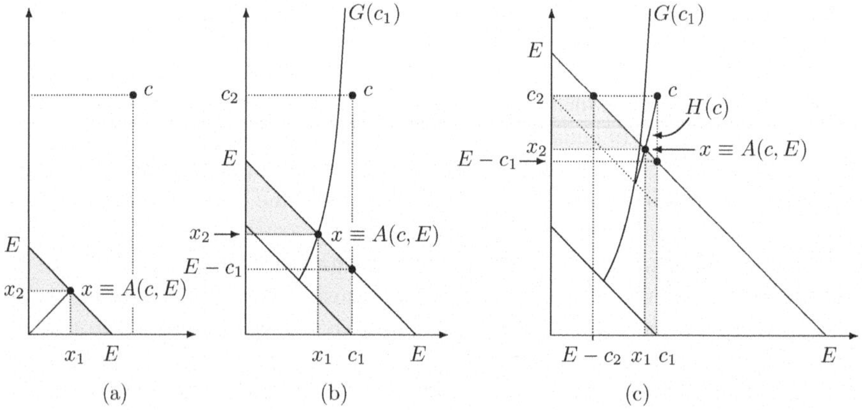

Fig. 2 Constructing a path of awards of the equal area rule. Here, $c_1 < c_2$. The path has three parts, each corresponding to one of the three intervals into which the range of variations in the endowment can be partitioned. (**a**) When $E \leq c_1$. (**b**) When $c_1 \leq E \leq c_2$. (**c**) When $E \geq c_2$

Equal Area Rule, A: Let $N \equiv \{1, 2\}$ and $(c, E) \in \mathcal{C}^N$. Then, $A(c, E)$ is the awards vector x with the property that among the points that are dominated by c and lie below the budget line, the area of those points whose abscissa is greater than x_1 is equal to the area of those points whose ordinate is greater than x_2.

Other Notation Given $a, b, c \in \mathbb{R}^N$, $\Delta(a, b, c)$ denotes the triangle with these points as vertices.

Because the bargaining problem associated with a claims problem is a rectangle truncated by a line of slope -1, the coordinates of its equal area point can be calculated explicitly. They are given in the following lemma. Let $c \in \mathbb{R}^N_+$. The lemma says that $p^A(c)$ has three parts, corresponding to a three-way partition of the set of possible values of the endowment given c. They are represented in the three panels of Fig. 2.

Lemma 1 (Ortells and Santos, 2011) *Let* $N \equiv \{1, 2\}$ *and* $(c, E) \in \mathcal{C}^N$ *be such that* $c_1 < c_2$, *say. The coordinates of its equal area awards vector are as follows:*
Case 1: $E \leq c_1$: $A(c, E) = (\frac{E}{2}, \frac{E}{2})$.
Case 2: $E \in [c_1, c_2]$: $A(c, E) = (c_1[1 - \frac{c_1}{2E}], E - c_1[1 - \frac{c_1}{2E}])$.
Case 3: $E \geq c_2$: $A(c, E) = (\frac{E}{2} + (c_1 - c_2)(1 - \frac{c_1+c_2}{2E}), E - \frac{E}{2} - (c_1 - c_2)(1 - \frac{c_1+c_2}{2E}))$.

In each of the three cases enumerated in the lemma, the coordinates of $A(c, E)$ are obtained by writing equality of

Case 1: the area of $\Delta(x, (x_1, 0), (E, 0))$ and the area of $\Delta(x, (0, x_2), (0, E))$ (panel (a)).

Case 2: the difference of the areas of $\Delta(x, (x_1, 0), (E, 0))$ and $\Delta((c_1, E - c_1), (c_1, 0), (E, 0))$, and the area of $\Delta(x, (0, x_2), (0, E))$ (panel (b)).

Case 3: the difference in the areas of $\Delta(x, (x_1, 0), (E, 0))$ and $\Delta((c_1, E - c_1), (c_1, 0), (E, 0))$, and the difference in the areas of $\Delta(x, (0, x_2), (0, E))$ and $\Delta((E - c_2, c_2), (0, c_2), (0, E))$, (panel (c)).

In Case 2, the coordinates of $A(c, E)$ do not depend on c_2. Given $c_0 \in \mathbb{R}_+$, let $G(c_0)$ be the locus of the point $(c_0[1 - \frac{c_0}{2E}], E - c_0[1 - \frac{c_0}{2E}])$ as E varies in $[c_0, \infty[$. Later on, we will consider claims vectors for the group $\{1, 3\}$ in which agent 1's claim is the smaller one, and for the group $\{2, 3\}$ in which agent 2's claim is the smaller one, and we will construct the paths of awards of the equal area rule for these claims vectors. Then, the notation $G(c_1)$ and $G(_2)$ will designate the copy of the curve we just defined in the spaces $\mathbb{R}^{\{1,3\}}$ and $\mathbb{R}^{\{2,3\}}$. For $p^A(c)$, we only need the part of it that corresponds to E varying in $[\min c_i, \max c_i]$.

In Case 3, the locus of $A(c, E)$ as E varies in $[\max c_i, c_1 + c_2]$ is a curve that we call $H(c)$.

3 Properties of the Equal Area Rule

In this section, we identify which of the basic properties of rules the equal area rule satisfies. These properties are as follows.

The $\frac{1}{|N|}$-truncated-claims lower bound on awards[8] says that each claimant should receive at least $\frac{1}{|N|}$th of his claim truncated at the endowment.

Order preservation says that, given two claimants, the award to the larger claimant should be at least as large as the award to the smaller claimant, and that

[8]The bound is introduced by Moreno-Ternero and Villar (2004) under the name of "securement." *Order preservation* is introduced by Aumann and Maschler (1985), and *order preservation under endowment variations* by Dagan et al. (1997) under the name of "supermodularity." *Linked claims-endowment monotonicity* appears in connection with a discussion of the duality operator in Thomson and Yeh (2008), and *bounded gain under claim increase* is introduced by Kasajima and Thomson (2012) together with a variety of other monotonicity properties. *Claims truncation invariance* is introduced by Curiel et al. (1987) and *minimal rights first* by the same authors under the name of the "minimal rights property." *Composition down* is introduced by Moulin (1987), *composition up* by Young (1988), and duality notions, including *self-duality*, by Aumann and Maschler (1985).

their losses should also be ordered in that way. This property obviously implies the common requirement that two claimants whose claims are equal be assigned equal amounts, **equal treatment of equals**. **Homogeneity** says that multiplying the data of a problem by any $\lambda > 0$ results in a new problem that is solved by rescaling by λ the awards vector chosen for the initial problem.

Endowment monotonicity says that if the endowment increases, each agent should receive at least as much as he did initially. **Order preservation under endowment variations** says that if the endowment increases, given two claimants, the award to the larger claimant should increase by at least as much as the award to the smaller claimant.

Claim monotonicity says that if an agent's claim increases, he should receive at least as much as he did initially. **Bounded award increase under claim increase** says that if an agent's claim increases, his award should not increase by more than his claim did. **Linked claim-endowment monotonicity** says that if an agent's claim and the endowment increase by equal amounts, that claimant's award should not increase by more than that amount.

Claims truncation invariance says that truncating a claim at the endowment should not affect the awards vector that is selected. **Minimal rights first** says a problem can be equivalently solved in either one of the following two ways: (1) directly; (2) in two steps, by first assigning to each claimant the difference between the endowment and the sum of the claims of the other claimants, or 0 if this difference is negative, and then the amount he would be assigned in the problem in which claims are reduced by these first-round awards and the endowment by their sum.

Composition down says that if the endowment decreases from some initial value, the awards vector for the new problem can be computed in either one of the following two ways: (1) directly; (2) by using as claims vector the awards vector calculated for the initial endowment. **Composition up** (Young, 1988) is a counter part of this invariance property that pertains to possible increases in the endowment.

Self-duality says that the awards vector selected by a rule for some problem is equal to the vector of losses implied by its choice in the "dual" problem, that is, the problem with the same claims vector but an endowment equal to the deficit in the initial problem.

When discussing *claims truncation invariance*, we will refer to the following characterization (Thomson, 2018):

Lemma 2 *For $|N| = 2$, say $N \equiv \{1, 2\}$. A rule S is claims truncation invariant if and only if it can be described in terms of the following networks of paths:*

(a) a path $F \subset \mathbb{R}^N_+$ that, for each $E \in \mathbb{R}_+$, meets the line of equation $x_1 + x_2 = E$ exactly once;

(b1) for each $c_2 \in \mathbb{R}_+$, a path $G(c_2) \subset \mathbb{R}^N_+$ that, for each $E \geq c_2$, meets the line of equation $x_1 + x_2 = E$ exactly once, and is bounded above by the line of equation $x_2 = c_2$;

(*b2*) *for each* $c_1 \in \mathbb{R}_+$, *a path* $G(c_1) \subset \mathbb{R}^N_+$ *that, for each* $E \geq c_1$, *meets the line of equation* $x_1 + x_2 = E$ *exactly once, and is bounded to the right by the line of equation* $x_1 = c_1$; *and*

(*c*) *for each* $c \in \mathbb{R}^N_+$ *a path* $H(c) \subset \mathbb{R}^N_+$ *that, for each* $E \in [\max\{c_i\}, c_1 + c_2]$, *meets the line of equation* $x_1 + x_2 = E$ *exactly once, and is bounded above by* c,

these paths being used as follows: for each $c \in \mathbb{R}^N_+$ *such that* $c_1 \geq c_2$, *the path for* c *follows* F *until the line of equation* $x_1 + x_2 = c_2$, *then follows* $G(c_2)$ *until the line of equation* $x_1 + x_2 = c_1$, *then follows* $H(c)$ *until* c; *also for each* $c \in \mathbb{R}^N_+$ *such that* $c_1 \leq c_2$, *the path for* c *follows* F *until the line of equation* $x_1 + x_2 = c_1$, *follows* $G(c_1)$ *until the line of equation* $x_1 + x_2 = c_2$, *then follows* $H(c)$ *until* c.

If in addition to *claims truncation invariance*, a rule satisfies *equal treatment of equals*, the path F is the 45° line.

Theorem 1 *The equal area rule satisfies the following properties: The* $\frac{1}{|N|}$*-truncated-claims lower bound on awards, order preservation, homogeneity, endowment monotonicity, order preservation under endowment variations, claims monotonicity, bounded gain under claim increase, linked claims-resource monotonicity, and claim truncation invariance.*

It violates minimal rights first, composition down, composition up, and self-duality.

Proof The proofs of most of these statements can be obtained from Lemma 1 by straightforward calculations that we omit.

- The $\frac{1}{|N|}$*-truncated-claims lower bound on awards*. For two claimants, meeting this bound requires each path of awards to contain the segment from the origin to the point whose coordinates are equal to half of the smaller claim. This is what is described under Case 1 of Lemma 1.
- *Order preservation*. Assuming $c_1 \leq c_2$ (and symmetrically if $c_2 \leq c_1$), the path of awards for each $c \in \mathbb{R}^N_+$ should lie on or above the 45° line and on or below the line of slope 1 passing through c. This is easily verified for the equal area rule.
- *Homogeneity*. Again, this property follows directly from the definition of the equal area rule.
- *Endowment monotonicity*. This means that paths of awards should be monotone curves. This is the case for the equal area rule. In fact, the rule satisfies the strict version of this property, which says that as the endowment increases, any claimant whose claim is positive should be assigned more.
- *Order preservation under endowment variation*. Let $c \in \mathbb{R}^N_+$. For a rule whose paths of awards are differentiable curves, this means that if $c_1 < c_2$, the slope of $p^A(c)$ is at least 1. Here, differentiability holds at every point except when the endowment is equal to c_2, and this slope requirement is easily verified.
- *Claim monotonicity*. The equal area rule satisfies this property but not its strict version, which says that, if the endowment is positive, a claimant whose claim

increases should be assigned more. Indeed, each of its paths of awards starts with a segment of slope 1 that emanates from the origin and whose length is equal to half of the smaller claim (Case 1 of Lemma 1).

- *Bounded gain under claim increase.* Proving that the equal area rule satisfies this property requires more extensive calculations, but they are straightforward as well. We omit them.
- *Linked claims-endowment monotonicity.* Let $x \equiv A(c, E)$. Assuming that c_1 increases by δ, at the point $x+(\delta, 0)$, the sacrifice made by claimant 1 is the same as at x whereas that of claimant 2 is larger. To reestablish equality, claimant 1's award should increase by less than δ.
- *Claims truncation invariance.* This follows directly from the definition of the equal area rule. The curves in terms of which its paths of awards can be described and whose existence is stated in Lemma 2 are $(G(c_1))_{c_1 \in \mathbb{R}_+}$ and $(G(c_2))_{c_2 \in \mathbb{R}_+}$. Given $c \in \mathbb{R}^N_+$ with $c_1 \leq c_2$, the path for c follows the 45° line up to the point of coordinates $(\frac{c_1}{2}, \frac{c_1}{2})$, then it follows $G(c_1)$ until it meets the line of equation $x_1 + x_2 = c_2$. Figure 3 shows a few sample paths of awards.
- *Minimal rights first.* Let $(c, E) \in \mathcal{C}^N$ be given by $c \equiv (4, 8)$ and $E = 8$. Then $A(c, E) = (3, 5)$. The vector of minimal rights in (c, E) is $(8-8, 8-4) = (0, 4)$ and $A(c - (0, 4), 8 - (0 + 4)) = (2, 2)$. Since $A(c, E) \neq (0, 4) + (2, 2)$, the equal area rule violates the property.
- *Composition down.* Let $(c, E) \in \mathcal{C}^N$ be given by $c \equiv (4, 8)$ and $E = 8$. Then $x \equiv A(c, E) = (3, 5)$. Let $E' \equiv 4$. We have $A(c, E') = (2, 2)$. However, the path of the equal area rule for x contains seg$[(0, 0), (\frac{3}{2}, \frac{3}{2})]$ and continues with the portion of the curve $G(3)$ which lies above the 45° line. Thus $A(x, E') \neq A(c, E')$; the equal area rule violates the property.

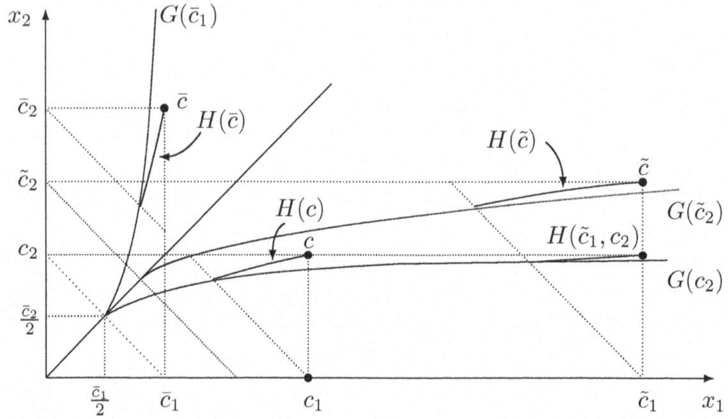

Fig. 3 Generating paths of awards of the equal area rule. Keeping agent 2's claim fixed at c_2 and $\tilde{c}_2$, we show the curves $G(c_2)$ and $G(\tilde{c}_2)$. The path for $\tilde{c} \equiv (\tilde{c}_1, \tilde{c}_2)$ consists of some initial segment of the 45° line, a piece of $G(\tilde{c}_2)$ and a curvi-linear segment $H(\tilde{c})$

- *Composition up*. Let $(c, E) \in \mathcal{C}^N$ be given by $c \equiv (4, 8)$ and $E = 4$. Then $x \equiv A(c, E) = (2, 2)$. Now, let $E' \equiv 8$. We have $A(c, E') = (3, 5)$. However, the path of A for $c - x = A(2, 6)$ contains seg[(0, 0), (1, 1)] and continues with the strictly monotone curve $G(2)$. Thus, $A(c, E') \neq A(c, E) + A(c - x, E'_E)$ and the equal area rule violates the property. We omit the straightforward derivation.
- *Self-duality*. This property implies that the path of awards for each $c \in \mathbb{R}^N_+$ pass through $\frac{c}{2}$. This is the case only if $c_1 = c_2$.

□

Two rules are dual if for each problem, one rule divides the endowment in the same way as the other divides the shortfall (the difference between the sum of the claims and the endowment) in the problem in which the claims vector is the same but the endowment is equal to the shortfall of the first problem. *Self-duality* is invariance under the duality operator.

It is clear that the equal area rule is not *self-dual*. Its dual is the rule that selects, for each problem $(c, E) \in \mathcal{C}^N$, the awards vector x with the property that among the points that are dominated by c and lie above the budget line, the area of those that are below the line of ordinate c_2 is equal to the area of those that are to the left of the line of abscissa c_1. When generalized to bargaining games in the obvious way (in the above statement, simply replace "lie above the budget line" by "lie above the boundary of the feasible set"), we obtain a solution proposed by Karagözoğlu and Rachmilevitch (2017).

4 Consistency

So far, we have only considered the two-claimant case. For more than two claimants, we begin by noting a difficulty that arises in extending the definition of the equal area rule. To illustrate, let us return to bargaining games. Let $N \equiv \{1, 2, 3\}$ and x be an efficient point of some $S \in \mathcal{B}^N$. In order to evaluate an agent's sacrifice at a proposed compromise, it appears natural to work with volumes. For each $i \in N$, let then $V_i(x, S)$ be the volume of the part of S of all points at which player i's utility is at least as large as x_i. The difficulty comes from the fact that $V_1(S, x)$ and $V_2(S, x)$ typically have a non-empty intersection. In Fig. 4, $V_1(S, x)$ is shown to consist of three regions, labeled $W_1(S, x)$, $W_{12}(S, x)$, and $W_{13}(S, x)$. At each point of $W_i(S, x)$, player i's utility is at least as large as at x and it is the opposite for players j and k. At each point of $W_{ij}(S, x)$, players i and j's utilities are at least as large as at x and it is the opposite for player k. Should we simply look for a point at which all $V_i(S, x)$ are equal? Would ignoring the region $W_{ij}(S, x)$ of overlap when defining the sacrifices made by players i and j at x be unfair to player k? Instead, should this common volume be somehow "shared" between players i and j? A discussion of these various options, and of their pros and cons, is in Thomson (1996).

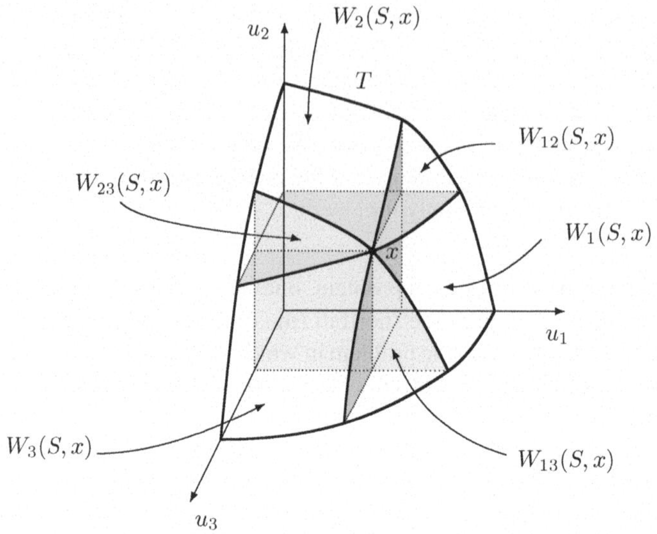

Fig. 4 Illustrating the difficulty in generalizing the equal area bargaining solution to more than two players. The region of points at which player i's payoff is at least as large as x_i is denoted $V_i(S,x)$. The regions of points at which two players's payoff are at least as large at a typical point x overlap. For instance, the intersection of $V_1(S,x)$ and $V_2(S,c)$ is $W_{12}(S,x)$

In our search for an extension of the equal area rule to more than two claimants, we will sidestep the difficulty just discussed and impose a property of coherence of rules across populations of different sizes. For that purpose, we need to generalize our framework of analysis. We imagine that there is an infinite set of "potential" claimants indexed by the natural numbers, $\mathbb{N}$. Let $\mathcal{N}$ be the family of finite subsets of $\mathbb{N}$; these are the populations that may be involved in a claims problem. A rule is now defined over $\bigcup_{N\in\mathcal{N}} \mathcal{C}^N$.

Consider the following property of such a rule. Having identified the awards vector it chooses for some problem, we imagine that some claimants leave the scene with their awards and we reevaluate the situation at this point. The amount available for the remaining claimants is equal to the endowment minus the sum of the awards to the claimants who left. Let us apply the rule to this "reduced" problem. **Consistency** says that the rule should choose the same award for each of the remaining claimants as it did initially. Formally, for each $N \in \mathcal{N}$, each $(c,E) \in \mathcal{C}^N$, and each $N' \subset N$, and—introducing $x \equiv S(c,E)$—we have $x_{N'} = S(c_{N'}, \sum_{N'} x_i) = S(c_{N'}, E - \sum_{N\setminus N'} x_i)$.

It will be convenient to rephrase this requirement by saying that if x belongs to the path of awards of the rule for c, its projection on any coordinate subspace belongs to its path for the projection of c onto the subspace. Thus, its path for c, when projected on that subspace, is a subset of its path for the projection of c. Moreover, if a rule is *endowment continuous*, which is the case for the equal area rule, the projection of its path for c is in fact equal to its path for the projection of c.

Similar questions have been asked about other two-claimant rules. One of them is the rule known as **concede-and-divide**. For each claims vector, this rule is defined by assigning to each claimant i the amount conceded by the other claimant j, namely the difference between the endowment and claimant j's claim, or 0 is that difference is negative, and in dividing the remainder equally. It turns out that concede-and-divide has a *consistent* extension, which is none other than the so-called Talmud rule. On the other hand, the rule obtained from the proportional rule by first truncating claims at the endowment has no such extension (Dagan and Volij, 1997).

A general technique to identify the consistent extension of a two-claimant rule when such an extension exists, or to prove that none does if that is the case, is developed in Thomson (2007). It exploits the projection implication of *consistency* just noted. This technique is particularly useful when paths of awards are piece-wise linear, as is often the case, but it has also helped address the question of existence of *consistent* extensions of rules whose paths of awards are not piece-wise linear. For example, it can be used to prove the non-existence mentioned above, of a *consistent* extension of the version of the proportional rule defined by truncating claims at the endowment first (Thomson, 2008). The proof of the negative result that we offer next follows the same logic.

Theorem 2 *The equal area rule has no consistent extension.*

Proof Let $N \equiv \{1, 2, 3\}$ and $c \in \mathbb{R}^N_+$ be such that $c_1 < c_2 < c_3$. Because $c_1 < c_2$, $p^A(c_1, c_2)$ includes seg$[(0, 0), (\frac{c_1}{2}, \frac{c_1}{2})]$ and the part C of the curve $G(c_1)$ in $\mathbb{R}^{\{1,2\}}_+$ that lies between the lines of equation $x_1 + x_2 = c_1$ and $x_1 + x_2 = c_2$.

Similarly, because $c_1 < c_3$, $p^A(c_1, c_3)$ includes seg$[(0, 0), (\frac{c_1}{2}, \frac{c_1}{2})]$, and the part D of the curve $G(c_1)$ in $\mathbb{R}^{\{1,3\}}$ that lies between the lines of equation $x_1 + x_3 = c_1$ and $x_1 + x_3 = c_3$.

Because A is *strictly endowment monotonic*, $p^A(c_1, c_2)$ and $p^A(c_1, c_3)$ are strictly monotone curves, and one can recover $p^A(c)$ from them as follows. Given $t \in [0, c_1]$, the plane P^t of equation $x_1 = t$ crosses $p^A(c_1, c_2)$ at a single point, x^t, and it crosses $p^A(c_1, c_3)$ at a single point, y^t. There is a unique point $z^t \in \mathbb{R}^N$ whose projections onto $\mathbb{R}^{\{1,2\}}$ and $\mathbb{R}^{\{1,3\}}$ are x^t and y^t, respectively. Because the same curve $G(c_1)$ is used to generate $C \subset p^A(c_1, c_2)$ and $D \subset p^A(c_1, c_3)$, it follows that up to an endowment equal to $c_2 = \min\{c_2, c_3\}$, C and D are the same curve (except that one lies in $\mathbb{R}^{\{1,2\}}_+$ and the other in $\mathbb{R}^{\{2,3\}}_+$), so that $x^t_2 = y^t_3$. Thus, the first two coordinates of z^t are equal, and by letting t run from $\frac{c_1}{2}$ to $c_1(1 - \frac{c_1}{2c_2})$, the abscissa of the topmost point of C, we deduce that the path for c of a *consistent* extension of A, if such an extension exists, contains, in addition to seg$[(0, 0, 0), (\frac{c_1}{2}, \frac{c_1}{2}, \frac{c_1}{2})]$, a monotone curve in $\mathbb{R}^N$ in the plane of equation $x_1 = x_3$ whose topmost point has second and third coordinates equal to $c_1(1 - \frac{c_1}{2c_2})$. (∗) The projection of these two objects onto $\mathbb{R}^{\{2,3\}}$ is seg$[(0, 0), (c_1(1 - \frac{c_1}{2c_2}), c_1(1 - \frac{c_1}{2c_2}))]$.

However, we also know that the path of awards of A for (c_2, c_3) consists of seg$[(0, 0), (\frac{c_2}{2}, \frac{c_2}{2})]$, and that it continues with the part of the curve $G(c_2)$ in $\mathbb{R}^{\{2,3\}}$

that lies between the lines of equation $x_2 + x_3 = c_2$ and $x_2 + x_3 = c_3$. Because $c_1(1 - \frac{c_1}{2c_2}) > \frac{c_2}{2}$, we obtain a contradiction to (∗). □

In the face of the negative result stated as Theorem 2, the question arises as to what to do for more than two claimants and preserve the spirit of the equal area rule. The notion of **average consistency** comes to our rescue. A rule satisfies this property if for each problem and each claimant, the award to this claimant is equal to the average of his awards in all of the two-claimant reduced problems associated with it involving him (Dagan and Volij, 1997). Formally, for each $N \in \mathcal{N}$, each $(c, E) \in \mathcal{C}^N$, and each $i \in N$, $x_i = \frac{1}{|N|-1} \sum_{j \in N \setminus \{i\}} S_i(c_i, c_j, x_i + x_j)$. Although the equal area rule has no *consistent* extension, we have the following existence and uniqueness result involving *average consistency*.

Theorem 3 *The equal area rule has a unique average consistent extension.*

Indeed, the only requirement for such an extension of a two-claimant rule to exist, and uniqueness is implied too, is that it be *endowment monotonic* (Dagan and Volij, 1997), and we have seen that the equal area rule enjoys this property.

The operator that associates with each two-claimant rule its *average consistent* extension preserves many of its properties. Included are *endowment monotonicity*, *anonymity* (Dagan and Volij, 1997), *claims monotonicity*, *claims continuity*, and *claims truncation invariance*. Thus, the *average consistent* extension of the two-claimant equal area rule satisfies each of the properties just enumerated.

5 Concluding Comments

In certain circumstances, one may decide that a particular claimant is more deserving than some other claimant, independently of the relative values of their claims. For example, one may give preferential treatment to a war veteran and to a single mother. To accommodate this possibility, one can assign weights to claimants and require that rules "respect" or "reflect" these weights. The most natural way to achieve this here is to select, for each problem, a point at which the areas appearing in the original definition, multiplied by the players' respective weights, are equal. For each claims vector, as the relative weights assigned to two claimants go to infinity, the path of awards for that claims vector approaches that of the sequential priority rule in which the claimant who is first is the one who is assigned the greater weight. All of the properties of the equal area rule are preserved under this generalization except, obviously, the $\frac{1}{|N|}$-*lower bound* and all *order preservation* properties. It is indeed the purpose of assigning different weights to claimants to inflect awards in their direction.

An alternative to the equal area bargaining solution that can also be understood as attempting to equate sacrifices among players and has been the object of some discussion is the solution that selects, for each bargaining problem, the point x for which the lengths of the curvi-linear segments in its boundary that connect x to the endpoints of the set of undominated payoff vectors are equal. This "equal length bargaining solution" can be applied to claims problems to generate an "equal length rule." It is an easy matter to check that this rule is none other than the well-studied "concede-and-divide" rule. The same comment applies to the Perles-Maschler bargaining solution (1981). Although these two bargaining solutions generally differ, the rules they induce for claims problem indeed coincide.[9] It is known that concede-and-divide has only one consistent extension, the "Talmud rule," so-called because it rationalizes resolutions proposed in the Talmud for particular numerical examples (Aumann and Maschler, 1985).

Finally, one may argue that instead of measuring agent i's sacrifice at a proposed compromise x by the area of the set of points y at which his award is greater than x_i, the difference between y_i and x_i be taken into consideration. A simple idea would be to measure the sacrifice imposed on claimant $i \in N$ by the integral over $t \in [x_i, c_i]$ of the product $(t - x_i)(E - t)$.

References

Anbarci, N. (1993). Non-cooperative foundations of the area monotonic solution. *Quarterly Journal of Economics, 108*, 245–258.

Anbarci, N., & Bigelow, J. F. (1994). The area monotonic solution to the cooperative bargaining problem. *Mathematical Social Sciences, 28*:133–142.

Aumann, R., & Maschler, M. (1985). Game theoretic analysis of a bankruptcy problem from the Talmud. *Journal of Economic Theory, 36*, 195–213

Calvo, E., & Peters, H. J. M. (2000). Dynamics and axiomatics of the equal area bargaining solution. *International Journal of Game Theory, 29*, 81–92.

Curiel, I., Maschler, M., & Tijs, S. H. (1987). Bankruptcy games. *Zeitschrift für Operations Research, 31*, A143–A159.

Dagan, N., Serrano, R., & Volij, O. (1997). A non-cooperative view of consistent bankruptcy rules. *Games and Economic Behavior, 18*, 55–72.

Dagan, N., & Volij, O. (1993). The bankruptcy problem: A cooperative bargaining approach. *Mathematical Social Sciences, 26*, 287–297.

Dagan, N., & Volij, O. (1997). Bilateral comparisons and consistent fair division rules in the context of bankruptcy problems. *International Journal of Game Theory, 26*, 11–25.

Karagözoğlu, E., & Rachmilevitch, S. (2017). Duality, area-considerations, and the Kalai-Smorodinsky solution. *Operations Research Letters, 45*, 30–33.

Kasajima, Y., & Thomson, W. (2012). Monotonicity properties of rules for the adjudication of conflicting claims. mimeo.

Moreno-Ternero, J., & Villar, A. (2004). The Talmud rule and the securement of agents' awards. *Mathematical Social Sciences, 47*, 245–257.

Moulin, H. (1987). Equal or proportional division of a surplus, and other methods. *International Journal of Game Theory, 16*, 161–186.

Nash, J. F. (1950). The bargaining problem. *Econometrica, 18*, 155–162.

[9]The typical path of awards of this rule for a claims vector $(c_1, c_2) > 0$ contains the same initial segment as the equal area rule, a segment that is symmetric with respect to the half claims vector, and a vertical or horizontal segment connecting these two objects, depending upon whether $c_1 \leq c_2$ or $c_2 \leq c_1$.

O'Neill, B. (1982). A problem of rights arbitration from the Talmud. *Mathematical Social Sciences, 2*, 345–371.
Ortells, T., & Santos, C. (2011). The pseudo-average rule: Bankruptcy, cost allocation and bargaining. *Mathematical Methods of Operations Research, 73*, 55–73.
Thomson, W. (1996). *Bargaining theory; the axiomatic approach* (forthcoming).
Thomson, W. (2003). Axiomatic and game-theoretic analysis of bankruptcy and taxation problems: A survey. *Mathematical Social Sciences, 45*, 249–297.
Thomson, W. (2007). On the existence of consistent rules to adjudicate conflicting claims: a geometric approach. *Review of Economic Design, 11*, 225–251.
Thomson, W. (2008). The two-agent claims-truncated proportional rule has no consistent extension: A constructive proof. *Economics Letters, 98*, 59–65.
Thomson, W. (2015). Axiomatic and game-theoretic analysis of bankruptcy and taxation problems: An update. *Mathematical Social Sciences, 74*, 41–59.
Thomson, W. (2018). *How to divide when there isn't enough; from Aristotle, the Talmud, and Maimonides to the axiomatics of resource allocation*. Cambridge: Cambridge University Press.
Thomson, W., & Yeh, C.-H. (2008). Operators for the adjudication of conflicting claims. *Journal of Economic Theory, 143*, 177–198.
Velez, R., & Thomson, W. (2012). Let them cheat! *Games and Economic Behavior, 75*, 948–963 (2012).
Young, P. (1987). Progressive taxation and the equal sacrifice principle. *Journal of Public Economics, 32*, 203–214.
Young, P. (1988). Distributive justice in taxation. *Journal of Economic Theory, 44*, 321–335.

Part V
Implementation

Recent Results on Implementation with Complete Information

Bhaskar Dutta

1 Introduction

The origins of the theory of mechanism design can be traced back to the debate on the relative merits of centralized planning and free markets. Lange (1936) and Lerner (1944) argued that centralized planning could emulate the achievements of free markets, and possibly do better since it could correct for market failure. On the other side of the debate were von Hayek (1944) and von Mises (1935) who argued that centralized planning could not possibly be as successful as free markets. However, it was Leo Hurwicz (1960, 1972) who provided a conceptual framework for what we now know as the theory of mechanism design. For instance, Hurwicz (1960) provided a formal definition of a mechanism as a system of communication in which agents send messages to each other or to a message center, and where a prespecified function or rule assigns an outcome to each profile of messages. He defined a mechanism as a communication system in which participants send messages to each other and/or to a "message center", and where a pre-specified rule assigns an outcome (such as an allocation of goods and services) for every collection of received messages. Hurwicz (1972) introduced the key notion of *incentive compatibility*. Hurwicz also formulated the basic implementation issue by asking whether there exist mechanisms through which the equilibrium interaction of self-interested agents yield the Walrasian equilibrium allocations.

B. Dutta (✉)
Ashoka University, Sonipat, India

University of Warwick, Coventry, UK
e-mail: b.dutta@warwick.ac.uk

W. Trockel (ed.), *Social Design*, Studies in Economic Design,
https://doi.org/10.1007/978-3-319-93809-7_15

In a seminal paper, Maskin (1999)[1] provided a very elegant answer to a more general question, by almost completely characterizing the correspondences that can be obtained as Nash equilibrium correspondences of various mechanisms in a very wide variety of settings. The Maskin framework studies mechanism design under complete information. The planner's goals are represented as a *social choice correspondence* that selects a set of feasible social outcomes for each state of the world. The state of the world is common knowledge amongst the set of individuals in the society, but is not known to the planner. The planner cannot simply ask the agent to announce the state of the world since the agents' preferences may not coincide with that of the planner. So, the planner has to design a mechanism that will induce agents to correctly reveal the state of the world. A mechanism along with a state of the world describes a complete information game. Maskin (1999) focussed on mechanisms where agents choose messages simultaneously. He used the conventional game-theoretic notion of *Nash equilibrium*, and provided an almost complete characterization of the class of social choice correspondences that are *implementable* in Nash equilibrium.

Maskin identified a condition that has come to be called *Maskin Monotonicity* as a necessary and almost sufficient condition for Nash implementation. Unfortunately, Monotonicity has a lot of bite—for instance, no non-dictatorial *social choice function* satisfies this condition if all logically possible states of the world are in the domain of the function. However, there has been a sudden resurgence of interest in implementation under complete information.[2] One particularly interesting implication of this literature is that a (small) modification of the original framework produces a dramatically different result. In this paper, we briefly discuss the main results when individual(s) have some preference for honesty as well as repeated Nash implementation.

2 Framework

Let N be a set of n individuals. The set of states of the world is represented by Θ, while X is the set of feasible social alternatives. A *social choice function* is a mapping $f : \Theta \to X$.

Each individual $i \in N$ has a utility function $u_i : X \times \Theta \to R$. So, $u_i(x, \theta)$ is the utility that i gets from x in state θ.[3]

[1] Although published in 1999, drafts of the paper have been in circulation since 1978.

[2] See, for instance, Bull and Watson (2007), Kartik and Tercieux (2012) on implementation with evidence, Dutta and Sen (2012), Matsushima (2008a,b), Ortner (2015), Saporiti (2014) and Lombardi and Yoshihara (2011), for the case when some individuals have a "small" preference for honesty, and Lee and Sabourian (2011), Mezzetti and Renou (2017) for repeated Nash implementation.

[3] Note that in the one-stage implementation problem, utility can be *ordinal*.

A one-stage *mechanism* is a pair $g = (M, \pi)$ where $M \equiv M_1 \times M_2 \times \ldots \times M_n$ and $\pi : M \to X$. Here, each M_i is the set of messages that i can send while π is the outcome space specifying an outcome for each n-tuple of messages.

Maskin (1999) asked the following question—what is the class of social choice functions such that there exists a mechanism g whose Nash equilibria coincide with the social choice function at each state of the world, and proceeded to give an almost complete answer.

Define the lower contour set for i at (x, θ) as follows. Let $L_i(x, \theta) = \{y \in X | u_i(x, \theta) \geq u_i(y, \theta)\}$. A fundamental condition for implementation is Monotonicity.

Definition 1 An scf satisfies *Monotonicity* if for all θ, θ', and x,

$$\left[f(\theta) = x, L_i(x, \theta) \subseteq L_i(x, \theta')\right] \to f(\theta') = x$$

This seems an intuitively appealing condition. It requires that if x is socially optimal in state θ, and the position of x vis-a-vis any alternative is no worse in state phi, then x should remain optimal in state θ'. However, *no* social choice function satisfies Monotonicity if the social choice function has full domain. As an example, consider the following.

Example 1 Let $N = \{1, 2, 3\}$, and $X = \{x, y, z\}$. The social choice function picks the pairwise majority winner if it exists and x otherwise. The utility profile is

- $u_1(x, \theta) > u_1(y, \theta) > u_1(z, \theta)$.
- $u_2(y, \theta) > u_2(z, \theta) > u_2(x, \theta)$.
- $u_3(z, \theta) > u_3(x, \theta) > u_3(x, \theta)$.
- $u_2(z, \phi) > u_2(y, \phi) > u_2(x, \phi)$.
- The utility functions of individuals 1 and 3 are identical in states θ and ϕ.

Notice that the utility profile in state θ corresponds to the voting paradox. Since there is no majority winner, x is the socially optimal the social optimal alternative according to the stipulated social choice function. However, z is the majority winner in state ϕ and hence socially optimal in that state. This is a violation of Monotonicity since $L_i(x, \theta) = L_i(x, \phi)$ for all individuals.

Definition 2 A scf satisfies *Absence of Veto* (AV) if $f(\theta) = x$ whenever at least $(n - 1)$ prefer x to any other alternative.

Unlike Monotonicity, AV is a very weak condition. For instance, it is trivially satisfied if there is a private good and no one is satiated with respect to this good.

The Maskin theorem follows.

Theorem 1 *If a social choice function f is implementable in Nash equilibrium, then it satisfies Monotonicity. Moreover, if $n \geq 3$, then f is implementable in Nash equilibrium if it satisfies Monotonicity and Absence of Veto.*

Dutta and Sen (1991) and Moore and Repullo (1990) deal with the two-person case. Moore and Repullo (1990) also have a complete characterization of the many-person case.[4] This is going to be the point of departure for what follows.

3 Preference for Honesty

The canonical mechanism for Nash implementation requires agents to announce (i) a state of the world θ in Θ (ii) an outcome $\in X$, (iii) and an integer $k \in \{1, 2, \ldots, n\}$. Let us write $M_i \equiv M_i^1 \times M_i^2$ where $M_i^1 = \Theta$.

In order to define an intrinsic preference for truthtelling—however small this preference may be—one needs to extend an individual's preferences over outcomes to one over messages. So, extend each individual i's utility function u_i over $\Theta \times X$ to $\tilde{u}_i$ over $\Theta \times M_i$.

Dutta and Sen (2012)[5] define the concept of *partial honesty*.

Definition 3 An individual is partially honest if for all $\theta \in \Theta$, for all $m_i, m_i' \in M_i$ and $m_{-i} \in M_{-i}$, $\tilde{u}_i(\theta, (m_i, m_{-i})) > \tilde{u}_i(\theta, (m_i', m_{-i}))$ if (i) $u_i(\theta, \pi((m_i, m_i)) = u_i(\theta, \pi((m_i', m_i))$, (ii) $m_i^1 = \theta$ and $m_i'^1 \neq \theta$.

The implication of partial honesty is that if m_i and m_i' are both best responses to m_{-i} in terms of u_i, but m_i involves declaration of the true state of the world, while m_i' involves a false state of the world, then i has a *strict* preference for telling the truth. In all other cases, partial honesty imposes no restriction on preferences. So, an individual is said to have a preference for honesty if she prefers to announce the true state of the world whenever a lie does not change the outcome given the messages announced by the others. Notice that this is a very weak preference for honesty since an "honest" individual may prefer to lie whenever the lie allows the individual to obtain a more preferred outcome. An alternative way of describing an honest individual's preference for honesty is that the preference ordering is lexicographic in the sense that the preference for honesty becomes operational only if the individual is indifferent on the outcome dimension, that is in terms of the utility function u_i.

This makes the following theorem, due to Dutta and Sen (2012),[6]

Theorem 2 *Suppose $n \geq 3$, and at least one agent is partially honest. Then, every social choice function satisfying AV is implementable in Nash equilibrium.*

Dutta and Sen also show that an additional condition on the social choice function is sufficient for Nash implementation in the two-person case. Importantly,

---

[4]The reader is referred to Jackson (2001) and Maskin and Sjostrom (2002) for very comprehensive surveys of the literature.

[5]Matsushima (2008a,b), Kartik and Tercieux (2012), Lombardi and Yoshihara (2011), Ortner (2015) also assume that some individuals have a "small" preference for honesty.

[6]Kartik and Tercieux (2012) also prove essentially the same result in a cardinal framework.

Monotonicity which has such strong implications is no longer a necessary condition for implementation. Hence, even a *small* preference for honesty has a large consequence for implementation!

Rather than presenting the proof of the theorem, I will discuss the role of Monotonicity in Maskin's original theorem and how the presence of one partially honest individual is sufficient to get rid of Monotonicity.

Suppose the social choice function is implementable in Nash equilibrium by means of the canonical mechanism (M, π). Then, for any state θ, $f(\theta)$ can be supported as a Nash equilibrium if everyone announces (θ, m_i^2). The canonical mechanism is designed so that a unilateral deviation by individual i either does not change the outcome or the outcome is in the lower contour set of i with respect to θ. Notice that Monotonicity plays no role here. Monotonicity is essential in establishing the difficult part of implementation—that no socially suboptimal outcome can be supported as a Nash equilibrium. Suppose that in state θ, everyone reports $m = (\phi, m^2)$ and that m is a Nash equilibrium in state ϕ. If $f(\theta) \neq f(\phi)$, then m must not be a Nash equilibrium in state θ. Monotonicity ensures the existence of some individual i and social outcome x such that $x \in L_i(f(\phi), \phi)$ but $x \notin L_i(f(\phi), \theta)$. The canonical mechanism gives individual i the power to unilaterally deviate to some message m_i' and get outcome x. Notice that since $x \in L_i(f(\phi)$, i will not want to deviate to m_i' when the true state is actually ϕ. But, since $x \notin L_i(f(\phi), \theta)$, individual i does want to deviate to m_i' when the true state is θ. This ensures that m is not a Nash equilibrium in state θ.

How does Partial Honesty help? Consider a modification of the canonical mechanism. Let (M, π) be a mechanism such that $\pi(m) = f(\theta)$ whenever at least $(n-1)$ individuals announce $m_i^1 = \theta$. As in the canonical mechanism, the modulo game is used to choose the outcome when no more than $(n-2)$ individuals agree on the first component of their messages. Now, suppose the true state is ϕ but everyone announces θ in the first component. Unanimous falsehood is no longer a Nash equilibrium. Consider an individual i who is partially honest. Individual i can deviate and announce ϕ. The outcome does not change since $(n-1)$ individuals continue to announce θ. However, individual i gains in terms of $\tilde{u}_i$ because she is now announcing the true state of the world.

The mechanism used in Dutta and Sen (2012) uses a *modulo game* as part of the mechanism in order to rule out unwanted equilibria when individuals make hybrid announcements—that is, less than $(n-1)$ individuals announce the same state. In a modulo game, each individual announces an integer between 1 and n. The winner of the modulo game is individual i if i equals the sum of the announced integers modulo n. The winner is then allowed to choose any outcome from X. Notice that given the integer announcements of the other $(n-1)$ individuals, individual i has a best response that allows her to be the winner of the modulo game. So, in the region of the message space where hybrid announcements are made, there can be a pure strategy Nash equilibrium only if all individuals share a common top outcome. But, then by AV, this outcome must be socially optimal. While modulo games do not have pure strategy equilibria, they do have mixed strategy equilibria which are not taken into account. To the extent that the outcomes associated with these mixed strategy equilibria are socially suboptimal, one can question whether this is a proper implementation result. An alternative to the use of modulo games is the so-called integer game in which the agent announcing the highest integer is the winner and gets to pick the outcome in the case of hybrid announcements. These mechanisms suffer from the Jackson (1992) critique—the mechanisms are unbounded[7] and casts into doubt the predictive value of Nash equilibrium. Unfortunately, to the best of my knowledge, all papers in Nash implementation where the social choice function has unrestricted domain of preferences employ either an integer game or a modulo game.

As I have mentioned earlier, the difficult part of Nash implementation is that the Nash equilibrium correspondence tends to be too large and so there is the possibility that socially suboptimal outcomes can be supported as equilibrium outcomes. The role of integer and modulo games is to restrict the (pure strategy) Nash equilibrium correspondence. Of course, the same purpose is served simply by refining the concept of equilibrium since this too reduces the size of the equilibrium correspondence. Another possibility is to seek appropriate restricted domains of preferences that will also eliminate unwanted equilibria. Both approaches have been tried in the literature.

Holden et al. (2014) restrict the domain of preferences to one allowing *separable punishments*. Strictly speaking, their separability restriction is not just on the domain of preferences since the permissible domain is *specific* to the social choice function under consideration. Their permissible domain has the following feature—there are two individuals i and j such that for each pair of states θ, θ', there is an alternative $a(\theta')$ which is distinct from $f(\theta')$, such that in any state θ, individual j is indifferent between $a(\theta')$ and $f(\theta')$ but individual i finds $a(\theta')$ to be strictly worse than $f(\theta')$. Formally, the condition is the following.

Definition 4 There is separable punishment if for each state $\theta' \in \Theta$, there is $a(\theta')$ and individuals i and j such that for all $\theta \in \Theta$, $u_j(a(\theta'), \theta) = u_j(f(\theta'), \theta)$ and $u_i(a(\theta'), \theta) < u_i(f(\theta'), \theta)$.

This restriction is particularly appropriate in economic environments with private goods. Such environments allow the possibility of selective punishment—one individual or group can be punished without punishing others. Domain restrictions in a similar spirit have been used in earlier literature. Perhaps the earliest was Jackson et al. (1994) who used a similar but not identical restriction. They characterized the class of social choice correspondences that can be implemented in undominated mechanisms in separable environments by bounded mechanisms which have the additional feature of there being no mixed strategy equilibrium.

[7]A mechanism is *bounded* if any weakly dominated strategy is weakly dominated by a strategy which is itself undominated. Jackson et al. (1994) characterize the class of social choice correspondences that are implemented in undominated Nash equilibrium by bounded mechanisms.

Holden, Kartik and Tercieux then go on to show that in the case of a separable punishment domain, if individuals are partially honest,[8] then any social choice function can be implemented by means of a direct mechanism in two rounds of iterative elimination of strictly dominated strategies. The use of a direct mechanism obviously avoids the Jackson critique. Moreover, since implementation is achieved in elimination of strictly dominated strategies, there are no unaccounted mixed strategy equilibria. Another feature of their result is that the two-person case no longer needs separate treatment.

The example below (Table 1 in their paper) illustrates the main idea underlying their proof. There are two states θ and θ' and two individuals 1 and 2, with the latter being partially honest. The mechanism designer can also levy small fines t_i on individual i, though these are not required on the equilibrium path. The outcome from X depends only on individual 1's announcement. A fine is levied on 1 if her announcement does not match that of 2. Assume that there are t_i, t_i' such that

$$u_1(f(\theta'), t_1', \theta) < u_1(f(\theta), \theta) \tag{1}$$

$$u_1(f(\theta), t_1, \theta') < u_1(f(\theta'), \theta') \tag{2}$$

Equation 1 requires that there is a fine t_1' such that 1 prefers $f(\theta)$ to $(f(\theta'), t_1')$ when the true state is θ, while Eq. (2) requires the existence of a large enough fine t_1 such that 1 prefers $f(\theta')$ to $(f(\theta), t_1)$ when the true state is θ'. The table below exhibits these outcomes.

	θ	θ'
θ	$f(\theta)$	$((f(\theta), t_1), f(\theta))$
θ'	$((f(\theta'), t_1'), f(\theta'))$	$f(\theta')$

Notice that it is a strictly dominant strategy for individual 2 to declare the true state of the world—her announcement has no effect on the outcome but she gets a bonus from telling the truth. Knowing that 2 will tell the truth, lying about the state is dominated by truthtelling for 1 in view of Eqs. (1) and (2).

Dutta and Sen (2012) have a related result on separable domains. Their concept of a separable domain assumes the existence of a reference alternative w with the property that for any alternative $x \in X$ and subset of agents J, there is an alternative a^J such that agents in J are indifferent to w while agents not in J are indifferent to a in all states of the world. An example of a separable domain is the pure exchange economy. The alternative w is the allocation $(0, \ldots, 0)$ where all agents get zero amounts of all goods. For any other allocation a and set of agents J, the allocation

[8] Their result goes through even if there is just one partially honest individual. But, then the construction of the mechanism would depend upon the identity of this individual and so the result will not be detail-free.

a^J is the one where agents in J get zero amounts of all goods while individuals not in J get the same consumption bundle that they are assigned in a. This definition has the advantage that it does not depend on the social choice function. On the other hand, its weakness is that it cannot handle the "standard" case of public goods economies with quasilinear preferences because a^J cannot depend on preferences—it has to be indifferent to w for all agents in J and indifferent to a for all agents outside J for *all* states.

Assuming that all agents are partially honest and that there are at least three individuals, Dutta and Sen prove a strong result—every social choice function is implementable in *strongly dominant* strategies. That is, declaring the true state of the world is a strongly dominant strategy for all agents in the direct mechanism. Notice that this does not contradict, for instance, the negative results of Barbera and Jackson (on strategyproofness of allocations in pure exchange economies because Dutta and Sen assume a complete information environment where individuals have to announce the *state of the world*, whereas the strategyproofness literature assumes that individuals only know their own preferences).

I conclude this section with a brief discussion of Ortner (2015), who uses two refinements of Nash equilibrium along with partial honesty. One refinement is that of Kandori et al. (1993) and Young (1993) *stochastically stable equilibrium*. The other refinement, labeled *fault tolerant equilibrium* while related to k-fault tolerant Nash equilibrium due to Eliaz (2002), is distinct. It incorporates the possibility that each player does not know whether other players are irrational and so chooses strategies to insure herself against irrational behavior of others. Ortner assumes that there are at least five individuals, and that all of them are partially honest. Moreover, there is a distinguished outcome a^* which is not in the range of the social choice function. Armed with these assumptions, he shows that any social choice function is implementable by means of a simple direct mechanism in either refinement of equilibrium.

4 Repeated Implementation

Two recent papers by Lee and Sabourian (2011) and Mezzetti and Renou (2017) extend the original Maskin framework in a new direction by considering the repeated implementation problem. In their setting, the same set of infinitely lived agents interact repeatedly over time, either a finite or infinite number of times. At each discrete point of time, a state of the world is drawn. Players learn the state so that there is complete information. The implementation problem arises because the planner does not observe the state in any period. Neither does the planner observe past states of the world. However, at each date, the planner does learn the outcome of past mechanisms.The planner's objective is to repeatedly implement an scf in each period after every possible history. He commits to a mechanism for each period, but the mechanism in period t can be conditioned on the past history of outcomes in periods $1, \ldots, t-1$. Players discount the future by δ.

Notice that although players interact repeatedly over time, this is not a repeated game since both the state of the world as well as the mechanism change over time. However, it does have the whiff of a repeated game and so one's first impression may be that the folk-theorem type results will enlarge the set of equilibria and hence make the implementation problem harder. Even if folk-type results do not hold, players can coordinate on past histories and generate additional equilibria in the repeated context. This implies that social choice functions satisfying Monotonicity and AV may not be repeatedly implementable. This intuition is confirmed in the following example.

	θ			θ'		
	$i=1$	$i=2$	$i=3$	$i=1$	$i=2$	$i=3$
a	4	2	2	3	1	2
b	0	3	3	0	4	4
c	0	4	4	0	2	3

Let $f(\theta) = a$, $f(\theta') = b$. Then, f is efficient, monotonic and satisfies AV. However, the repeated use of the Maskin canonical mechanism does not implement f. Consider the unanimous report of $(\theta', b, 0)$ in each state. Player 1 would prefer to deviate since he prefers a to b in both states. But, the mechanism does not allow him to change the outcome. Player 2 does not want to deviate—she is getting her best outcome in each state. Any deviation of player 3 is punished by playing the stage game equilibrium in all subsequent stages. He can deviate in state θ and obtain ? instead of b, but this would be met by punishment in which his continuation payoff is a convex combination of 2 in θ and 4 in θ'. This is less than the equilibrium payoff provided the player is patient enough.

This example illustrates a well-known phenomenon in repeated games. Unless the minmax payoffs of the game lie on the efficiency frontier, there will typically be many equilibrium paths along which unwanted outcomes are implemented if players are patient. Notice that in this example, the equilibrium payoffs are less than the minmax payoff of player 1. The minmax payoff of player 1 is 0 arising from the equilibrium strategies of 2 and 3 in state θ'. All this suggests that the conditions that guarantee one-shot implementation may not be sufficient for repeated implementation. A key result of Lee and Sabourian is that they may not be necessary either.

In order to derive their characterization conditions, Lee and Sabourian depend heavily upon insights from results in repeated games. For instance, let $v(f)$ be the payoff vector associated with f. Lee and Sabourian show that there cannot be any payoff vector in the convex hull of all payoffs that can be constructed from the range of f that strictly dominates $v(f)$. This is because if this condition is not met, then there can be a collusive equilibrium in which all agents earn the higher payoff. This then gives a *necessary* condition for implementation—if $v(f)$ is strictly Pareto dominated by another vector $v' \in CO(V(f))$ (where $V(f)$ is the set of payoff vectors in the range of f), then if players are sufficiently patient, f cannot be repeatedly implementable from any period t onwards. So, *weak efficiency* in the range is a necessary condition for repeated implementation. Lee and Sabourian then go on to prove a partial converse—under some mild conditions, if there are at least three individuals, then *strong efficiency* in the range of f is a sufficient condition for repeated Nash implementation if players are patient enough and from period two onwards.

Weak and Strong Efficiency are quite far removed from Monotonicity, and there are no apparent connections between them. However, Mezzetti and Renou (2017) actually establish a close relationship. They study repeated implementation in its full generality by considering repeated implementation both over *infinite* and *finite* periods, so that the standard one-shot implementation problem becomes a special case. They introduce a condition *dynamic monotonicity* and show that it is necessary and almost sufficient in *all* repeated implementation problems, including the case of a finite number of interactions and the case of infinitely repeated interactions with general discount factors. Moreover, dynamic monotonicity is equivalent to Maskin monotonicity in the one-shot case. In infinitely repeated problems with an arbitrarily high enough discount factor, dynamic monotonicity is closely related to weak efficiency in the range.

5 Conclusion

There has been a resurgence of interest in complete information implementation theory. In this paper, I have briefly discussed two strands of the recent literature, one on *partial honesty* being close to behavioral economics. This is in keeping with developments in economic theory which is now modelling individual behavior that does not necessarily conform to the neoclassical paradigm of maximization of a preference ordering. The recent literature in Behavioral economics focuses on departures from classical notions of choice behavior determined by preference maximization. Choice behavior can be influenced by a variety of phenomena such as menu dependence, framing, temptation and self-control. In multi-person settings, considerations of *reciprocity* mean that one agent's choice behavior or optimal action(s) may be influenced by other agents' actions. These phenomena then mean that choice behavior may no longer be rationalizable by means of a preference ordering defined over X.

Obviously, implementation theory has to change if it is to keep up with developments in behavioral economics. In an important paper, de Clippel (2014) takes a big step in incorporating such behavioral concerns into complete information implementation. He assumes that each individual's choice behavior is described by a choice correspondence $C_i(-, \theta)$ which describes the set of alternatives that individual i would choose from any set $S \subseteq X$. This explicitly allows for menu dependence since the choice out of a set S need not have any relationship with choice out of a set T in any state of the world. de Clippel goes on to derive necessary and sufficient conditions for Nash implementability in this setting. But, these conditions are far from a complete characterization, at least partly because of the generality of his framework—no restriction is imposed at all on the choice correspondences $\{C_i\}_{i \in N}$.

However, a common approach in behavioral economics is to impose *some* restrictions in the form of axioms on individual choice behavior.[9] For example, one axiom—WWARP—is a weakening of the well-known Weak axiom of Revealed Preference that is necessary for full rationalizability of choice behavior. Manzini and Mariotti (2007) use WWARP and another relatively weak axiom to characterize a class of choice correspondences that can be described as short-list method. Given any set S, the individual first shortlists candidates according to one criterion and then chooses one candidate out of the set of candidates according to a (common) preference ordering defined over X. Thus, choice behavior follows a well-defined procedure providing some structure to the choice correspondences $\{C_i\}_{i \in N}$. This suggests the possibility of using the de Clippel framework, but borrowing from the axiomatic approach in behavioral economics to develop a theory of behavioral implementation.

Another interesting and unexplored area again arises from the empirical observation that individual preferences may be endogenous and may in fact be influenced by the institutional mechanism itself. Bowles and Polania-Reyes discuss some literature which describes how incentives can alter individuals' social preferences for adhering to social norms of doing the right thing—the introduction of overtime allowances resulting in shorter hours worked being an example. More generally, the provision of incentives may actually be counterproductive. This obviously renders problematic standard design approach of constructing a mechanism which would operate on *fixed* individual preferences to produce desirable outcomes in equilibrium.

References

Bull, J., & Watson, J. (2007). Hard evidence and mechanism design. *Games and Economic Behavior, 58*, 75–93.

de Clippel, G. (2014). Behavioral implementation. *American Economic Review, 104*, 2975–3002.

Dutta, B., & Sen, A. (1991). A necessary and sufficient condition for two-person Nash implementation. *Review of Economic Studies, 58*, 121–128.

Dutta, B., & Sen, A. (2012). Nash implementation with partially honest individuals. *Games and Economic Behavior, 74*, 154–169.

Eliaz, K. (2002). Fault tolerant implementation. *Review of Economic Studies, 69*, 589–610.

Holden, R., Kartik, N., & Tercieux, O. (2014). *Games and Economic Behavior, 83*, 284–290.

[9]See, for instance, Manzini and Mariotti (2007).

Hurwicz, L. (1960). Optimality and informational efficiency in resource allocation processes. In K. Arrow, S. Karlin, & P. Suppes (Eds.), *Mathematical methods in social sciences* (pp. 27–46). Stanford: Stanford University Press.
Hurwicz, L. (1972). On Informationally decentralized systems. In C. McGuire & R. Radner, (Eds.), *Decision and Organization* (pp. 297–336). Amsterdam: North-Holland.
Jackson, M. O. (1992). Implementation in undominated.strategies: A look at bounded mechanisms. *Review of Economic Studies, 59*, 757–775.
Jackson, M. O. (2001). A crash course in implementation theory. *Social Choice and Welfare, 18*, 655–708.
Jackson, M. O., Palfrey, T. R., & Srivastava, S. (1994). Undominated Nash implementation in bounded mechanisms. *Games and Economic Behavior, 6*, 474– 501.
Kandori, M., Mailath, G., & Rob, R. (1993). Learning, mutations and long run equilibria in games. *Econometrica, 61*, 29–56.
Kartik, N., & Tercieux, O. (2012). Implementation with evidence? *Theoretical Economics, 7*, 323–355.
Lombardi, M., & Yoshihara, N. (2011). Partially-honest Nash implementation: Characterization results. http://dx.doi.org/10.2139/ssrn.1759924
Lange, O. (1936). On the economic theory of socialism. *Review of Economic Studies, 4*, 53–71.
Lee, J., & Sabourian, H. (2011). Efficient repeated implementation. *Econometrica, 79*, 1967–1994.
Lerner, A. (1944). *The economics of control*. New York: McMillan.
Manzini, P., & Mariotti, M. (2007). Sequentially rationalizable choice. *American Economic Review, 97*, 1824–1839.
Maskin, E. (1999). Nash equilibrium and welfare optimality. *Review of Economic Studies, 66*, 23–38.
Maskin, E., & Sjostrom, T. (2002). Implementation theory. In K. J. Arrow, A. K. Sen, & K. Suzumura (Eds.), *Handbook of social choice and welfare* (pp. 237–288). North Holland, Amsterdam.
Matsushima, H. (2008a). Behavioral aspects of implementation theory. *Economics Letters, 100*, 161–164.
Matsushima, H. (2008b). Role of honesty in full implementation. *Journal of Economic Theory, 139*, 353–359.
Mezzetti, C., & Renou, L. (2017). Repeated Nash implementation. *Theoretical Economics, 12*, 249–285.
Moore, J., & Repullo, R. (1990). Nash implementation: A full characterization. *Econometrica, 58*, 1083– 1099.
Ortner, J. (2015). Direct implementation with minimally honest individuals. *Games and Economic Behavior, 90*, 1–16.
Saporiti, A. (2014). Securely implementable social choice rules with partially honest agents. *Journal of Economic Theory, 154*, 216–228.
von Hayek, F. (1944). *The road to serfdom*. London: Routledge.
von Mises, L. (1935). Die Wirtschaftsrechnung im Sozialistischen Gemeinwesen. In F. von Hayek (Ed.), *Collectivist economic planning*. London: Routledge.
Young, P. (1993). The evolution of conventions. *Econometrica, 61*, 57–84.

Unrestricted Domain Extensions of Dominant Strategy Implementable Allocation Functions

Paul H. Edelman and John A. Weymark

1 Introduction

A mechanism consists of an allocation function and a payment function that, respectively, determine the alternative that is chosen and the payment that must be made by each individual as a function of their reported types. It is well known that for a dominant strategy incentive compatible mechanism, there is no loss of generality if attention is restricted to a one-person mechanism in which the types of all but one individual are fixed. We show that any one-person dominant strategy implementable allocation function g on a restricted domain of types can be extended to the unrestricted domain in such a way that dominant strategy implementability is preserved when utility is quasilinear. We identify a sufficient condition for which this extension is essentially unique in a sense made precise below. Much is known about the properties of dominant strategy implementable allocation functions and their implementing payment functions on an unrestricted domain (see, e.g., Cuff et al., 2012; Vohra, 2011). Because g is the restriction of any of its unrestricted domain extensions, the properties of g's extensions can be used to analyze the properties of g itself, particularly when the extension is essentially unique.

For an arbitrary type space, Rochet (1987) identifies a necessary and sufficient condition for an allocation function to be dominant strategy implementable. Gui et al. (2004) show that Rochet's conditions are equivalent to all cycles in the corresponding allocation graph having nonnegative length. The allocation graph is a graph derived from the allocation function whose nodes are the alternatives. Gui

P. H. Edelman
Department of Mathematics and the Law School, Vanderbilt University, Nashville, TN, USA
e-mail: paul.edelman@law.vanderbilt.edu

J. A. Weymark (✉)
Department of Economics, Vanderbilt University, Nashville, TN, USA
e-mail: john.weymark@vanderbilt.edu

W. Trockel (ed.), *Social Design*, Studies in Economic Design,
https://doi.org/10.1007/978-3-319-93809-7_16

et al. (2004) also show that the partition of the type space into the sets of types that are assigned the same alternative by the allocation function can be identified using polyhedra known as difference sets that are defined using the lengths of the arcs in the allocation graph.[1] Our arguments draw on the analysis by Edelman and Weymark (2017) of the geometric structure of this partition when the cycle lengths in the allocation graph are all zero. They also draw on an alternative characterization of dominant strategy implementability in terms of node potentials due to Heydenreich et al. (2009).

In Sect. 2, we describe the model. Section 3 introduces allocation graphs and states Rochet's Theorem. Difference sets and the zero 2-cycle condition are considered in Sect. 4. Node potentials are introduced in Sect. 5. The existence of an unrestricted domain extension of a dominant strategy implementable allocation function is established in Sect. 6 and a sufficient condition for this extension to be essentially unique is provided in Sect. 7. Examples illustrating our results are presented in Sect. 8. In Sect. 9, we offer some concluding remarks.[2]

2 Preliminaries

As noted in Sect. 1, there is no loss of generality in restricting attention to one-person mechanisms. The set of alternatives is $A = \{a_1, \ldots, a_m\}$, where $m \geq 2$. An alternative is sometimes referred to by the integer $i \in M = \{1, \ldots, m\}$ that indexes it. The individual's *type* is a vector $v = (v_1, \ldots, v_m) = (v(a_1), \ldots, v(a_m))$, where $v_i = v(a_i)$ is his valuation of the ith alternative. The *type space* (the set of possible types) is V, where $|V| \geq 2$. The type space is *unrestricted* if $V = \mathbb{R}^m$.

The mechanism designer knows that the individual's type is in V, but does not know which type in V it is. He designs a *mechanism* (g, π), where $g\colon V \to A$ is an *allocation function* and $\pi : V \to \mathbb{R}$ is a *payment function*. These functions specify the alternative that is chosen and the individual's payment (subsidy, if negative) as a function of his reported type. The type space V is the *domain* of the mechanism.

The individual's utility is his valuation minus his payment, and so is quasilinear. Formally, given the mechanism (g, π), his *utility* is given by

$$v(g(\tilde{v})) - \pi(\tilde{v}) \tag{1}$$

when v is his true type and $\tilde{v}$ is his reported type. The individual reports a type that maximizes his utility, which need not be his true type.

[1] The main results in Gui et al. (2004) also appear in Vohra (2011).

[2] Further details about the material discussed in Sects. 2–5 and 9 may be found in Edelman and Weymark (2017), Heydenreich et al. (2009), and Vohra (2011).

A mechanism (g, π) is *dominant strategy incentive compatible* if

$$v(g(v)) - \pi(v) \geq v(g(\tilde{v})) - \pi(\tilde{v}), \quad \forall v, \tilde{v} \in V. \tag{2}$$

For such a mechanism, the individual has an incentive to report his true type whatever it is. The allocation function g is *dominant strategy implementable* if there exists a payment function π such that (g, π) is *dominant strategy incentive compatible*. We only consider dominant strategy incentive compatible mechanisms.

Dominant strategy implementability has two implications that allow for some simplification. First, the allocation and payment functions only depend on the valuations of the alternatives that are ever chosen, so, as in Edelman and Weymark (2017), we can reinterpret A as being this set of alternatives. With this interpretation of A, g is surjective. Second, payments must be the same for types that are allocated the same alternative, so a payment function that implements the allocation function g can be equivalently described by a function $\rho_g \colon M \to \mathbb{R}$, where $\rho_g(i)$ is the payment if the ith alternative is chosen. That is, using ρ_g, $g(v)$ solves the following affine maximization problem:

$$g(v) = a_j \text{ for some } j \in \arg\max_{i \in M}\{v_i - \rho_g(i)\}, \quad \forall v \in V. \tag{3}$$

The fact that g can be implemented by payments that only depend on the chosen alternative is known as the *taxation principle*.

The ith *alternative preimage* is

$$R_i = \{v \in V | g(v) = a_i\}, \quad \forall i \in M. \tag{4}$$

That is, R_i is the set of types that are assigned the ith alternative by g. By assumption, g is surjective, so each of these sets is nonempty.

3 Allocation Graphs and Rochet's Theorem

The *allocation graph* Γ_g corresponding to g is the complete directed graph whose nodes are the set M viewed as labels for the m alternatives. The *length* (which could be negative) of the directed arc from node i to node j is

$$l_{ij} = \inf_{v \in R_j} \left[v_j - v_i\right]. \tag{5}$$

By definition, $l_{ii} = 0$ for all $i \in M$. Provided that g is dominant strategy implementable, all of these lengths are finite. Let

$$\bar{l}_i = \frac{1}{m}\sum_j l_{ji}, \quad \forall i \in M, \tag{6}$$

denote the average length of the arcs in Γ_g that terminate at node i.

For any pair of nodes i and j in Γ_g, a *path* is a sequence of directed arcs connecting i to j and a *k-cycle* is a path from i to i with k arcs, where k is any positive integer. The allocation function g satisfies the *k-cycle nonnegativity condition* if all k-cycles in Γ_g have nonnegative length and it satisfies the *zero k-cycle condition* if all k-cycles in Γ_g have zero length.

For an arbitrary type space, Rochet (1987) identifies a necessary and sufficient condition for an allocation function to be dominant strategy implementable. Theorem 1 provides a statement of Rochet's Theorem in terms of cycles in the allocation graph Γ_g.

Theorem 1 (Rochet (1987)) *The following conditions for the allocation function* $g: V \to A$ *are equivalent:*

1. *g is dominant strategy implementable.*
2. *For every integer $k \geq 2$, the k-cycle nonnegativity condition is satisfied.*

4 Difference Sets and the Zero 2-Cycle Condition

Our analysis exploits the geometric structure of the partition of the type space V provided by the m alternative preimages. This structure is identified using polyhedra defined on all of $\mathbb{R}^m$. In the following, we let int$S$ denote the interior of the set S and $\mathbf{1}$ denote the vector whose components are all equal to 1.

For all distinct $i, j \in M$, the *pairwise difference set for* the ordered pair of alternatives (a_i, a_j) is

$$\overline{H}_{ij} = \{v \in \mathbb{R}^m \,|\, v_i - v_j \geq l_{ji}\} \tag{7}$$

and its boundary is

$$H_{ij} = \{v \in \mathbb{R}^m \,|\, v_i - v_j = l_{ji}\}. \tag{8}$$

Each of these pairwise difference sets is a closed halfspace in $\mathbb{R}^m$. It is convenient to let $H_{ii} = \overline{H}_{ii} = \mathbb{R}^m$. For all $i \in M$, the *difference set for* a_i is the polyhedron

$$P_i = \bigcap_{j=1}^{m} \overline{H}_{ij}. \tag{9}$$

As Theorem 2 demonstrates, except for possibly on its boundary, the intersection of the difference set P_i with the type space V is the set of types that are assigned the ith alternative by g.

Theorem 2 (Gui et al. (2004)) *For the allocation function $g\colon V \to A$, for any alternative $a_i \in A$:*

1. *For any type $v \in R_i$, $v \in P_i \cap V$.*
2. *If g satisfies the 2-cycle nonnegativity condition, then for any type $v \in \mathrm{int} P_i \cap V$, $v \in R_i$.*

An implication of Theorem 2 is that if $v \in V$ but $v \notin P_i$, then $g(v) \neq a_i$. If $H_{ij} = H_{ji}$, then P_i and P_j have a facet in common and $l_{ij} + l_{ji} = 0$. Dominant strategy implementation implies that the difference sets for distinct alternatives have no interior points in common. As a consequence, if $H_{ij} \neq H_{ji}$, then $l_{ij} + l_{ji} > 0$.

A further implication of Theorem 2 is that if $g(v) = a_i$ and $v' = v + c \cdot \mathbf{1}$, then $g(v') = a_i$ except possibly when v (and, hence v') is on the boundary of P_i. The latter observation permits us to normalize the type vectors so that their components sum to 0 or, equivalently, that they lie in the subspace $\mathbf{1}^{\perp}$ of $\mathbb{R}^m$ orthogonal to $\mathbf{1}$.

For all $i \in M$, the *normalized difference set for a_i* is

$$\hat{P}_i = P_i \cap \mathbf{1}^{\perp}. \tag{10}$$

Theorem 3 shows that this set is a pointed cone with vertex p^i whose jth component is the average length of the arcs in Γ_g that terminate at node i minus the length of the arc that goes from node j to node i.

Theorem 3 (Edelman and Weymark (2017)) *For all $i \in M$, $\hat{P}_i$ is a pointed cone with vertex p^i whose jth component is*

$$p^i_j = \bar{l}_i - l_{ji}, \quad \forall j \in M. \tag{11}$$

If the allocation function g is dominant strategy implementable and all of the 2-cycles in Γ_g have zero length, then all cycles in Γ_g have zero length (see Cuff et al., 2012). The relationship between zero cycle lengths and the vertices of the normalized difference sets is provided in Theorem 4.

Theorem 4 (Edelman and Weymark (2017)) *If the allocation function $g\colon V \to A$ is dominant strategy implementable, then the following conditions are equivalent:*

1. *The vertices $\{p^i\}$ of the normalized difference sets $\{\hat{P}_i\}$ coincide.*
2. *g satisfies the zero 2-cycle condition.*

Restrictions on the type space for which the conditions in Theorem 4 are satisfied when the allocation function is dominant strategy implementable have been identified by Cuff et al. (2012) and Edelman and Weymark (2017). For example, they hold if the type space is unrestricted.

Because P_i is a cone, $\hat{P}_i$ is the orthogonal projection of P_i onto $\mathbf{1}^{\perp}$. The orthogonal projection of the type space V onto $\mathbf{1}^{\perp}$ (the *projected type space*) is also of interest. This projection is denoted by $\hat{V}$.

5 Implementability and Node Potentials

An alternative characterization of dominant strategy implementability to that provided by Rochet's Theorem can be obtained using node potentials. The function $\rho_g : M \to \mathbb{R}$ is a *node potential* for the allocation function $g : V \to A$ if

$$\rho_g(j) \leq \rho_g(i) + l_{ij}, \quad \forall i, j \in M. \tag{12}$$

That is, a node potential assigns a scalar to each node in the graph Γ_g in such a way that (12) holds.

The payment function $\pi : V \to \mathbb{R}$ *corresponds to the node potential* ρ_g if for all $i \in M$ and all $v \in R_i$, $\pi(v) = \rho_g(i)$. In other words, the payment required by the payment function π for any type $v \in V$ that the allocation function g assigns a_i is the value assigned to the ith node in Γ_g by the node potential ρ_g. Theorem 5 provides a characterization of dominant strategy incentive compatibility in terms of node potentials.

Theorem 5 (Heydenreich et al. (2009)) *For the allocation function* $g : V \to A$ *and payment function* $\pi : V \to \mathbb{R}$, (g, π) *is dominant strategy incentive compatible if and only if* π *corresponds to a node potential* $\rho_g : M \to \mathbb{R}$.

The node potential ρ_g thus provides a set of implementing payments for the m alternatives. Using Theorems 3 and 4, Edelman and Weymark (2017) show that when the zero 2-cycle condition is satisfied, the common vertex p of the normalized difference sets are implementing payments. By (11), the payment for the ith alternative is then $\bar{l}_i$ (the average length of the arcs that terminate at node i in the allocation graph Γ_g) because $l_{ii} = 0$ for all $i \in M$.

6 Extending the Domain

The allocation function $g^+ : \mathbb{R}^m \to A$ is an *unrestricted domain extension* of the allocation function $g : V \to A$ if $g^+(v) = g(v)$ for all $v \in V$. We are interested in universal domain extensions that preserve dominant strategy implementability. Theorem 6 shows that any dominant strategy implementable allocation function on a restricted type space has such an extension.

Theorem 6 *If the allocation function $g: V \to A$ is dominant strategy implementable, then g has a unrestricted domain extension $g^+: \mathbb{R}^m \to A$ that is dominant strategy implementable.*

We give two proofs for Theorem 6 that provide different insights about the nature of the extension. The first proof combines a revealed preference argument with the taxation principle's optimization problem in (3).

Proof (Version 1) Because g is dominant strategy implementable, there exists a payment function $\pi: V \to \mathbb{R}$ that implements it. By the taxation principle, this payment function can be written as a function $\rho_g: M \to \mathbb{R}$ because types that are assigned the same alternative have the same payment. Let

$$\mathcal{O} = \{(a_i, \rho_g(i)) \mid i \in M\} \tag{13}$$

be the set of all combinations of an alternative and its corresponding payment for the mechanism (g, π).

For $v \in V$, let $g^+(v) = g(v)$. For all $v \in \mathbb{R}^m \setminus V$, let

$$g^+(v) = a_i \text{ for some } i \in \arg\max_{i \in M}\{v(a_i) - \rho_g(i)\}. \tag{14}$$

Because there are a finite number of alternatives, $g^+(v)$ is well defined. Thus, when the type v is not in the domain V, the individual gets to choose any one of the alternatives and pays the amount associated with it in the original mechanism. By construction, when the individual is of type v, he is choosing a combination of an alternative and a payment from $\mathcal{O}$ that is utility maximal for him. As a consequence, because $g^+(v) = g(v)$ for $v \in V$, g^+ is an extension of g that is dominant strategy implementable. □

This proof of Theorem 6 is quite simple and highlights the importance of the taxation principle for the construction of the extension of g. However, it does not exploit the geometric structure that is provided by the difference sets and the lengths in the allocation graph that are used to define them. Our second proof of Theorem 6 does.

Consider any dominant strategy implementable allocation function g and let π be a payment function that implements it. By Theorem 5, π corresponds to some node potential ρ_g. Let

$$l^+_{ij} = \rho_g(j) - \rho_g(i), \quad \forall i, j \in M. \tag{15}$$

The value l^+_{ij} is the increment in the payment required if a_j is chosen instead of a_i by the allocation function g using the payment function π corresponding to the node potential ρ_g. The *node potential allocation graph* Γ^+_g is defined to be the complete directed graph with node set M for which the length of the directed arc from node i to node j is l^+_{ij}.

It follows immediately from (15) that every cycle in Γ_g^+ has zero length.

Lemma 1 *If $\rho_g\colon M \to \mathbb{R}$ is a node potential for the dominant strategy implementable allocation function $g\colon V \to A$, then for every integer $k \geq 2$, any k-cycle in the node potential allocation graph Γ_g^+ has zero length.*

Lemma 2 shows that the length of any arc in the allocation graph Γ_g is at least as large as the length of the corresponding arc in the node potential allocation graph Γ_g^+ and that these arc lengths coincide when an arc is part of a zero length 2-cycle of Γ_g.

Lemma 2 *If $\rho_g\colon M \to \mathbb{R}$ is a node potential for the dominant strategy implementable allocation function $g\colon V \to A$, then for all $i, j \in M$,*

$$l_{ij} \geq l_{ij}^+. \tag{16}$$

and for all $i, j \in M$ for which $l_{ij} + l_{ji} = 0$,

$$l_{ij}^+ = l_{ij}. \tag{17}$$

Proof Because ρ_g is a node potential for g, (16) follows from (12) and (15). Consider any $i, j \in M$ for which $l_{ij}+l_{ji} = 0$. Because $l_{ij}+l_{ji} = 0$ and $l_{ij}^+ + l_{ji}^+ = 0$, if $l_{ij} > l_{ij}^+$, we would have

$$0 = l_{ij} + l_{ji} > l_{ij}^+ + l_{ji}^+ = 0,$$

which is impossible. Hence, because (16) holds, (17) does as well. □

For the allocation function $g\colon V \to A$, the *zero 2-cycle graph* is the graph Γ_g^2 with node set M that has an edge between nodes i and j, denoted $i \sim j$, if and only if $l_{ij} + l_{ji} = 0$. This graph is undirected and only has an edge between two nodes if the length of the 2-cycle formed by the arcs connecting these nodes in Γ_g is zero.

For all $i \in M$, let P_i^+ be the difference set for a_i defined as in (9) but using the lengths $\{l_{ij}^+\}$ instead of the lengths $\{l_{ij}\}$ when defining the analogues of the pairwise difference sets in (7). Also let $\hat{P}_i^+ \subseteq \mathbf{1}^\perp$ be the corresponding normalized difference set for a_i. An implication of Lemma 2 is that for all $i \in M$, $P_i \subseteq P_i^+$ and $\hat{P}_i \subseteq \hat{P}_i^+$. In moving from P_i to P_i^+, any facet of P_i that is defined using an alternative whose node forms a 2-cycle of Γ_g^2 with node i is unchanged, whereas any facet of P_i that is defined using an alternative whose node does not form a 2-cycle of Γ_g^2 with node i is moved parallel so as to increase the size of this difference set. We use these observations in our second proof of Theorem 6.

Proof (Version 2) Because g is dominant strategy implementable, by Theorem 5, there exists a node potential $\rho_g\colon M \to \mathbb{R}$ and a payment function $\pi\colon V \to A$ corresponding to it that implements g. By Lemma 2, $l_{ij}^+ = l_{ij}$ and $l_{ji}^+ = l_{ji}$ for any pair of nodes i and j for which $i \sim j$ in the 2-cycle graph Γ_g^2. For any pair of nodes i and j for which $i \not\sim j$, by (16), $l_{ij} > l_{ij}^+$ and $l_{ji} > l_{ji}^+$. Hence, by the definitions of P_i and P_i^+,

$$P_i \subseteq P_i^+, \quad \forall i \in M. \tag{18}$$

We now show that

$$\cup_{i \in M} P_i^+ = \mathbb{R}^m. \tag{19}$$

On the contrary, suppose that there exists a $v \in \mathbb{R}^m$ for which $v \notin P_i^+$ for any $i \in M$. Using the lengths $\{l_{ij}^+\}$ instead of the lengths $\{l_{ij}\}$ in (7) and (9), it then follows that for all $i \in M$, there exists an $i_j \in M$ such that

$$v_i - v_{i_j} < l_{i_j i}^+. \tag{20}$$

Because the number of nodes is finite, there exists a k-cycle for some $k \in \{2, \ldots, M\}$ in which each arc is the arc from i to i_j for some i. Let E be the set of the arcs in this cycle with the arc that starts at node i denoted by ii_j. By (20),

$$0 = \sum_{ii_j \in E} [v_i - v_{i_j}] < \sum_{ii_j \in E} l_{ii_j}^+. \tag{21}$$

By Lemma 1, every cycle in the complete directed graph Γ_g^+ has zero length, which contradicts (21). Hence, (19) holds.

We now construct the allocation function $g^+ \colon \mathbb{R}^m \to A$. For all $v \in V$, we let $g^+(v) = g(v)$ so that g^+ is an unrestricted domain extension of g. By construction, $\mathrm{int}P_i^+ \cap \mathrm{int}P_j^+ = \varnothing$ for all $i, j \in M$. For all $i \in M$, let $g^+(v) = a_i$ for any $v \in \mathrm{int}P_i^+ \setminus V$. For any other $v \in \mathbb{R}^m$, there exists a maximal subset $\mathscr{I} \subseteq M$ for which $v \in \cap_{I \in \mathscr{I}} P_i^+$. For such a v, let $g^+(v) = a_i$ for some $i \in \mathscr{I}$. By construction, the allocation function g^+ satisfies the conditions in Theorem 2 reinterpreted so as to apply to g^+.

By Lemma 1, all cycles in Γ_g^+ have zero length. Hence, by Rochet's Theorem (Theorem 1), g^+ is dominant strategy implementable. □

An implication of Theorem 6 is that Γ_g^+ is the allocation graph for the allocation function g^+. Because all 2-cycles in this graph have zero length and g^+ is dominant strategy implementable, it follows from Theorem 4 that the normalized difference sets $\{\hat{P}_i^+\}$ have a common vertex, which we denote by p^+.

7 Essential Uniqueness of an Unrestricted Domain Extension

Two allocation functions g and g' that have the same domain are *essentially equivalent* if their difference sets are identical. By Theorem 2, both of these functions assign the same alternative to any type in their common domain that is in the interior of any of the difference sets. It is only when v is on the boundaries of two or more difference sets that $g(v)$ and $g'(v)$ can differ. An unrestricted domain extension g^+ of an allocation function g is *essentially unique* if any other unrestricted domain extension of g is essentially equivalent to g^+.

In Theorem 7, we show that a dominant strategy implementable allocation function g has an essentially unique unrestricted domain extension if the zero 2-cycle graph Γ_g^2 is connected. This graph need not have any cycles, but as Lemma 3 establishes, if there are any, they must have zero length. This observation is used to help prove our uniqueness result.

Lemma 3 *If the allocation function $g: V \to A$ is dominant strategy implementable, then any cycle of the zero 2-cycle graph Γ_g^2 has zero length.*

Proof By Lemma 2, for any $i, j \in M$ for which $i \sim j$ in Γ_g^2, $l_{ij}^+ = l_{ij}$. Because Γ_g^+ is complete and all of its cycles have zero length, it follows that any cycle of Γ_g^2 must have zero length. □

Theorem 7 demonstrates that connectedness of the zero 2-cycle graph is sufficient for the uniqueness of an unrestricted domain extension.

Theorem 7 *If the allocation function $g: V \to A$ is dominant strategy implementable and the zero 2-cycle graph Γ_g^2 is connected, then g has an essentially unique unrestricted domain extension $g^+: \mathbb{R}^m \to A$.*

Proof Consider any three nodes $i, j, k \in M$ of Γ_g^2 for which $i \sim j$ and $j \sim k$, but $i \not\sim k$. By Lemma 3, the length of the path from node i to node k via node j is the negative of the reverse path. Adding the arc from node k to node i to the first path results in a cycle. Moreover, there is a unique arc length l_{ki}^* that results in this cycle having zero length. The reverse cycle only has zero length if the arc from node i to node k has length $-l_{ki}^*$. The graph Γ_g^2 is connected, and so by assigning lengths in this way, we have uniquely extended Γ_g^2 to a graph for which all three cycles exist and have zero length. A simple induction argument shows that this way of assigning lengths to arcs that are not in Γ_g^2 uniquely extends Γ_g^2 to a complete graph Γ_g^* all of whose cycles have zero length. Lemmas 1 and 2 and Theorem 6 then imply that Γ_g^* coincides with the node potential allocation graph Γ_g^+. The difference sets for any unrestricted domain extension g^+ of g are uniquely determined by the lengths of the arcs in Γ_g^+. Hence, any unrestricted domain extension of g must have the same difference sets and, therefore, there is an essentially unique unrestricted domain extension of g. □

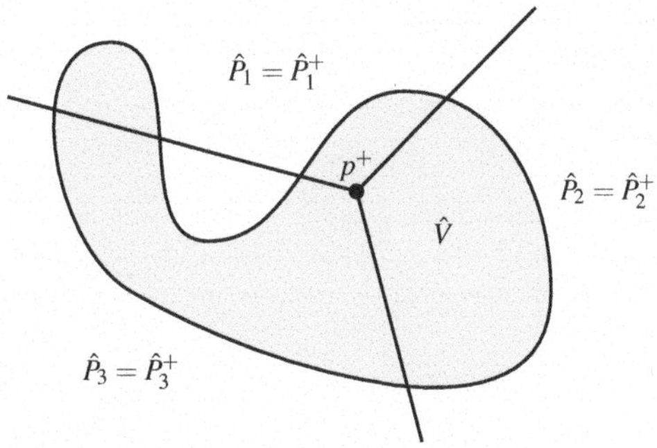

Fig. 1 Illustration of Example 1

8 Examples

We provide three examples to illustrate how to construct an unrestricted domain extension of an allocation function whose domain is not all of $\mathbb{R}^m$. Edelman and Weymark (2017) use the allocation functions in the first two examples to illustrate Theorem 4, but they do not consider domain extensions.

In each of our examples, there are three alternatives. When this is the case, $\mathbf{1}^\perp$ is a plane, which facilitates the use of diagrams. In our diagrams, the orientation is chosen so that $\mathbf{1}^\perp$ lies flat in the page. Each of the three normalized difference sets $\hat{P}_1$, $\hat{P}_2$, and $\hat{P}_3$ lies in this plane. These sets are pointed cones whose bounding rays form a 120° angle. Because the allocation function g is surjective, each of the normalized difference sets must have a nonempty intersection with the projected type space $\hat{V}$ and each type in $\hat{V}$ must be in at least one of them.

Example 1 A situation in which the conditions in Theorem 4 are satisfied is illustrated in Fig. 1. Each pair of normalized difference sets shares a common facet, and so all 2-cycles (and, hence, all cycles) have zero length. By Theorem 4, this is only possible if $\hat{P}_1$, $\hat{P}_2$, and $\hat{P}_3$ share a common vertex. As we have seen in Sect. 5, the ith component of this vertex is the average length $\bar{l}_i$ of the arcs in Γ_g that terminate at node i.

To define the allocation function g^+ that extends g to all of $\mathbb{R}^3$, we must, of course, let $g^+(v) = g(v)$ for all $v \in V$. For $v \notin V$, for all $i, j \in M$, g^+ assigns alternative a_i to any $v \in \text{int}P_i$, a_i or a_j to any $v \in P_i \cap P_j$, and a_1, a_2, or a_3 to any $v \in P_1 \cap P_2 \cap P_3$.

In Fig. 1, the union of the three normalized difference sets $\{\hat{P}_i\}$ is all of $\mathbf{1}^\perp$ and, hence, the union of the corresponding difference sets $\{P_i\}$ is all of $\mathbb{R}^m$. As a consequence, for each $i \in M$, the normalized difference set $\hat{P}_i^+$ for g^+ coincides with the corresponding normalized difference set $\hat{P}_i$ for g and, hence, their common vertex p^+ is also the common vertex of $\hat{P}_1$, $\hat{P}_2$, and $\hat{P}_3$.

Example 2 A situation in which the conditions in Theorem 4 are not satisfied is illustrated in Fig. 2. The vertex p^2 of $\hat{P}_2$ lies outside of $\hat{V}$ and differs from the vertices p^1 of $\hat{P}_1$ and p^3 of $\hat{P}_3$. Because the type space V is connected and $m = 3$,

Fig. 2 Illustration of Example 2

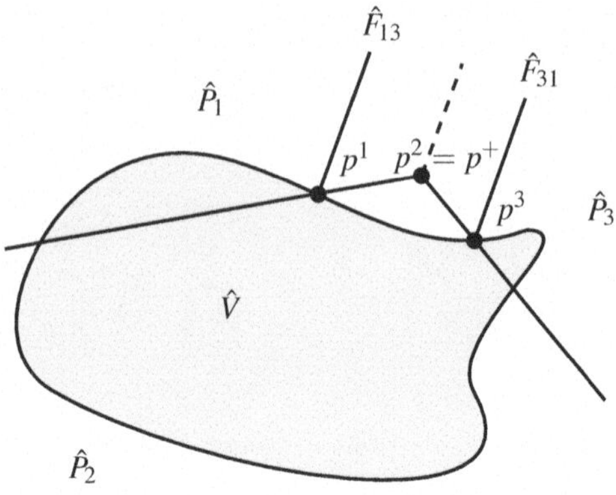

there must be at least two zero length 2-cycles (see Edelman and Weymark, 2017; Vohra, 2011). Because the vertices of the normalized difference sets are not all the same, it then follows from Theorem 4 that exactly one of the two cycles has positive length. Here, it is the 2-cycle for a_1 and a_3. This 2-cycle has positive length because $\hat{P}_1$ and $\hat{P}_3$ have no type in common. In contrast, each of the other two pairs of normalized difference sets share a common facet, and so the other 2-cycles have zero length.

There are points in $\mathbb{R}^3$ that are not in any of the normalized difference sets. The allocation function g^+ that extends g to all of $\mathbb{R}^3$ is defined by first constructing difference sets P_1^+, P_2^+, and P_3^+ for which (i) $P_i \subseteq P_i^+$ for all $i \in M$ and (ii) $\cup_{i \in M} P_i = \mathbb{R}^3$. This is done by constructing normalized difference sets $\hat{P}_1^+$, $\hat{P}_2^+$, and $\hat{P}_3^+$ for which (i) $\hat{P}_i \subseteq P_i^+$ for all $i \in M$ and (ii) $\cup_{i \in M} \hat{P}_i = \mathbf{1}^\perp$. The only way to do this is to make p^2 the common vertex of $\hat{P}_1^+$, $\hat{P}_2^+$, and $\hat{P}_3^+$.

By (8), for each $i, j \in M$, $v_i - v_j = l_{ji}$ on the line $H_{ij} \cap \mathbf{1}^\perp$. Hence, any normalized difference set $\hat{P}_i$ has a facet whose slope is the same as one of the facets of $\hat{P}_j$ for $j \neq i$. In Fig. 2, $\hat{F}_{13}$ and $\hat{F}_{31}$ are the parallel facets of $\hat{P}_1$ and $\hat{P}_3$, respectively. The sets $\hat{P}_1^+$ and $\hat{P}_3^+$ are obtained from $\hat{P}_1$ and $\hat{P}_3$ by moving these facets so that they coincide with the dashed line in the figure. The set $\hat{P}_2^+$ is set equal to $\hat{P}_2$.[3] The three normalized difference sets constructed in this way have p^2 as their common vertex p^+. For each $i \in M$, $P_i^+ = \{v \in \mathbb{R}^3 \mid v = \tilde{v} + c \cdot \mathbf{1}$ for some $\tilde{v} \in \hat{P}_i^+\}$.

The allocation function g^+ that extends g to all of $\mathbb{R}^3$ is now defined as in Example 1. That is, $g^+(v) = g(v)$ for all $v \in V$ and for all other $v \in \mathbb{R}^3$, for all $i, j \in M$, g^+ assigns alternative a_i to any $v \in \text{int} P_i$, a_i or a_j to any $v \in P_i \cap P_j$, and a_1, a_2, or a_3 to any $v \in P_1 \cap P_2 \cap P_3$. All 2-cycles in the corresponding allocation graph have zero length.

[3]In Figs. 2 and 3, we do not label these three normalized difference sets. However, they are easily identified by our descriptions of their construction.

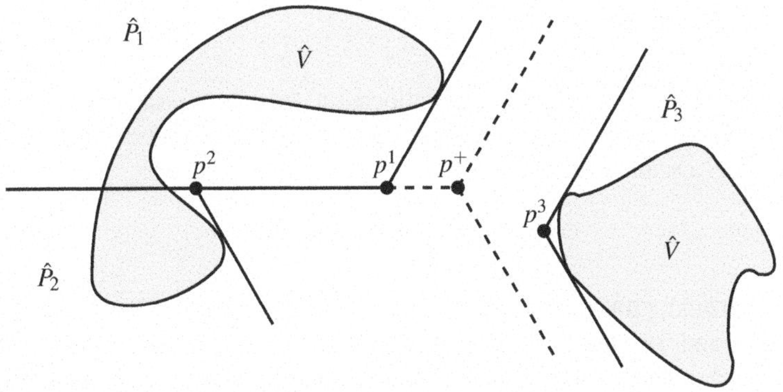

Fig. 3 Illustration of Example 3

In both Examples 1 and 2, that allocation function g has an essentially unique unrestricted domain extension g^+. Moreover, the common vertex p^+ of the normalized difference sets $\hat{P}_1^+$, $\hat{P}_2^+$, and $\hat{P}_3^+$ for g^+ coincides with some of the vertices of the normalized difference sets $\hat{P}_1$, $\hat{P}_2$, and $\hat{P}_3$ for g. In Example 3, the allocation function g does not have an essentially unique unrestricted domain extension. For the extension g^+ considered in this example, p^+ does not coincide with a vertex of any of the normalized difference sets for g.

Example 3 The projected type space $\hat{V}$ and the three normalized difference sets $\hat{P}_1$, $\hat{P}_2$, and $\hat{P}_3$ for the allocation function g are as illustrated in Fig. 3. Because $\hat{V}$ is not connected, V is not connected either. Because $l_{12} + l_{21} = 0$, the common vertex p^+ of the three normalized difference sets $\hat{P}_1^+$, $\hat{P}_2^+$, and $\hat{P}_3^+$ for the extension g^+ must lie on the line through p^1 and p^2. It must also lie on a line that is parallel to the upward sloping facets of $\hat{P}_1$ and $\hat{P}_3$ and on a line that is parallel to the downward sloping facets of $\hat{P}_2$ and $\hat{P}_3$. Furthermore, it must lie weakly to the right of $\hat{P}_1$ and weakly to the left of $\hat{P}_3$. It is because these constraints leave some freedom about where to locate p^+ that there is not an essentially unique unrestricted domain extension of g. The exact location of p^+ (subject to these constraints) depends on which payment function is used to implement g or, equivalently, what node potential is used.

The rays that originate at p^+ are the facets of the normalized difference sets for g^+. These sets are used as in Examples 1 and 2 to specify the alternative assigned by g^+ for types that are not in V. All 2-cycles in the allocation graph for g^+ have zero length.

9 Concluding Remarks

A dominant strategy implementable allocation function g satisfies the *revenue equivalence* property if for any two payment functions π and π' that implement it, there exists a scalar c such that

$$\pi'(v) = \pi(v) + c, \quad \forall v \in V. \tag{22}$$

Heydenreich et al. (2009) show that revenue equivalence holds if and only if for any two nodes i and j in the allocation graph Γ_g, the length of the shortest path from i to j is the negative of the length of the shortest path from j to i. An implication of this result is that the length of the cycle formed by the shortest paths from node i to j and from node j to i is zero. This cycle need not be a 2-cycle because these paths need not be the direct paths between these two nodes. However, Edelman and Weymark (2017) show that when the zero 2-cycle condition is satisfied, the shortest path between two nodes is the direct path. As a consequence, revenue equivalence is implied by the zero 2-cycle condition when g is dominant strategy implementable. In general, g need not satisfy either the revenue equivalence property or the zero 2-cycle condition. Nevertheless, any unrestricted domain extension of g must satisfy the zero 2-cycle condition because the domain is unrestricted (Cuff et al., 2012) and, hence, it satisfies the revenue equivalence property.

When the zero 2-cycle condition is satisfied by a dominant strategy implementable allocation function g, the normalized difference sets for it and for any unrestricted domain extension are the same. As we have seen, their common vertex p is a set of implementing payments for the alternatives and, hence, the payment function π that corresponds to it is an implementing payment function (as a function of the type). Because this is a situation in which revenue equivalence holds, the set of all payment functions that implement g is the set of all π' that satisfy (22) for some scalar c for the payment function π identified in this way.

It is an open question whether there is a simple way to characterize all of the implementing payment functions for a dominant strategy implementable allocation function g when revenue equivalence does not hold. Such a characterization can be obtained if there exists a simple characterization of the normalized difference sets for all of the unrestricted domain extensions of g when there is not an essentially unique extension. Using the vertices of these normalized difference sets, implementing payments can be identified as is done here for the case in which an unrestricted domain extension is essentially unique.

Acknowledgements We are grateful to a referee of Edelman and Weymark (2017) for suggesting the argument used in the first proof of Theorem 6. We are also grateful to Alexey Kushnir for a conversation about unrestricted domain extensions and to Walter Trockel for comments on a previous draft.

References

Cuff, K., Hong, S., Schwartz, J. A., Wen, Q., & Weymark, J. A. (2012). Dominant strategy implementation with a convex product set of valuations. *Social Choice and Welfare, 39*, 567–597.

Edelman, P. H., & Weymark, J. A. (2017). *Dominant Strategy Implementability and Zero Length Cycles*. Working paper No. 2972177, SSRN.

Gui, H., Müller, R., & Vohra, R. V. (2004). *Characterizing Dominant Strategy Mechanisms with Multi-Dimensional Types*. Discussion Paper No. 1392, Center for Mathematical Studies in Economics and Management Science, Northwestern University.

Heydenreich, B., Müller, R., Uetz, M., & Vohra, R. V. (2009). Characterization of revenue equivalence. *Econometrica, 77*, 307–316.

Rochet, J. C. (1987). A necessary and sufficient condition for rationalizability in a quasi-linear context. *Journal of Mathematical Economics, 16*, 191–200.

Vohra, R. V. (2011). *Mechanism design: A linear programming approach*. Cambridge: Cambridge University Press.

Self-implementation of Social Choice Correspondences in Strong Equilibrium

Bezalel Peleg and Hans Peters

1 Introduction

A social choice correspondence chooses alternatives based on the preferences of the agents. Generally speaking, one looks for social choice correspondences with desirable properties, such as anonymity, Pareto optimality, and many more. The problem, as already studied in Hurwicz (1972), is that preferences may be private knowledge or, more generally, agents are entitled to report any preferences they wish, resulting in alternatives chosen on the basis of the wrong information, and thus in the desired properties of the social choice correspondence being violated. Requiring strategy-proofness of a social choice function, meaning that no agent can ever benefit from not reporting truthfully, is in general too strong and results in dictatorship (Gibbard, 1973; Satterthwaite, 1975).

Implementation theory is concerned with finding game forms (mechanisms, decentralized systems) of which the equilibrium (Nash, strong, etc.) alternatives in the game with the true preferences coincide with the alternatives assigned to those preferences by the social choice correspondence under consideration. In particular since the work of Hurwicz (1972) there is a large literature on necessary and/or sufficient conditions for implementation of social choice correspondences under various equilibrium concepts, with Maskin (1999) as one of the basic contributions. For an overview of this literature up to the current millennium, see Jackson (2001).

B. Peleg
The Federmann Center for the Study of Rationality and the Institute of Mathematics, The Hebrew University of Jerusalem, Jerusalem, Israel
e-mail: pelegba@math.huji.ac.il

H. Peters (✉)
Department of Quantitative Economics, Maastricht University, Maastricht, The Netherlands
e-mail: h.peters@maastrichtuniversity.nl

W. Trockel (ed.), *Social Design*, Studies in Economic Design,
https://doi.org/10.1007/978-3-319-93809-7_17

A well-recognized drawback of many of the game forms or mechanisms employed in implementation theory is that they tend to be fairly complicated and not easy to use in practice. For instance, they may require agents to report not just preferences but complete preference profiles, to report integer numbers, etc. In the present paper, we therefore ask what is still feasible by using what we call 'self-implementation': this means implementation by a game form that is simply a selection (social choice function) from the correspondence under consideration and, thus, requires the agents just to report their own preferences and nothing else. Apart from the simplicity of such a mechanism its use is also defendable in the sense that it is close to the social choice correspondence that is deemed desirable. Specifically, we ask the following question: which social choice correspondences are self-implementable in strong equilibrium (that is, strategy-profiles such that no coalition can gain by deviating, as introduced in Aumann, 1959)?

It turns out that under some natural additional conditions we are able to give a precise answer to this question: if the number of agents is not too small and the social choice function that selects from the correspondence and implements it is anonymous and satisfies 'no veto power', then the correspondence must result from so-called feasible elimination, as already introduced in Peleg (1978). The number of agents being not too small will be made precise and, together with the no veto power property boils down to this number being at least as large as twice the number of alternatives minus one—a condition satisfied in most (political) elections. No veto power means that no agent on its own is able to exclude any alternative from being chosen—again a natural condition in larger elections. This result is quite involved: its proof can be based on a selection from existing results in the literature, as we will indicate; nevertheless, for the convenience of the reader and in order to avoid having to introduce many additional concepts, we present a completely self-contained proof.

As already mentioned, the concept of feasible elimination was introduced by Peleg (1978), in order to construct the so-called exactly and strongly consistent social choice functions: for such social choice functions there is for every profile of (true) preferences a strong equilibrium profile resulting in the truthful alternative. What we explicitly add in the present paper is not only that social choice functions that select from a feasible elimination social choice correspondence implement this correspondence in strong equilibrium, but also that under the additional conditions mentioned above, the feasible elimination correspondence is the unique correspondence for which this can be done.

Section 2 introduces the main concepts and Sect. 3 presents the main result. Most parts of the proof are shifted to the Appendix. Section 4 concludes.

Notations The following basic notations are used throughout. For a set D, $|D|$ denotes the cardinality of D, $P(D)$ the power set, i.e., the set of all subsets of D, and $P_0(D)$ the set of all nonempty subsets of D.

2 Self-implementation in Strong Equilibrium

Let A be the set of m *alternatives*, $m \geq 2$, and let $N = \{1, \ldots, n\}$, $n \geq 2$, be the set of *voters*. Subsets of N are called *coalitions*. Let L be the set of all *preferences*, i.e., complete, antisymmetric and transitive binary relations, on A. Then L^N is the set of all (preference) *profiles*. A *social choice correspondence* (SCC) is a function $H : L^N \to P_0(A)$. A *social choice function* (SCF) is a function $F : L^N \to A$. A social choice function F is a *selection* from a social choice correspondence H if $F(R^N) \in H(R^N)$ for every $R^N \in L^N$.

A *game form* is an $(n+1)$-tuple $g = (\Sigma^1, \ldots, \Sigma^n, \pi)$, where Σ^i is the *strategy set* of player (voter) $i \in N$, and $\pi : \Pi_{i=1}^n \Sigma^i \to A$ is the *outcome function*. For every $R^N \in L^N$ the pair (g, R^N) is a(n ordinal) game. A strategy profile $\sigma \in \Pi_{i=1}^n \Sigma^i$ is a *strong equilibrium* (Aumann, 1959) in the game (g, R^N) if there are no $S \in P_0(N)$ and $\widetilde{\sigma}^S \in \Pi_{i \in S} \Sigma^i$ such that $\pi(\widetilde{\sigma}^S, \sigma^{N \setminus S}) \neq \pi(\sigma)$ and $\pi(\widetilde{\sigma}^S, \sigma^{N \setminus S}) R^i \pi(\sigma)$ for all $i \in S$.[1]

A social choice correspondence H is *strong equilibrium implementable* if there is a game form $g = (\Sigma^1, \ldots, \Sigma^n, \pi)$ such that for every $R^N \in L^N$ we have

$$H(R^N) = \{\pi(\sigma) : \sigma \text{ is a strong equilibrium in } (g, R^N)\} .$$

In this case we also say that the game form g implements the SCC H in strong equilibrium.

A social choice function F can be identified with the game form in which the strategy set of each voter is the set L and the outcome function is F, i.e., to each strategy profile (preference profile) $Q^N \in L^N$ the outcome (alternative) $F(Q^N)$ is assigned. We denote this game form simply by F. Then (F, R^N) is a game for every $R^N \in L^N$.

Let H be a social choice correspondence. We call H *strong self-implementable* if there is a social choice function F such that

(i) $F(R^N) \in H(R^N)$ for every $R^N \in L^N$, and
(ii) $H(R^N) = \{F(Q^N) : Q^N \text{ is a strong equilibrium in } (F, R^N)\}$.

In words, the selection F from H implements H in strong equilibrium.

We assume that every SCC H (including every SCF, since this can be viewed as a single-valued SCC) occurring in the rest of the paper is *non-imposed*, i.e., for every $x \in A$ there is an $R^N \in L^N$ such that $H(R^N) = \{x\}$.

A well-known necessary condition (Maskin, 1999; see also Jackson, 2001) for H to be (self-)implementable is the following.

Maskin Monotonicity For all $R^N = (R^1, \ldots, R^n)$, $Q^N = (Q^1, \ldots, Q^n) \in L^N$, and $x \in H(Q^N)$, if $x Q^i y$ implies $x R^i y$ for all $y \in A$ and $i \in N$, then $x \in H(R^N)$.

[1] Here, $\sigma^{N \setminus S}$ denotes the restriction of σ to $N \setminus S$. Similar notation will be used throughout the paper.

3 Main Result

The purpose of this section is to characterize all social choice correspondences H that are self-implementable in strong equilibrium if the number of voters is relatively large and the selection that implements H satisfies two natural properties, namely anonymity and no-veto power. The latter means that no voter on his own is able to exclude any alternative from being chosen. We arrive at this theorem by combining a number of existing results in the literature, but our proof will be self-contained.

We start with the following concept, introduced by Peleg (1978). A social choice function F is *exactly and strongly consistent* (ESC) if for every $R^N \in L^N$ the game (F, R^N) has a strong equilibrium $Q^N \in L^N$ such that $F(Q^N) = F(R^N)$. We now immediately have the following result.

Lemma 3.1 *Let the selection F from the social choice correspondence H implement H in strong equilibrium. Then F is ESC.*

Proof Let $R^N \in L^N$ and $x = F(R^N)$. Then $x \in H(R^N)$ and therefore there is a strong equilibrium Q^N of the game (F, R^N) such that $F(Q^N) = x$. Hence, $F(Q^N) = F(R^N)$. □

The SCCs of interest in this section are based on the so-called feasible elimination procedures, defined for the case where $n + 1 \geq m$. Informally, first, assign weights $\beta(x) \in \mathbb{N}$ to the alternatives $x \in A$ such that the sum of these weights is equal to $n+1$. Consider a preference profile and take an alternative x that is bottom ranked at least $\beta(x)$ times. Eliminate $\beta(x)$ preferences where x is bottom ranked, and next eliminate x everywhere in the remaining profile. Repeat this procedure until one alternative remains.

Formally, we have the following definition. Let $n+1 \geq m$. A function $\beta : A \to \mathbb{N}$ such that $\sum_{x \in A} \beta(x) = n + 1$ is called a *weight function*.

Definition 3.2 Let β be a weight function. Let $R^N \in L^N$. A (β-)*feasible elimination procedure* ((β-)f.e.p.) for R^N is a sequence $(x_1, C_1; \ldots; x_{m-1}, C_{m-1}; x_m)$ such that

(a) $A = \{x_1, \ldots, x_m\}$,
(b) $C_1, \ldots, C_{m-1}$ are pairwise disjoint subsets of N and $|C_j| = \beta(x_j)$ for all $j = 1, \ldots, m-1$,
(c) $x_k R^i x_j$ for all $j = 1, \ldots, m-1$, $k = j+1, \ldots, m$, and $i \in C_j$.

Thus, in a feasible elimination procedure[2] $(x_1, C_1; \ldots; x_{m-1}, C_{m-1}; x_m)$, by condition (c) alternative x_1 is bottom ranked for all voters in C_1 and by condition (b), $|C_1| = \beta(x_1)$. Now eliminate the preferences of the voters in C_1, and eliminate x_1 from the preferences of the remaining voters. In the remaining profile, x_2 is bottom ranked for all voters in C_2 by condition (c), and by condition (b), $|C_2| = \beta(x_2)$, so

[2]Dependence on β is often not mentioned when confusion is unlikely.

that the preferences of the voters in C_2 can be eliminated and x_2 can be eliminated from the remaining profile. And so on and so forth. Observe that after eliminating x_1 there are $n-\beta(x_1)$ voters left, after eliminating x_2 there are $n-\beta(x_1)-\beta(x_2)$ voters left, and after eliminating x_{m-1} there are $n-\beta(x_1)-\ldots-\beta(x_{m-1}) = \beta(x_m)-1 \geq 0$ voters left.

An important observation about f.e.p.s. is the following. Suppose an alternative x is bottom ranked by (at least) the voters in some coalition S with $|S| = \beta(x)$, in a profile $R^N \in L^N$. Then x must be eliminated in every f.e.p. for R^N. To see this suppose there is an f.e.p. in which x is not eliminated and let y be the alternative eliminated last, say via coalition T. Then the finally left voters form a coalition S' containing S. We have $\beta(y) + \beta(x) = |T| + |S'| + 1$ by the foregoing, but also $|T| + |S'| \geq \beta(y) + \beta(x)$, a contradiction.

It is not difficult to see that there exists always at least one f.e.p. under the assumptions in the definition. If every alternative x_j is bottom ranked less than $\beta(x_j)$ times, then the total number of voters is at most $\sum_{j=1}^{m} \beta(x_j) - m$, which is equal to $n + 1 - m$ and therefore strictly smaller than n. A similar argument can be made after elimination of each alternative $x_1, \ldots, x_{m-2}$.

Let β be a weight function. An alternative x is R^N*-maximal for* β if there exists a β-f.e.p. $(x_1, C_1; \ldots; x_{m-1}, C_{m-1}; x)$. We denote

$$M_\beta(R^N) = \{x \in A : x \text{ is } R^N\text{-maximal for } \beta\}.$$

The following lemma repeats the known result that M_β is Maskin monotonic. For completeness, a proof can be found in the appendix, where also references to the literature are provided. For a weight function β as in Definition 3.2 we use the notation $\beta(B) = \sum_{x \in B} \beta(x)$ for $B \subseteq A$.

Lemma 3.3 *Let β be a weight function. Then M_β is Maskin monotonic.*

Next, we provide a characterization of maximal alternatives. Again, see the appendix for references and a proof.

Lemma 3.4 *Let β be a weight function. Let $x \in A$ and $R^N \in L^N$. The following statements are equivalent.*

(i) $x \in M_\beta(R^N)$.
(ii) *There are no $S \in P_0(N)$ and $B \in P_0(A)$ such that $|S| \geq \beta(A \setminus B)$, $x \in A \setminus B$, and $y\ R^i\ x$ for all $i \in S$ and $y \in B$.*

The following result says that M_β is self-implementable in strong equilibrium by any selection from it.

Proposition 3.5 *Let β be a weight function and let F be a selection from M_β. Then F implements M_β in strong equilibrium.*

Proof

(a) Let $R^N \in L^N$ and $x \in M_\beta(R^N)$. We show that there is a strong equilibrium Q^N of (F, R^N) such that $F(Q^N) = x$. Let $(x_1, C_1; \ldots; x_{m-1}, C_{m-1}; x)$ be an f.e.p. for R^N and consider the profile $Q^N \in L^N$ obtained from R^N by lowering x_j to the last position in the preferences of the voters in C_j, $j = 1, \ldots, m-1$, leaving everything else in tact. Then $M_\beta(Q^N) = \{x\}$, hence $F(Q^N) = x$. Also, Q^N is a strong equilibrium of (F, R^N). Indeed assume on the contrary that there exist $S \in P_0(N)$ and $P^S \in L^S$ such that $F(P^S, Q^{N\setminus S}) = z \neq x$ and $zR^i x$ for all $i \in S$. Then $z = x_j$ for some $1 \leq j \leq m-1$. By the definition of an f.e.p., $xR^i z$ for all $i \in C_j$, hence $S \cap C_j = \emptyset$. Since $|C_j| = \beta(z)$ and z is the last ranked alternative of Q^ℓ for all $\ell \in C_j$, we have that $z \notin M_\beta(P^S, Q^{N\setminus S})$, contradicting $F(P^S, Q^{N\setminus S}) = z$.

(b) Let Q^N be strong equilibrium of (F, R^N) with $F(Q^N) = x$. We show that $x \in M_\beta(R^N)$. It is sufficient to show that (ii) of Lemma 3.4 holds for x. Suppose not. Then there is an $S \in P_0(N)$ and $B \in P_0(A)$, $x \notin B$, such that $yR^i x$ for all $y \in B$ and $i \in S$, and $|S| \geq \beta(A \setminus B)$. Consider a profile $P^S \in L^S$ with $A \setminus B$ at bottom for all voters in S. Then by the remarks following Definition 3.2, all elements of $A \setminus B$ will be eliminated in any f.e.p. for $(P^S, Q^{N\setminus S})$, so that $M_\beta(P^S, Q^{N\setminus S}) \subseteq B$, hence S has an improvement, a contradiction to the assumption that Q^N is strong equilibrium of (F, R^N). □

Before turning to a converse of Proposition 3.5 we introduce two additional possible properties of a social choice correspondence H. Of course, these properties also apply for a social choice function F, since a social choice function can be identified with a single-valued social choice correspondence.

Anonymity For all $R^N \in L^N$ and for all permutations π of N, $H(R^1, \ldots, R^n) = H(R^{\pi(1)}, \ldots, R^{\pi(n)})$.

No Veto Power For all $x \in A$ and $i \in N$, there is no $R^i \in L$ such that $x \notin H(R^i, R^{N\setminus\{i\}})$ for all $R^{N\setminus\{i\}} \in L^{N\setminus\{i\}}$.

Proposition 3.6 *Let social choice function F be ESC, anonymous, and satisfy No Veto Power, and let $n+1 \geq m$. Then there is a weight function β such that F is a selection from M_β.*

Also this proposition can be deduced from earlier results in the literature, but for completeness we provide a self-contained proof in the appendix. The following theorem is a corollary to Propositions 3.5 and 3.6 and the main result of this section.

Theorem 3.7 *Let $n+1 \geq m$ and let the social choice function H be implementable in strong equilibrium by a selection F which is anonymous and satisfies No Veto Power. Then $H = M_\beta$ for some weight function β.*

Proof By Lemma 3.1 and Proposition 3.6 it follows that there is a weight function β such that $F(R^N) \in M_\beta(R^N)$ for all $R^N \in L^N$. By Proposition 3.5, F implements M_β in strong equilibrium. Hence,

$$H(R^N) = \{F(Q^N) : Q^N \text{ is a strong equilibrium in } (F, R^N)\} = M_\beta(R^N)$$

for all $R^N \in L^N$, which completes the proof. □

Theorem 3.7 says, roughly, that if the number of voters is relatively large, then the only social choice correspondences which are self-implementable in a reasonable way in strong equilibrium are the correspondences M_β. Typically, in political elections the constraint $n + 1 \geq m$ is satisfied and the conditions of Anonymity and No Veto Power for a final selection of a candidate are natural if not compelling.

The conditions of Anonymity and No Veto Power in the theorem are on the selection F. In general we can make the following observations. It is possible that H is anonymous but F is not: let H assign to every profile the set of all top-ranked alternatives, and let F select from that set the top-ranked alternative of agent 1. Also, F can be anonymous but H not: fix an alternative $a \in A$ and let H assign all top-ranked alternatives, but leave out a if this is top-ranked by agent 1 alone, and let F select from H the alternative that is ranked maximally according to a fixed preference Q which has a last. Further, if F satisfies No Veto Power, then also H does, but the converse is not necessarily true: fix an alternative a, let H assign all top-ranked alternatives, and let F select from that according to a fixed ordering Q, but leave out a as a possible choice if it is last ranked by agent 1. Then H satisfies No Veto Power but F does not.

Since, by the preceding remarks, M_β in the theorem satisfies No Veto Power, it follows by the definition of a β-f.e.p. that $\beta(x) \geq 2$ for all $x \in A$ and, thus, that the number of agents is at least as large as twice the number of alternatives minus one.

4 Concluding Remarks

Clearly, the approach in this paper leaves many open questions. We mention two of these. First, which social choice correspondences are self-implementable in strong equilibrium if the number of agents is relatively small—for instance, a small group of people in a restaurant has to make some common choices from a large menu of dishes? Second, what can be said about self-implementation in Nash equilibrium?

Appendix: Remaining Proofs

Proofs of Lemmas 3.3 and 3.4

Proof of Lemma 3.3[3] Let Q^N and R^N be as in the definition of Maskin monotonicity, and $x \in M_\beta(Q^N)$. Without loss of generality we assume that there is a voter v such that $Q^{N\setminus\{v\}} = R^{N\setminus\{v\}}$. Let $f^* = (x_1, C_1; \ldots; x_{m-1}, C_{m-1}; x)$ be an f.e.p. for Q^N, where $A = \{x_1, \ldots, x_{m-1}, x\}$. If $v \notin C_1 \cup \ldots \cup C_{m-1}$, then it is easy to see that f^* is still an f.e.p. for R^N, so that $x \in M_\beta(R^N)$. Now assume $v \in C_1 \cup \ldots \cup C_{m-1}$. If $v \in C_j$ with $j > 1$, then we may eliminate $x_1, \ldots, x_{j-1}$ and all voters in $C_1 \cup \ldots \cup C_{j-1}$ first, and next continue the argument with the remaining profile, where now all voters in C_j have x_j bottom ranked according to Q^{C_j}. So, without loss of generality, let $v \in C_1$.

The rest of the proof is based on a three-step algorithm.

Step 1 If the bottom alternative of R^v is equal to x_1, then f^* is still an f.e.p. for R^N and we are done. Otherwise, go to Step 2.

Step 2 Let the bottom alternative of R^v be $x_\ell \neq x_1$, so $\ell \in \{2, \ldots, m-1\}$. If all voters in C_ℓ have x_ℓ as bottom alternative in R^N, then we can first eliminate x_ℓ via C_ℓ and go back to Step 1 for the reduced profile. Otherwise, go to Step 3.

Step 3 Take $\hat{v} \in C_\ell$ with x_ℓ not as bottom alternative and note that the bottom alternative of $R^{\hat{v}} = Q^{\hat{v}}$ is some x_j with $j < \ell$ (since x_j must be eliminated before x_ℓ in f^*). Then modify C_ℓ to $\hat{C}_\ell = (C_\ell \cup \{v\}) \setminus \{\hat{v}\}$ and modify C_1 to $\hat{C}_1 = (C_1 \cup \{\hat{v}\}) \setminus \{v\}$. (In words, we switch v and $\hat{v}$.) Go back to Step 1.

Repeat this procedure until the final substitute of v in the modified C_1 has x_1 at bottom. Then we can apply an f.e.p. resulting in x, so that $x \in M_\beta(R^N)$. □

Proof of Lemma 3.4[4] For the implication $(i) \Rightarrow (ii)$, let $x \in M_\beta(R^N)$ and let $(x_1, C_1; \ldots; x_{m-1}, C_{m-1}; x)$ be an f.e.p. for R^N. Suppose there were S and B as in (ii). Write $B = \{x_{i_1}, \ldots, x_{i_{|B|}}\} \subseteq \{x_1, \ldots, x_{m-1}\}$, then $\left(\cup_{j=1}^{|B|} C_{i_j}\right) \cap S = \emptyset$ by definition of an f.e.p., and $|\cup_{j=1}^{|B|} C_{i_j}| = \beta(B)$. Hence $|S| + |\cup_{j=1}^{|B|} C_{i_j}| \geq \beta(A \setminus B) + \beta(B) = n + 1$, a contradiction.

We prove the implication $(ii) \Rightarrow (i)$ by induction on the number of alternatives m. Let $x \in A$ and assume that (ii) holds.

If $m = 2$, say $A = \{x, y\}$, then there is no $S \in P_0(N)$ such that $|S| \geq \beta(x)$ and $yR^i x$ for all $i \in S$, so that $M_\beta(R^N) = \{x\}$.

---

[3]See Lemma 5.3.5 in Peleg (1984); or Remark 9.3.7 in Peleg and Peters (2010), based on Theorem 9.3.6 in the same source. In turn, the latter result goes back to Polishchuk (1978). More generally, Lemma 3.7 in Peleg and Peters (2017b) shows Maskin monotonicity of an extension of M_β.

[4]Also this result can be deduced from Theorem 9.3.6 in Peleg and Peters (2010). It is included as Lemma 3.5 in Peleg and Peters (2017a).

Now suppose that $m > 2$ and that the implication $(ii) \Rightarrow (i)$ holds if there are less than m alternatives. For every $B \in P_0(A \setminus \{x\})$ denote $S_B = \{i \in N : yR^i x \text{ for all } y \in B\}$. Then (ii) is equivalent to

$$|S_B| < \beta(A \setminus B) \text{ for all } B \in P_0(A \setminus \{x\}) \tag{1}$$

hence to

$$|N \setminus S_B| \geq \beta(B) \text{ for all } B \in P_0(A \setminus \{x\}). \tag{2}$$

We consider two cases.

Case 1 There exists $\widetilde{B} \in P_0(A \setminus \{x\})$ with $|\widetilde{B}| \leq m - 2$ and $|N \setminus S_{\widetilde{B}}| = \beta(\widetilde{B})$.

For this case we consider the two following subproblems:

- $N_1 = N \setminus S_{\widetilde{B}}$, $A_1 = \widetilde{B} \cup \{x\}$, $\beta_1(y) = \beta(y)$ for all $y \in \widetilde{B}$, $\beta_1(x) = 1$, and $R^i_1 = R^i_{|A_1}$ for all $i \in N_1$.[5]
- $N_2 = S_{\widetilde{B}}$, $A_2 = A \setminus \widetilde{B}$, $\beta_2(y) = \beta(y)$ for all $y \in A_2$, and $R^i_2 = R^i_{|A_2}$ for all $i \in N_2$.

We next show that (1) holds for the first subproblem. If not, then there is a $B \in P_0(\widetilde{B})$ such that $|T| \geq \beta_1(A_1 \setminus B)$, where $T = \{i \in N_1 : yR^i_1 x \text{ for all } y \in B\}$. Then $|T \cup S_{\widetilde{B}}| = |T| + |S_{\widetilde{B}}| \geq [\beta_1(x) + \beta(\widetilde{B}) - \beta(B)] + [n - \beta(\widetilde{B})] = \beta(A \setminus B)$, hence $|S_B| \geq \beta(A \setminus B)$, which is a violation of (1) for the original problem. Therefore, (1) must hold for the first subproblem, implying that $x \in M_{\beta_1}(R_1^{N_1})$ by induction.

Similarly, suppose that (1) does not hold for the second subproblem. Then there is a $B \in P_0(A \setminus (\widetilde{B} \cup \{x\}))$ such that $|T| \geq \beta_2(A_2 \setminus B)$, where now $T = \{i \in S_{\widetilde{B}} : yR^i_2 x \text{ for all } y \in B\}$. Then $|T \cup (N \setminus S_{\widetilde{B}})| = |T| + |N \setminus S_{\widetilde{B}}| \geq [\beta(A) - \beta(B) - \beta(\widetilde{B})] + \beta(\widetilde{B}) = \beta(A \setminus B)$, which is a violation of (1) for the original problem. We conclude that (1) must hold for the second subproblem as well, so that $x \in M_{\beta_2}(R_2^{N_2})$ by induction.

Now let $(z_1, C_1; \ldots; z_{|\widetilde{B}|}, C_{|\widetilde{B}|}; x)$ be an f.e.p. for the first subproblem and let $(u_1, D_1; \ldots; u_{m-1-|\widetilde{B}|}, D_{m-1-|\widetilde{B}|}; x)$ be an f.e.p. for the second subproblem. Since, in particular, $yR^i x$ for all $y \in \widetilde{B}$ and $i \in N_2 = S_{\widetilde{B}}$, it follows that

$$(u_1, D_1; \ldots; u_{m-1-|\widetilde{B}|}, D_{m-1-|\widetilde{B}|}; z_1, C_1; \ldots; z_{|\widetilde{B}|}, C_{|\widetilde{B}|}; x)$$

is an f.e.p. for the original problem, implying that in this case we have $x \in M_\beta(R^N)$.

Case 2 For all $\widetilde{B} \in P_0(A \setminus \{x\})$ with $|\widetilde{B}| \leq m - 2$ we have $|N \setminus S_{\widetilde{B}}| > \beta(\widetilde{B})$.

Suppose there is an $\ell \in N$ such that x is not ranked at the last or second last position in R^ℓ, and let $\widehat{y}$ be the alternative ranked right below x. We switch x and $\widehat{y}$ in voter ℓ's preference to obtain a new preference $\widehat{R}^\ell$ and a new preference profile

[5] $R^i_{|B}$ denotes the restriction of R^i to B.

$\widehat{R}^N = (R^1, \ldots, R^{\ell-1}, \widehat{R}^\ell, R^{\ell+1}, \ldots, R^N)$ that still satisfies (2): for any set B with $|B| \leq m-2$ this holds because of the strict inequality in Case 2, and for $B = A \setminus \{x\}$ this holds since x is not ranked last in $\widehat{R}^\ell$.

If Case 1 applies to $\widehat{R}^N$, then $x \in M_\beta(\widehat{R}^N)$. Thus, by Lemma 3.3, $x \in M_\beta(R^N)$. If Case 1 does not apply to $\widehat{R}^N$, then we repeat this step for some voter $\ell' \in N$ with x not ranked last or second last at $\widehat{R}^{\ell'}$, and so on, until either Case 1 applies or there is no voter left with x not ranked at the last or second last position.

In the latter case, we have a profile, say $\widetilde{R}^N$, for which still (2) holds and with x ranked last or second last for each voter $i \in N$. Observe that y is last ranked for all voters in $N \setminus S_{\{y\}}$ for all $y \in A \setminus \{x\}$. Also, by (2), $|N \setminus S_{\{y\}}| \geq \beta(y)$ for all $y \in A \setminus \{x\}$. It follows that in any f.e.p. for $\widetilde{R}^N$ every $y \in A \setminus \{x\}$ is bottom ranked by at least $\beta(y)$ voters and therefore eliminated, so that $M_\beta(\widetilde{R}^N) = \{x\}$. By Lemma 3.3 again, $x \in M(R^N)$.

By (2), Cases 1 and 2 are exhaustive, which completes the proof of the lemma. □

Proof of Proposition 3.6

We now turn to the proof of Proposition 3.6.[6] It will be convenient to introduce some terminology related to effectivity functions.[7] Let F be a social choice function and let $S \subseteq N$ and $B \subseteq A$. Then S is (F-)*effective for* B is there is $R^S \in L^S$ such that $F(R^S, Q^{N \setminus S}) \in B$ for all $Q^{N \setminus S} \in L^{N \setminus S}$. For every $x \in A$ define the integer $b(x)$ (the 'blocking coefficient' of x) by

$$b(x) = \min\{|S| : S \subseteq N \text{ is effective for } A \setminus \{x\}\}\,.$$

By non-imposition of F, we have $1 \leq b(x) \leq n$ for all $x \in A$. We write $b(B)$ for $\sum_{x \in B} b(x)$, $B \subseteq A$. Of course, $b(\cdot)$ depends on F but this will be suppressed from notation if confusion is unlikely.

We start with three useful observations.[8]

Lemma A.1 *Let the SCF F be anonymous. Let $S \subseteq N$ and $B \subseteq A$ such that $|S| \geq b(A \setminus B)$. Then S is effective for B.*

[6] Alternatively, a proof can be deduced from Theorem 5.5.3 in Peleg (1984), which in turn is based on Holzman (1986). We include a proof here for completeness, and additionally to avoid introduction of more definitions and concepts.

[7] These functions have been first formally introduced in Moulin and Peleg (1982). Here we just use some of the associated terminology.

[8] Many of the arguments in this part are based on Chapter 10 in Peleg and Peters (2010) and the references therein, in particular Holzman (1986).

Proof Write $A \setminus B = \{x_1, \ldots, x_k\}$, where $k \geq 0$. Let $S_1, \ldots, S_k$ be a partition of S such that $|S_j| \geq b(x_j)$ for each $j = 1, \ldots, k$, and let $R^{S_j} \in L^{S_j}$ such that $F(R^{S_j}, Q^{N \setminus S_j}) \in A \setminus \{x_j\}$ for each $j = 1, \ldots, k$ and $Q^{N \setminus S_j} \in L^{N \setminus S_j}$. Then $F(R^S, Q^{N \setminus S}) \in B$ for all $Q^{N \setminus S} \in L^{N \setminus S}$. So S is effective for B. □

Lemma A.2 *Let the SCF F be ESC and let $S \subseteq N$ be effective for $B \subseteq A$. Let $R^N \in L^N$ and $x \in A \setminus B$ such that $yR^i x$ for all $y \in B$ and $i \in S$. Then $F(R^N) \neq x$.*

Proof Suppose on the contrary that $F(R^N) = x$ and let Q^N be a strong equilibrium in (F, R^N) with $F(Q^N) = x$. Since S is effective for B, there is $P^S \in L^S$ such that $F(P^S, Q^{N \setminus S}) \in B$, contradicting that Q^N is a strong equilibrium in (F, R^N). □

Lemma A.3 *Let the SCF F be ESC and anonymous, and assume that $b(A) = n+1$. Then F is a selection from M_b.*

Proof Let $R^N \in L^N$ and $x = F(R^N)$. Let $B \subseteq A$, $S \subseteq N$, $|S| \geq b(A \setminus B)$, and $x \in A \setminus B$. In order to prove that $x \in M_b(R^N)$, it is by Lemma 3.4 sufficient to prove that we do not have $yR^i x$ for all $y \in B$ and $i \in S$.

On the contrary, suppose that $yR^i x$ for all $y \in B$ and $i \in S$. Since $|S| \geq b(A \setminus B)$, Lemma A.1 implies that S is effective for B. Then Lemma A.2 implies that $F(R^N) \neq x$, a contradiction. □

Notice that in order to obtain Proposition 3.6 we may try and derive the condition $b(A) = n + 1$ in Lemma A.3. This is, essentially, what is done in the remainder of the proof.

Lemma A.4 *Let the SCF F be ESC, $S \subseteq N$, $B \subseteq A$, and suppose that for every $Q^{N \setminus S} \in L^{N \setminus S}$ there is $P^S \in L^S$ such that $F(P^S, Q^{N \setminus S}) \in B$. Then S is effective for B.*[9]

Proof On the contrary, suppose that for every $Q^S \in L^S$ there is $P^{N \setminus S} \in L^{N \setminus S}$ such that $F(Q^S, P^{N \setminus S}) \in A \setminus B$. Consider a profile $R^N \in L^N$ such that $xR^i y$ and $yR^j x$ for every $i \in S$, $j \in N \setminus S$, $x \in B$, and $y \in A \setminus B$. Let $z = F(R^N)$ and let Q^N be a strong equilibrium of (F, R^N) with $F(Q^N) = z$. If $z \in A \setminus B$, then S can improve by a profile P^S as in the statement of the lemma. If $z \in B$, then $N \setminus S$ can improve by a profile $P^{N \setminus S}$ as above. □

In what follows we will use the notion of a generalized partition or *g-partition* of a set, which is a partition in which some elements may be empty.

Lemma A.5 *Let the SCF F be ESC. Then there are no $p \geq 2$, partition $B_1, \ldots, B_p$ of A and g-partition $S_1, \ldots, S_p$ of N such that $N \setminus S_i$ is effective for B_i, for every $i = 1, \ldots, p$.*

[9] This lemma states that the effectivity function associated with F is 'maximal'. See Moulin and Peleg (1982) or Peleg (1984).

Proof Suppose not, so (g-)partitions as in the lemma exist. Consider a profile R^N as in the following table:

S_1	S_2	$\cdots$	S_p
B_2	B_3	$\cdots$	B_1
$\vdots$	$\vdots$		$\vdots$
B_p	B_1	$\cdots$	B_{p-1}
B_1	B_2	$\cdots$	B_p

(meaning that every member of coalition S_1 prefers all alternatives of B_2 over all alternatives of B_3, all alternatives of B_3 over all alternatives of B_4, and so on and so forth). Now by Lemma A.2, $F(R^N) \notin B_i$ for every $i = 1, \ldots, p$. Since $\cup_{i=1}^{p} B_i = A$, this is a contradiction. □

Lemma A.6 *Let the SCF F be ESC and satisfy NVP. Then there are no partition $\{x\}, B_1, B_2$ of A and g-partition S, T_1, T_2 of N such that $|S| = b(x)$ and $N \setminus T_j$ is effective for B_j for $j = 1, 2$.*

Proof Suppose not, so (g-)partitions as in the lemma exist.

First, suppose $S = N$. Then for every $i \in N$, $|N \setminus \{i\}| < |S| = b(x)$. Therefore, for every $Q^{N\setminus\{i\}} \in L^{N\setminus\{i\}}$ there is $P^i \in L$ such that $F(P^i, Q^{N\setminus\{i\}}) = x$, so that by Lemma A.4, $\{i\}$ is effective for x. Since $|A| \geq 2$ this violates NVP of F. Thus, $S \neq N$ and $b(x) < n$. By NVP, also $b(x) > 1$. So $|S| \geq 2$ and $T_1 \cup T_2 \neq \emptyset$.

Let now S_1, S_2 be a partition of S and consider a profile R^N as in the following table:

S_1	S_2	T_1	T_2
B_2	B_1	$\{x\}$	$\{x\}$
B_1	B_2	B_2	B_1
$\{x\}$	$\{x\}$	B_1	B_2

Since $S = S_1 \cup S_2$ is effective for $A \setminus \{x\} = B_1 \cup B_2$ we have by Lemma A.2 that $F(R^N) \neq x$. Without loss of generality we assume that $F(R^N) \in B_1$. Let Q^N be a strong equilibrium in (F, R^N) with $F(Q^N) = F(R^N)$, hence $F(Q^N) \neq x$.

Case 1 $xQ^i y$ for some $i \in S$, without loss of generality $i \in S_1$, and $y \in A \setminus \{x\}$.

In this case consider the partition $\{x\}, \{y\}, A \setminus \{x, y\}$ of A and the g-partition $S \setminus \{i\}, \{i\}, T_1 \cup T_2$ of N. Since $|S \setminus \{i\}| < b(x)$ we have that $N \setminus (S \setminus \{i\})$ is effective for $\{x\}$ by Lemma A.4. By NVP and Lemma A.4, $N \setminus \{i\}$ is effective for $\{y\}$. Hence, by Lemma A.5, $N \setminus (T_1 \cup T_2)$ is not effective for $A \setminus \{x, y\}$. In turn, again by Lemma A.4, this implies that $T_1 \cup T_2$ is effective for $\{x, y\}$. Consider a profile $P^{T_1 \cup T_2} \in L^{T_1 \cup T_2}$ such that $xP^j yP^j z$ for all $j \in T_1 \cup T_2$ and $z \in A \setminus \{x, y\}$. Then by Lemma A.2, $F(P^{T_1 \cup T_2}, Q^S) \in \{x, y\}$. Since $xQ^i y$ and since $T_1 \cup T_2 \cup \{i\} = N \setminus (S \setminus \{i\})$ is effective for $\{x\}$, again by Lemma A.2, $F(P^{T_1 \cup T_2}, Q^S) \neq y$. Hence, $F(P^{T_1 \cup T_2}, Q^S) = x$. This contradicts that Q^N is a strong equilibrium in (F, R^N).

Case 2 yQ^ix for all $i \in S$ and $y \in A \setminus \{x\}$.

In this case, consider the partition $\{x\}, B_1, B_2$ of A and the g-partition $S_2, S_1 \cup T_1, T_2$ of N. Since $|S_2| < b(x)$ we have by Lemma A.4 that $N \setminus S_2$ is effective for $\{x\}$. By assumption, $N \setminus T_2$ is effective for B_2. Hence by Lemma A.5, $N \setminus (S_1 \cup T_1)$ is not effective for B_1, which in turn by Lemma A.4 implies that $S_1 \cup T_1$ is effective for $A \setminus B_1$. Consider a profile $P^{S_1 \cup T_1} \in L^{S_1 \cup T_1}$ such that yP^jxP^jz for all $j \in S_1 \cup T_1$, $y \in B_2$, and $z \in B_1$. By Lemma A.2, $F(P^{S_1 \cup T_1}, Q^{S_2 \cup T_2}) \notin B_1$. Since by assumption $S_1 \cup S_2 \cup T_1$ is effective for B_2, by Case 2 yQ^ix for all $y \in B_2$ and $i \in S$, and $N \setminus T_2$ is effective for B_2, we have by Lemma A.2 that $F(P^{S_1 \cup T_1}, Q^{S_2 \cup T_2}) \neq x$. Hence $F(P^{S_1 \cup T_1}, Q^{S_2 \cup T_2}) \in B_2$. Since $F(Q^N) = F(R^N) \in B_1$, $S_1 \cup T_1$ has an improvement, contradicting that Q^N is a strong equilibrium of (F, R^N). □

Lemma A.7 *Let the SCF F be ESC and satisfy NVP, and* $1 \leq k \leq m - 2$. *Then there are no partition* $\{x_1\}, \ldots, \{x_k\}, B_1, B_2$ *of A and g-partition* $S_1, \ldots, S_k, T_1, T_2$ *of N such that* $|S_i| = b(x_i)$ *for each* $i = 1, \ldots, k$, $N \setminus T_1$ *is effective for* B_1, *and* $N \setminus T_2$ *is effective for* B_2.

Proof The proof is by induction on k. For $k = 1$ this is Lemma A.6. Let $2 \leq k \leq m - 2$ and suppose that the statement in the lemma holds for $k - 1$. Suppose, on the contrary, that the statement does not hold for k, and let $\{x_1\}, \ldots, \{x_k\}, B_1, B_2$ and $S_1, \ldots, S_k, T_1, T_2$ be as in the lemma. Since $S_i \neq \emptyset$ for every $i = 1, \ldots, k$, we have $\emptyset \neq S_k \cup T_1 \neq N$. By Lemma A.4, either $S_k \cup T_1$ is effective for $A \setminus (\{x_k\} \cup B_1)$ or $N \setminus (S_k \cup T_1)$ is effective for $\{x_k\} \cup B_1$. In the first case, Lemma A.6 is violated for the partition $\{x_k\}, B_1, A \setminus (\{x_k\} \cup B_1)$ of A and the g-partition $S_k, T_1, N \setminus (S_k \cup T_1)$ of N. In the second case, the induction hypothesis is violated for the partition $\{x_1\}, \ldots, \{x_{k-1}\}, \{x_k\} \cup B_1, B_2$ of A and the g-partition $S_1, \ldots, S_{k-1}, S_k \cup T_1, T_2$ of N. □

The next lemma says that an ESC social choice function is 'subadditive'.[10]

Lemma A.8 *Let the SCF F be ESC, let* $S_1 \subseteq N$ *be effective for* $B_1 \subseteq A$ *and let* $S_2 \subseteq N$ *be effective for* $B_2 \subseteq A$, *such that* $B_1 \cap B_2 = \emptyset$. *Then* $S_1 \cap S_2$ *is effective for* $B_1 \cup B_2$.

Proof

(a) Say that coalition S is *s-effective* for a set of alternatives B if there is a partition $B_1, \ldots, B_k$ of B and there are coalitions $S_1, \ldots, S_k$ such that S_j is effective for B_j, $j = 1, \ldots, k$, and $S = \cap_{j=1}^k S_j$. Clearly, if S is effective for B, then S is also s-effective for B by taking $k = 1$, $S_1 = S$, $B_1 = B$. We will prove the converse, which will imply the lemma.
(b) We first prove that if S is s-effective for B, then $N \setminus S$ is not s-effective for $A \setminus B$. Suppose the latter were not the case, i.e., both S is s-effective for B and $N \setminus S$ is s-effective for $A \setminus B$. Let $B_1, \ldots, B_k$ and $C_1, \ldots, C_\ell$ be the associated partitions of B and $A \setminus B$, and let $S_1, \ldots, S_k$ and $T_1, \ldots, T_\ell$ be

[10]Cf. Moulin (1983).

the associated coalitions, hence $S = \cap_{j=1}^{k} S_j$ and $N \setminus S = \cap_{h=1}^{\ell} T_h$. List $S_1, \ldots, S_k, T_1, \ldots, T_\ell$ as $V_1, \ldots, V_p$ and list the associated sets of alternatives as $D_1, \ldots, D_p$ (where $p = k+\ell$). Then for every $i \in N$ there is $q \in \{1, \ldots, p\}$ such that $i \notin V_q$. Consider a preference profile R^N such that for every $i \in N$, $D_{q+1} R^i D_{q+2} R^i \ldots R^i D_p R^i D_1 R^i \ldots R^i D_q$. Let $x \in A$. If $x \in D_q$ for some $q > 1$, then $D_{q-1} R^i x$ for all $i \in V_{q-1}$, so that by Lemma A.2 we have $F(R^N) \neq x$. If $x \in D_1$, then $D_p R^i x$ for all $i \in V_p$, so that again by Lemma A.2 we have $F(R^N) \neq x$. This is not possible, hence we have that if S is s-effective for B, then $N \setminus S$ is not s-effective for $A \setminus B$.

(c) Now, finally, assume that S is s-effective for B. Then by part (b), $N \setminus S$ is not s-effective for $A \setminus B$, hence by part (a), $N \setminus S$ is not effective for $A \setminus B$. By Lemma A.4, S is effective for B. This concludes the proof of the lemma. □

The final lemma we need is the following.

Lemma A.9 *Let the SCF F be ESC and satisfy NVP. Let* $0 \leq k \leq m-2$. *Then there are no partition* $\{x_1\}, \ldots, \{x_m\}$ *of A and g-partition* $S_1, \ldots, S_m$ *of N such that* $|S_j| = b(x_j)$ *for* $j = 1, \ldots, k$ *and* $|N \setminus S_j|$ *is effective for* $\{x_j\}$ *for* $j = k+1, \ldots, m$.

Proof For $k = 0$ this follows from Lemma A.5. Now let $k > 0$. Suppose on the contrary that we had $\{x_1\}, \ldots, \{x_m\}$ and $S_1, \ldots, S_m$ as in the lemma. By repeated application of Lemma A.8 we have that $N \setminus (S_{k+1} \cup \ldots \cup S_{m-1})$ is effective for $\{x_{k+1}, \ldots, x_{m-1}\}$. Now the partition $\{x_1\}, \ldots, \{x_k\}, \{x_{k+1}, \ldots, x_{m-1}\}, \{x_m\}$ and g-partition $S_1, \ldots, S_k, T_1, T_2$ with $T_1 = S_{k+1} \cup \ldots \cup S_{m-1}$ and $T_2 = S_m$ violate Lemma A.7. □

Proof of Proposition 3.6 In view of Lemma A.3, it is sufficient to prove that $b(A) = n+1$. Clearly, $b(A) \geq n+1$, otherwise N would have some profile R^N such that $F(R^N) \notin A$, which is clearly impossible. Write $A = \{x_1, \ldots, x_m\}$. We distinguish two cases.

Case 1 $b(A) \geq n+m$. Then $n \leq b(A) - m = \sum_{j=1}^{m}(b(x_j) - 1)$, so that there is a g-partition $S_1, \ldots, S_m$ of N with $|S_j| \leq b(x_j) - 1$ for every $j = 1, \ldots, m$, which by using Lemma A.4 violates Lemma A.9 for $k = 0$.

Case 2 $b(A) = n + (m-k)$ for some $k \in \{1, \ldots, m-2\}$. In this case, let S_j, $j = 1, \ldots, k$, be coalitions with $|S_j| = b(x_j)$. Since

$$\begin{aligned}\sum_{j=1}^{k} |S_j| &= b(A) - (b(x_{k+1} + \ldots + b(x_m)) \\ &= n + (m-k) - (b(x_{k+1} + \ldots + b(x_m)) \\ &\leq n + (m-k) - (m-k) \\ &= n\end{aligned}$$

the S_j can be chosen disjoint. Also,

$$\begin{aligned} n - \sum_{j=1}^{k} |S_j| &= n - (b(A) - \sum_{j=k+1}^{m} b(x_j)) \\ &= n - n - (m-k) + \sum_{j=k+1}^{m} b(x_j)) \\ &= \sum_{j=k+1}^{m} (b(x_j) - 1) \end{aligned}$$

so that we can find disjoint $S_{k+1}, \ldots, S_m$ with $|S_j| = b(x_j) - 1$ for all $j = k+1, \ldots, m$, hence, by Lemma A.4, $N \setminus S_j$ is effective for $\{x_j\}$. This is again a violation of Lemma A.9.

Thus, $b(A) = n + 1$, which concludes the proof. □

References

Aumann, R. J. (1959) Acceptable points in general cooperative n-person games. In *Contributions to the theory of games. Annals of mathematic studies no. 40* (Vol. IV). Princeton, NJ: Princeton University Press.

Gibbard, A. (1973). Manipulation of voting schemes: A general result. *Econometrica, 41*, 587–602.

Holzman, R. (1986). On strong representations of games by social choice functions. *Journal of Mathematical Economics, 15*, 39–57.

Hurwicz, L. (1972). On informationally decentralized systems. In C. B. Mcguire & R. Radner (Eds.), *Decision and organization*. Amsterdam: North-Holland.

Jackson, M. O. (2001). A crash course in implementation theory. *Social Choice and Welfare, 18*, 655–708.

Maskin, E. (1999). Nash equilibrium and welfare optimality. *The Review of Economic Studies, 66*, 23–38.

Moulin, H. (1983). *The strategy of social choice*. Amsterdam: North-Holland.

Moulin, H., & Peleg, B. (1982). Cores of effectivity functions and implementation theory. *Journal of Mathematical Economics, 10*, 115–145.

Peleg, B. (1978). Consistent voting systems. *Econometrica, 46*, 153–161.

Peleg, B. (1984). *Game theoretic analysis of voting in committees*. Cambridge: Cambridge University Press.

Peleg, B., & Peters, H. (2010). *Strategic social choice: Stable representations of constitutions*. Heidelberg: Springer.

Peleg, B., & Peters, H. (2017a). Feasible elimination procedures in social choice: An axiomatic characterization. *Research in Economics, 71*, 43–50.

Peleg, B., & Peters, H. (2017b). Choosing k from m: Feasible elimination procedures revisited. *Games and Economic Behavior, 103*, 254–261.

Polishchuk, I. (1978). *Monotonicity and uniqueness of consistent voting systems*. Center for Research in Mathematical Economics and Game Theory, Hebrew University of Jerusalem.

Satterthwaite, M. A. (1975). Strategy-proofness and Arrow's conditions: Existence and correspondence theorems for voting procedures and social welfare functions. *Journal of Economic Theory, 10*, 187–207.

Part VI
New Directions in Design

Domains Admitting Ex Post Incentive Compatible and Respectful Mechanisms: A Characterization for the Two-Alternative Case

Salvador Barberà, Dolors Berga, and Bernardo Moreno

1 Introduction

We consider collective decision-making problems when only two outcomes are possible. Finding satisfactory mechanisms in this case is an easy task in private values environments, where each agent's preferences are fully determined by their type. Then, majority voting (with tiebreaking) and its generalizations are among the many voting rules that satisfy good incentive and efficiency conditions, among other properties.[1]

The situation is more complicated in the case of interdependent values, where the type of each agent does not fully determine that agent's preferences, which may also depend on the type of other agents. This is, for example, the case of deliberative juries, where each agent is endowed with information that she may share with others

[1]See, for example, May (1952), Austen-Smith and Banks (1999), and Fishburn (1973) Part I.

S. Barberà
MOVE, Universitat Autònoma de Barcelona, Barcelona, Spain

Barcelona GSE, Barcelona, Spain

Departament d'Economia i d'Història Econòmica, Universitat Autònoma de Barcelona, Bellaterra, Spain
e-mail: salvador.barbera@uab.cat

D. Berga
Departament d'Economia, Universitat de Girona, Girona, Spain
e-mail: dolors.berga@udg.edu

B. Moreno (✉)
Departamento de Teoría e Historia Económica, Facultad de Ciencias Económicas y Empresariales, Universidad de Málaga, Málaga, Spain
e-mail: bernardo@uma.es

W. Trockel (ed.), *Social Design*, Studies in Economic Design,
https://doi.org/10.1007/978-3-319-93809-7_18

strategically, and each one forms preferences on the basis of signals she gets from nature and from other jurors. Each profile of revealed types, to which mechanisms must assign an outcome, will still fully determine societies' preference profiles. But finding mechanisms with good incentives and other criteria is no longer a simple task, even if only two alternatives are at stake.

In a preceding paper (Barberà et al. 2018), we have identified domains of type profiles such that no mechanisms defined on them, other than the constant ones, can be ex post incentive compatible and respectful.[2] And we also provide examples of alternative domains where satisfactory mechanisms can be designed.

This chapter uses essentially the same framework, notation, and definitions as our preceding work. However, by focusing on situations where only two alternatives are possible, we arrive at a sharp characterization result: a condition that is only necessary to obtain positive results in the general case becomes sufficient as well as in the special case we consider here. Given the general difficulties that one encounters in designing mechanisms in the presence of interdependent types, and the relevance of the two-alternative case for applications, we think that the sharp result we present is of independent interest.

We work on a purely ordinal framework. This leads us to a somewhat nonstandard formulation of the model. Part of this formulation involves the distinction between types, as a holistic representation of agents, and preference functions, which are the rules used by agents to form the preferences that will guide their voting behavior. Properties of type domains are expressed in terms of the types themselves and of the preferences they give rise to.

The chapter proceeds as follows. In Sect. 2, we present the general framework and define the domain restrictions and the type of mechanisms we shall concentrate on. Section 3 contains the result and its proof, and an application. We conclude in Sect. 4.

2 The Model

Let $N = \{1, 2, \ldots, n\}$ be a finite set of *agents* with $n \geqslant 2$. Let $A = \{a, b\}$ be the set of alternatives. Let $R_i \in \mathcal{R}$ denote agent *i's preference* over A, where P_i and I_i are the strict and the indifference part of R_i, respectively, where $\mathcal{R}$ contains three individual preferences over A: aP_ib meaning that agent i (strictly) prefers a to b, bP_ia meaning that i (strictly) prefers b to a, and aI_ib meaning that i is indifferent between a and b. A *preference profile*, denoted by $R_N = (R_1, \ldots, R_n)$, is an element of $\mathcal{R}^n$.

It will be useful to pay attention to the relationship between pairs of preferences.

Definition 1 *Let $x \in A$. We say that $R'_i \in \mathcal{R}$ is an x-monotonic transform of $R_i \in \mathcal{R}$ if for $z \in A\backslash\{x\}$, either xI_iz and xR'_iz, or xP_iz and xP'_iz, or zP_ix.*

[2]See the definitions of properties in Sect. 2.

In words: R_i' is an x-monotonic transform of R_i if the position of x relative to z has weakly improved when going from R_i to R_i'. A special case of x-monotonic transforms of preferences is when $R_i = R_i'$.

Each agent $i \in N$ is endowed with a type θ_i belonging to a set Θ_i. Each θ_i includes all the information in the hands of i. We denote by $\Theta = \times_{i \in N} \Theta_i$ the set of type profiles. A *type profile* is an n-tuple $\theta = (\theta_1, \ldots, \theta_n)$.

The information about agent's preferences is already contained in each type profile. But we find it useful to develop a language that allows us to explicitly differentiate between the overall information contained in the types and the specific information that refers to preferences. This is often achieved in the literature by predicating that agents are endowed with a utility function that depends on the profile of types. But since we work with ordinal preferences, we formalize this dependence by means of rules of the form $R : \Theta \to \mathcal{R}^n$, which assign a preference profile to each type profile and that we call *preference functions.* We will refer to $R(\theta)$ as the preference profile induced by θ. $R_i(\theta)$ will stand for the induced preferences of agent i at type profile θ.

Notice that the domain of R is a Cartesian product including all possible type profiles. Its range is a set of preference profiles. We exclude the trivial case where all type profiles lead to the same preference profile and assume that the range is not a singleton. Also notice that this range may be a non-Cartesian strict subset of $\mathcal{R}^n$.

Following the standard use, we will call *private values environments* those where each agent's component of the preference function only depends on her type. That is, $R_i(\theta) = R_i(\theta_i)$ for each agent $i \in N$ and $\theta \in \Theta$. Otherwise, we are in *interdependent values environments*.

We now introduce an example which adapts in ordinal terms and for two alternatives the one proposed by Bergemann and Morris (2005) as their Example 1.

Example 1 An interdependent values example with a non-Cartesian range.

Let $N = \{1, 2\}$ *and* $A = \{a, b\}$. *Each agent i has two possible types:* $\Theta_i = \{\underline{\theta}_i, \overline{\theta}_i\}$. *The preference function R is defined in Table 1. We write, in each cell, the preferences of both agents for a given type profile, where preferences are represented by an ordered list from better to worse, with parenthesis in case of indifferences. Observe that* $aP_1(\underline{\theta}_1, \underline{\theta}_2)b$ *while* $bP_1(\underline{\theta}_1, \overline{\theta}_2)a$; *hence agent 1's preferences over b and a depend on agent 2's type and we are in an interdependent values environment.*

To show that the range of R is not a Cartesian product, note that $\mathcal{R}_1 = \{ab, ba, (ab)\}$ *and* $\mathcal{R}_2 = \{ab, ba\}$, *but the preference profile (ba, ba) is not in the range of the preference function R.*

Table 1 Preference function for Example 1

R	$\underline{\theta}_2$		$\overline{\theta}_2$	
$\underline{\theta}_1$	$R_1(\underline{\theta}_1, \underline{\theta}_2)$	$R_2(\underline{\theta}_1, \underline{\theta}_2)$	$R_1(\underline{\theta}_1, \overline{\theta}_2)$	$R_2(\underline{\theta}_1, \overline{\theta}_2)$
	ab	*ba*	*ba*	*ab*
$\overline{\theta}_1$	$R_1(\overline{\theta}_1, \underline{\theta}_2)$	$R_2(\overline{\theta}_1, \underline{\theta}_2)$	$R_1(\overline{\theta}_1, \overline{\theta}_2)$	$R_2(\overline{\theta}_1, \overline{\theta}_2)$
	(ab)	*ab*	*(ab)*	*ab*

Our result focuses on direct mechanisms. In fact, the properties we discuss are best analyzed with reference to the direct mechanism associated with any general one that might be described in terms of different message spaces and outcome functions.

A *direct mechanism on* Θ is a function $f : \Theta \to A$ such that $f(\theta) \in A$ for each $\theta \in \Theta$. From now on, we drop the term "direct" and refer to mechanisms, without danger of ambiguity.

Notice that, by letting Θ be the domain of f, we implicitly assume that all type profiles within this set are considered to be feasible by the designer.

We are interested in the characteristics of the domains on which mechanisms are defined. We shall now identify an important condition on domains (see Definition 4) that may or may not be satisfied by given sets of type profiles. This condition start by considering sequences of type profiles that result from changing the type of individual agents, one at a time. These sequences are identified in detail in Definitions 2 and 3.

Let $S = \left\{\theta^S_{i(S,1)}, \ldots, \theta^S_{i(S,t_S)}\right\} \in \prod_{h=1}^{t_S} \Theta_{i(S,h)}$ be a sequence of individual types of length t_S.

The sequence of agents whose types appear in S is denoted by $I(S) = (i(S, 1), \ldots, i(S, t_S))$, where $i(S, h)$ is the agent in position h in S. Notice that agents may appear in that sequence several times or not at all.

Given $\theta \in \Theta$ and $S = \left\{\theta^S_{i(S,1)}, \ldots, \theta^S_{i(S,t_S)}\right\} \in \prod_{h=1}^{t_S} \Theta_{i(S,h)}$, we consider the sequence of type profiles $m^h(\theta, S)$ that results from changing one at a time the types of agents according to S, starting from θ. Formally, $m^h(\theta, S) \in \Theta$ is defined recursively so that $m^0(\theta, S) = \theta$ and for each $h \in \{1, \ldots, t_S\}$, $m^h(\theta, S) = \left(\left(m^{h-1}(\theta, S)\right)_{N \setminus i(S,h)}, \theta^S_{i(S,h)}\right)$.

Definition 2 *Let* $\theta \in \Theta, S = \left\{\theta^S_{i(S,1)}, \ldots, \theta^S_{i(S,t_S)}\right\} \in \prod_{h=1}^{t_S} \Theta_{i(S,h)}$. *We call the sequence of type profiles* $\left\{m^h(\theta, S)\right\}_{h=0}^{t_S}$ *the passage from* θ *to* θ' *through* S *if* $m^{t_s}(\theta, S) = \theta'$ *for* $\theta' \in \Theta$.

More informally, we say that θ leads to θ' through S.

Notice that a given passage from θ to θ' through S induces a corresponding sequence of preference profiles, $R^h(\theta, S)$ for $h \in \{0, 1, \ldots, t_S\}$, where for each agent $i \in N$, $R^h_i(\theta, S) \equiv R_i(m^h(\theta, S))$.

We can now establish a condition on the connection between sequences of changes in type profiles and the changes in preferences profiles that they induce.

Definition 3 *Let* $x \in A$, $\theta, \theta' \in \Theta$. *We say that the passage from* θ *to* θ' *through* S *is x-satisfactory if for each* $h \in \{1, \ldots, t_S\}$, $R^h_{i(S,h)}(\theta, S)$ *is an x-monotonic transform of* $R^{h-1}_{i(S,h)}(\theta, S)$.

Example 1 (continued) Satisfactory and nonsatisfactory passages.

Let $x = a$, $\theta = (\underline{\theta}_1, \underline{\theta}_2), \theta' = (\overline{\theta}_1, \underline{\theta}_2)$, *and* $S = \{\overline{\theta}_2, \overline{\theta}_1, \underline{\theta}_2\}$ *a sequence of individual types. Note that* $I(S) = \{2, 1, 2\}$ *and* $t_S = 3$. *We claim that the passage from* θ *to* θ' *through* S *is* a*-satisfactory. To show it, we have to check that for each* $h \in \{1, 2, t_S = 3\}$, $R^h_{i(S,h)}(\theta, S)$ *is an* a*-monotonic transform of* $R^{h-1}_{i(S,h)}(\theta, S)$.

For that, observe first that $R^0_{i(S,1)}(\theta, S) = R_2(\underline{\theta}_1, \underline{\theta}_2)$, $R^1_{i(S,1)}(\theta, S) = R_2(\underline{\theta}_1, \overline{\theta}_2)$, $R^1_{i(S,2)}(\theta, S) = R_1(\underline{\theta}_1, \overline{\theta}_2)$, $R^2_{i(S,2)}(\theta, S) = R_1(\overline{\theta}_1, \overline{\theta}_2)$, $R^2_{i(S,2)}(\theta, S) = R_2(\overline{\theta}_1, \overline{\theta}_2)$, *and* $R^3_{i(S,2)}(\theta, S) = R_2(\overline{\theta}_1, \underline{\theta}_2)$. Then, using the table in Example 1, note that the following three facts hold: $R_2(\underline{\theta}_1, \overline{\theta}_2)$ is an a-monotonic transform of $R_2(\underline{\theta}_1, \underline{\theta}_2)$. Moreover, $R_1(\overline{\theta}_1, \overline{\theta}_2)$ is an a-monotonic transform of $R_1(\underline{\theta}_1, \overline{\theta}_2)$. And, $R_2(\overline{\theta}_1, \underline{\theta}_2) = R_2(\overline{\theta}_1, \overline{\theta}_2)$.

Let $x = a, \theta = (\underline{\theta}_1, \underline{\theta}_2), \theta' = (\overline{\theta}_1, \underline{\theta}_2)$, and $S = \{\overline{\theta}_1\}$ a sequence of individual types. Note that, $I(S) = \{1\}$ and $t_S = 1$. We claim that the passage from θ to θ' through S is not a-satisfactory. To show it, observe that for $R^1_{i(S,1)}(\theta, S) = (ab)$ is not an a-monotonic transform of $R^0_{i(S,1)}(\theta, S) = R_{i(S,1)}(\theta) = ab$.

Notice that in the case of private values the order of individuals in S could be changed and the new sequence would still serve the same purpose. This is because the changes in the type of each agent only induce changes in the preferences of this agent. By contrast, the precise order of agents $I(S)$ may be crucial in the case of interdependent values. We say that x is the reference alternative when going from θ to θ'.

Armed with these definitions we now state the concept of knit domains.

Definition 4 We say that Θ is *knit* if for any two pairs formed by an alternative and a type profile each, $(x, \theta), (z, \widetilde{\theta}) \in A \times \Theta, \theta \neq \widetilde{\theta}, x \neq z$, there exist $\theta' \in \Theta$ and sequences of types S and $\widetilde{S}$, such that the passage from θ to θ' through S is x-satisfactory and the passage from θ to $\widetilde{\theta}$ through $\widetilde{S}$ is z-satisfactory.

Although the above definition is general, we want to remark that no domain is knit in private values environments.

Proposition 1 *No domain* Θ *in a private values environment is knit.*[3]

Proof of Proposition 1 Since we are assuming that the range of the preference function is not a singleton, we can choose $i \in N, \theta_i, \widetilde{\theta}_i \in \Theta_i, \theta_i \neq \widetilde{\theta}_i$ to be such that $R(\theta_i) \neq R(\widetilde{\theta}_i)$. There will be a pair of alternatives, say x and z, such that $xP_i(\theta_i)z$ and $zR_i(\widetilde{\theta}_i)x$ (otherwise, for $\theta_i, \widetilde{\theta}_i \in \Theta_i$, $R(\theta_i) = R(\widetilde{\theta}_i)$). To show that the set of types Θ is not knit, we prove that for the two pairs $(x, (\theta_i, \theta_{-i}))$, and $(z, (\widetilde{\theta}_i, \theta_{-i}))$ for some θ_{-i}, there does not exist any θ', S, and $\widetilde{S}$ such that the passage from θ to θ'

[3]The argument used here is the same as in the proof of Proposition 1 in Barberà et al. (2018). We include it here for completeness and to illustrate a case where knitness fails.

through S be x-satisfactory and the passage from $\widetilde{\theta}$ to θ' through $\widetilde{S}$ be z-satisfactory. We prove it by contradiction. Suppose otherwise that there exist θ^*, S^*, $\widetilde{S}^*$, such that the passages $\left\{m^h(\theta, S^*)\right\}_{h=0}^{t_{S^*}}$ and $\left\{m^h(\theta, \widetilde{S}^*)\right\}_{h=0}^{t_{\widetilde{S}^*}}$ from θ to θ^* through S^* and $\widetilde{\theta}$ to θ^* through $\widetilde{S}^*$ are x and z-satisfactory, respectively.

Since we are in a private values environment, changes in the type of agent j never affect the induced preferences of other agents and in particular never affect i's induced preferences if $j \neq i$. Moreover, we know that $xP_i(\theta_i)z$ and $zR_i(\widetilde{\theta}_i)x$. These two observations imply that agent i must belong to $I(S^*) \cup I(\widetilde{S}^*)$. That is, i will appear in at least one of these two sequences.

We concentrate on the steps of the passage where agent i changes her type and we show that there is no θ^* compatible with the existence of x-satisfactory and z-satisfactory passages from θ to θ^* and from $\widetilde{\theta}$ to θ^*.

Without loss of generality, by the remark just after Definition 3, we can assume that all types of agent i in S^* and $\widetilde{S}^*$ appear in the first positions in these sequences. Let's define $I_{S^*,i} \equiv \{h \in \{1, 2, \ldots, i_{S^*}\} : i(S^*, h) = i\}$ and $I_{\widetilde{S}^*,i} = \left\{h \in \{1, 2, \ldots, i_{\widetilde{S}^*}\} : i(\widetilde{S}^*, h) = i\right\}$.

Take $1 \in I_{S^*,i}$. Since $R_i^1(\theta, S^*)$ is an x-monotonic transform of $R_i(\theta_i)$, we have that $xP_i(m_i^1(\theta, S^*))z$. By repeating the same argument for each $h \in I_{S^*,i}$ we finally obtain that $xP_i(m_i^{i_{S^*}}(\theta, S^*))z$ where $m_i^{i_{S^*}}(\theta, S^*) = \theta_i^*$.

Now, take $1 \in I_{\widetilde{S}^*,i}$. Since $R_i^1(\widetilde{\theta}^*, \widetilde{S}^*)$ is a z-monotonic transform of $R_i(\widetilde{\theta}_i^*)$, we have that $zR_i(m_i^1(\widetilde{\theta}^*, \widetilde{S}^*))x$. By repeating the same argument for each $h \in I_{\widetilde{S}^*,i}$ we finally obtain that $zR_i(m_i^{i_{\widetilde{S}^*}}(\theta, \widetilde{S}^*))z$ where $m_i^{i_{\widetilde{S}^*}}(\theta, \widetilde{S}^*) = \theta_i^*$.

As mentioned above, changes in the types of agents different from i will not change agent i's preferences. Thus, we have obtained the desired contradiction: on the one hand that $xP_i(\theta^*)z$ and on the other hand that $zR_i(\theta^*)x$. ■

We now turn attention to some properties of the mechanisms themselves.

We first look at incentives. *Ex post incentive compatibility* requires, for all agents to prefer truthtelling at a given type profile θ, if all the other agents also report truthfully.[4]

Definition 5 A mechanism f is *ex post incentive compatible* on Θ if, for any agent $i \in N$, $\theta \in \Theta$, and $\theta_i' \in \Theta_i$, $f(\theta)R_i(\theta)f(\theta_i', \theta_{N\setminus\{i\}})$.

We say that an agent $i \in N$ *can ex post profitably deviate under mechanism f at* $\theta \in \Theta$ if there exists $\theta_i' \in \Theta_i$ such that $f(\theta_i', \theta_{N\setminus\{i\}})P_i(\theta)f(\theta)$. Note that ex post incentive compatibility requires that no agent can ex post profitably deviate at any type profile.

Finally, we shall require our mechanisms to satisfy a condition that we call *respectfulness*. It is a relatively weak requirement since it only applies to some

[4]This property is called uniform incentive compatibility by Holmstrom and Myerson (1983). See also Bergemann and Morris (2005).

limited changes in type profiles, and has no bite in some important cases (e.g., in public good economies where agents' preferences are strict). The condition essentially requires that for those limited changes in type profiles, no agent can affect the outcome (for her and for others) unless she changes her own level of satisfaction.

Definition 6 A mechanism f is (*outcome*) *respectful* on Θ if

$$f(\theta)I_i(\theta)f(\theta_i', \theta_{N\setminus\{i\}}) \text{ implies } f(\theta) = f(\theta_i', \theta_{N\setminus\{i\}}),$$

for each $i \in N$, $\theta \in \Theta$, and $\theta_i' \in \Theta_i$ such that $R_i(\theta_i', \theta_{N\setminus\{i\}})$ is a $f(\theta)$-monotonic transform of $R_i(\theta)$.

3 The Main Result

Our Theorem 1 provides a full characterization of those domains that admit nonconstant, ex post incentive compatible, and respectful mechanisms in the two-alternative case that we are considering.

Theorem 1 *A domain Θ admits nonconstant, respectful, and ex post incentive compatible mechanism if and only if it is not knit.*

Proof Let us prove, by construction, that if a domain Θ is not knit then there exist nonconstant, respectful, and ex post incentive compatible mechanisms on Θ. In any domain that is not knit, there will be two pairs formed by an alternative and a type profile each, $(x, \theta), (z, \widetilde{\theta}) \in A \times \Theta, \theta \neq \widetilde{\theta}, x \neq z$ such that there do not exist $\theta' \in \Theta, S, \widetilde{S}$ where the passage from θ to θ' through S is x-satisfactory and the passage from $\widetilde{\theta}$ to θ' through $\widetilde{S}$ is z-satisfactory. Without loss of generality, assume that $x = a$ and $z = b$.

Before defining the desired mechanism, let us first propose the following partition of Θ:$\Theta_1 = \{\overline{\theta} \in \Theta$: there are S_1 and S_2 such that the passage from θ to $\overline{\theta}$ through S_1 is a-satisfactory and the passage from $\overline{\theta}$ to θ through S_2 is b-satisfactory$\}$, $\Theta_2 = \{\overline{\theta} \in \Theta$: there are $\widetilde{S}_1$ and $\widetilde{S}_2$ such that the passage from $\widetilde{\theta}$ to $\overline{\theta}$ through $\widetilde{S}_1$ is b-satisfactory and the passage from $\overline{\theta}$ to $\widetilde{\theta}$ through $\widetilde{S}_2$ is a-satisfactory$\}$, $\Theta_3 = \Theta\backslash(\Theta_1 \cup \Theta_2)$. Note that since the domain is not knit $\Theta_1 \cap \Theta_2 = \varnothing$.

We now define a mechanism f as follows: $f(\widehat{\theta}) = a$ if $\widehat{\theta} \in \Theta_1 \cup \Theta_3$ and $f(\widehat{\theta}) = b$, otherwise. Let us first check, by contradiction, that f is ex post incentive compatible.

Suppose that there exist $\overline{\theta} \in \Theta$, agent $i \in N, \theta_i' \in \Theta$ such that $f(\theta_i', \overline{\theta}_{N\setminus\{i\}}) P_i(\overline{\theta})f(\overline{\theta})$.

Case 1. $f(\overline{\theta}) = a$ and $f(\theta_i', \overline{\theta}_{N\setminus\{i\}}) = b$. Thus, by definition of f, either (1) $\overline{\theta} \in \Theta_1$ or (2) $\overline{\theta} \in \Theta\backslash(\Theta_1 \cup \Theta_2)$. However, $\left(\theta_i', \overline{\theta}_{N\setminus\{i\}}\right) \in \Theta_2$.

Observe that $R_i(\overline{\theta})$ is such that $bP_i(\overline{\theta})a$ and $R_i(\theta_i', \overline{\theta}_{N\setminus\{i\}})$ can be any preference. That is, $bR_i(\theta_i', \overline{\theta}_{N\setminus\{i\}})a$, $aR_i(\theta_i', \overline{\theta}_{N\setminus\{i\}})b$, or $aI_i(\theta_i', \overline{\theta}_{N\setminus\{i\}})b$ holds. For the three cases, observe that the passage from $\overline{\theta}$ to $(\theta_i', \overline{\theta}_{N\setminus\{i\}})$ through $S' = \{\theta_i'\}$ where $I(S') = (i)$ is a-satisfactory and, also, the passage from $(\theta_i', \overline{\theta}_{N\setminus\{i\}})$ to $\overline{\theta}$ through $\widehat{S}$ where $I(\widehat{S}) = (i)$ is b-satisfactory.

Subcase 1.1: $\overline{\theta} \in \Theta_1$. Since $\overline{\theta} \in \Theta_1$, we have that the passage from θ to $\overline{\theta}$ through S_1 is a-satisfactory and therefore the passage from θ to $(\theta_i', \overline{\theta}_{N\setminus\{i\}})$ through $S_1 \cup S'$ is a-satisfactory. Also since $\overline{\theta} \in \Theta_1$, we have that the passage from $\overline{\theta}$ to θ through S_2 is b-satisfactory and therefore the passage from $(\theta_i', \overline{\theta}_{N\setminus\{i\}})$ to θ through $\widehat{S} \cup S_2$ is b-direct satisfactory. Therefore, $(\theta_i', \overline{\theta}_{N\setminus\{i\}}) \in \Theta_1$ which is a contradiction to the assumption that $(\theta_i', \overline{\theta}_{N\setminus\{i\}}) \in \Theta_2$.

Subcase 1.2: $\overline{\theta} \in \Theta\setminus(\Theta_1 \cup \Theta_2) = \Theta_3$. Since $(\theta_i', \overline{\theta}_{N\setminus\{i\}}) \in \Theta_2$, we have that the passage from $\widetilde{\theta}$ to $(\theta_i', \overline{\theta}_{N\setminus\{i\}})$ through $\widetilde{S}_1$ is b-satisfactory, and therefore the passage from $\widetilde{\theta}$ to $\overline{\theta}$ through $\widetilde{S}_1 \cup \widehat{S}$ is b-satisfactory. Also, since $(\theta_i', \overline{\theta}_{N\setminus\{i\}}) \in \Theta_2$, we have that the passage from $(\theta_i', \overline{\theta}_{N\setminus\{i\}})$ to $\widetilde{\theta}$ through $\widetilde{S}_2$ is a-satisfactory and therefore the passage from $\overline{\theta}$ to $\widetilde{\theta}$ through $\widehat{S} \cup \widetilde{S}_2$ is a-satisfactory. Therefore, $\overline{\theta} \in \Theta_2$ which is a contradiction to the assumption that $\overline{\theta} \in \Theta\setminus(\Theta_1 \cup \Theta_2)$.

Case 2. $f(\theta_i', \overline{\theta}_{N\setminus\{i\}}) = a$ and $f(\overline{\theta}) = b$. Thus, by definition of f, either (1) $(\theta_i', \overline{\theta}_{N\setminus\{i\}}) \in \Theta_1$ or (2) $(\theta_i', \overline{\theta}_{N\setminus\{i\}}) \in \Theta\setminus(\Theta_1 \cup \Theta_2)$. However, $\overline{\theta} \in \Theta_2$.

Observe that $R_i(\overline{\theta})$ is such that $aP_i(\theta)b$ and $R_i(\theta_i', \overline{\theta}_{N\setminus\{i\}})$ can be any preference. That is, $bR_i(\theta_i', \overline{\theta}_{N\setminus\{i\}})a$, $aR_i(\theta_i', \overline{\theta}_{N\setminus\{i\}})b$, or $aI_i(\theta_i', \overline{\theta}_{N\setminus\{i\}})b$ holds. For the three cases, observe that the passage from $\overline{\theta}$ to $(\theta_i', \overline{\theta}_{N\setminus\{i\}})$ through $S' = \{\theta_i'\}$ where $I(S') = (i)$ is b-satisfactory, also, the passage from $(\theta_i', \overline{\theta}_{N\setminus\{i\}})$ to $\overline{\theta}$ through $\widehat{S}$ where $I(\widehat{S}) = (i)$ is a a-satisfactory.

Since $\overline{\theta} \in \Theta_2$, we have that the passage from $\widetilde{\theta}$ to $\overline{\theta}$ through $\widetilde{S}_1$ is b-satisfactory, and therefore the passage from $\widetilde{\theta}$ to $(\theta_i', \overline{\theta}_{N\setminus\{i\}})$ through $\widetilde{S}_1 \cup S'$ is b-satisfactory. Also, since $\overline{\theta} \in \Theta_2$, we have that the passage from $\overline{\theta}$ to $\widetilde{\theta}$ through $\widetilde{S}_2$ is a-satisfactory and therefore the passage from $(\theta_i', \overline{\theta}_{N\setminus\{i\}})$ to $\widetilde{\theta}$ through $\widehat{S} \cup \widetilde{S}_2$ is a-satisfactory. Therefore, $(\theta_i', \overline{\theta}_{N\setminus\{i\}}) \in \Theta_2$ which is a contradiction to the assumption that $(\theta_i', \overline{\theta}_{N\setminus\{i\}}) \in \Theta_1 \cup \Theta_3$.

Now, we show that f is respectful.

By contradiction suppose that there exist $\overline{\theta} \in \Theta$, agent $i \in N$, $\theta_i' \in \Theta$ such that $f(\theta_i', \overline{\theta}_{N\setminus\{i\}})I_i(\overline{\theta})f(\overline{\theta})$, $f(\theta_i', \overline{\theta}_{N\setminus\{i\}}) \neq f(\overline{\theta})$, and $R_i(\theta_i', \overline{\theta}_{N\setminus\{i\}})$ is a $f(\overline{\theta})$-monotonic transform of $R_i(\overline{\theta})$.

First, assume that $f(\overline{\theta}) = a$ and $f(\theta_i', \overline{\theta}_{N\setminus\{i\}}) = b$. Thus, by definition of f, either (1) $\overline{\theta} \in \Theta_1$ or (2) $\overline{\theta} \in \Theta\setminus(\Theta_1 \cup \Theta_2)$. However, $(\theta_i', \overline{\theta}_{N\setminus\{i\}}) \in \Theta_2$.

Observe that $R_i(\overline{\theta})$ is such that $bI_i(\overline{\theta})a$ and $R_i(\theta_i', \overline{\theta}_{N\setminus\{i\}})$ can be any preference. That is, $bR_i(\theta_i', \overline{\theta}_{N\setminus\{i\}})a$, $aR_i(\theta_i', \overline{\theta}_{N\setminus\{i\}})b$, or $aI_i(\theta_i', \overline{\theta}_{N\setminus\{i\}})b$ holds. For the three cases, observe that the passage from $\overline{\theta}$ to $(\theta_i', \overline{\theta}_{N\setminus\{i\}})$ through $S' = \{\theta_i'\}$ where $I(S') = (i)$ is a-satisfactory, also, the passage from $(\theta_i', \overline{\theta}_{N\setminus\{i\}})$ to $\overline{\theta}$ through $\widehat{S}$ where $I(\widehat{S}) = (i)$ is a b-satisfactory.

Repeating the same argument as in Cases 1 and 2 above we get the desired contradiction.

An identical argument as above holds if $f(\theta_i', \overline{\theta}_{N\setminus\{i\}}) = a$ and $f(\overline{\theta}) = b$.

To prove the second part of our result, we invoke the fact that on knit domains any respectful and ex post incentive compatible mechanism must be constant as stated in Theorem 1 in Barberà et al. (2018). ■

As an application of our framework and result we present an example of a deliberative jury, where jurors are endowed with preference functions that result from very specific reactions to information. The example is inspired in Austen-Smith and Feddersen (2006), but we extend it here to allow for agents to be indifferent between the two outcomes. We exhibit a family of interesting mechanisms satisfying our conditions and which in addition may reach efficient outcomes. Clearly, the domain of types in this example is not knit.

Example 2 A three-person jury $N = \{1, 2, 3\}$ must decide over two alternatives: whether to acquit (A) or to convict (C) a defendant under a given mechanism. The defendant is either guilty (g) or innocent (i). Each juror j gets a signal $s_j = g$ or $s_j = i$.

In this application, the type of each juror is the combination of the signals she has received and the idiosyncratic methods she uses to process her information and that of others into a preference over alternatives. In this and other applications, we find it useful to be explicit about the way in which agents form their preferences, once they know the type profile. We do so by distinguishing between two components of their type: one is the set of signals they get, s_i, and the other is the rule they use to form their preferences once they have the signals, b_i. Formally, a type for agent $i \in N$ can be written as $\theta_i = (b_i, s_i) \in \Theta_i$, where b_i is a function from type profiles to individual i's preferences. If we take those elements as the primitives of the model, we can then induce the relevant preference function. Let B_i be the set of possible preference formation rules for individual i.

Let S_i be the set of signals for agent i. We consider types in $\Theta_i \equiv B_i \times S_i$ and the preference function $R : \Theta \to \times_{i \in N} \mathcal{R}_i$ is such that for each $i \in N$, $R_i(\theta) = b_i(s)$ where $s = (s_1, \ldots, s_n)$. Thus, juror's preferences arise from combining the different signals they obtain from the deliberation, according to their particular preference formation rules. These are of two possible kinds, depending on the agents' tendency to convict in view of their observed signals and of those declared by others.

Each juror may now be either unswerving or median. Unswerving jurors (u) prefer to convict rather than acquit (this preference denoted by C) if they have observed the "guilty" signal and have also received such a signal from at least another juror. They prefer to acquit rather than to convict (denoted by A) if they have the "innocent" signal and have also received such a signal from at least another juror. Otherwise, they are indifferent. Median jurors (m) prefer to convict rather than to acquit if they have observed the "guilty" signal and have also received such a signal from at least another juror but also if they receive two "guilty" signals from other jurors. Otherwise, they prefer to acquit rather than convict (denoted by AC).

Formally, each agent can have two preference formation rules that are defined as follows: (1) $b_i^u(s) = C\ if\ s_i = g$ and $s_j = g$ for some $j \neq i$, $b_i^u(s) = A\ if\ s_i = i$ and $s_j = i$ for some $j \neq i$, and $b_i^u(s) = (AC)$, otherwise; (2) $b_i^m(s) = b^m(s)$ such that $b^m(s) = C$ *if* $\#\{i \in N : s_j = g\} \geq 2$ and $b^m(S) = A$, otherwise. Here $B_i = \{b_i^u, b^m\}$. The preference function R is such that for each agent $i \in N$, $R_i((b_i^u, s_i), \theta_{-i}) = b_i^u(s)$ and $R_i((b^m, s_i), \theta_{-i}) = b^m(s)$.[5]

We now define mechanisms that are nonconstant, respectful, and ex post incentive compatible on Θ. In view of Theorem 1, this is possible since the domain of definition is not knit. To prove that, consider any two pairs of alternatives and types such that unanimity of induced preferences holds in favor of the corresponding alternative. For example, let (C, θ) where $\theta = ((b_1^m, g), (b_2^m, g), (b_3^m, i))$ and $(A, \widetilde{\theta})$ where $\widetilde{\theta} = ((b_1^m, i), (b_2^m, i), (b_3^m, i))$ and observe that for each $i \in N$, $R_i(\theta) = C$ and $R_i(\widetilde{\theta}) = A$. Then, there are no θ' and sequences S and $\widetilde{S}$ such that the passage from θ to θ' through S is C-satisfactory and the passage from $\widetilde{\theta}$ to θ' through $\widetilde{S}$ is A-satisfactory.

Let $q \in \{1, 2, 3\}$. A mechanism f is voting by quota q if f chooses C for a type profile θ if and only if at least q agents have induced preferences from θ such that C is preferred to A.[6]

Formally, for each type profile $\theta = (b, s) \in \Theta$,

$$f(\theta) = C \mathit{if\ and\ only\ if} \# \{i \in N : b_i(s) = C\} \geq q.$$

In Table 2, we describe all possible results of voting by quota for different values of q. We have four matrices, one for each type of agent 3. In the rows of each matrix, we write the four types of agent 1 and in the columns the four types of agent 2. In each cell, we write each agent's best alternatives according to their preference at a given type profile, followed by the outcome of a quota mechanism. When two outcomes appear in a cell, the one in the left stands for the outcome of voting by quota 3 and the one in the right is the outcome for both quota 1 and 2, which in this example are always the same.

Given Table 2, it is easy to check that this rule is ex post incentive compatible and respectful. In addition, it also satisfies anonymity.

[5]Being unswerving for agents 1 and 2 is different. Suppose that both jurors are unswerving and the signal profile is (g, i, g). Then, juror 1 will prefer to convict rather than to acquit but juror 2 would not. Yet being both jurors median would induce the same preferences for both agents: they would prefer to convict rather than to acquit.

[6]See Austen-Smith and Feddersen (2006) and Barberà and Jackson (2004) for papers where these rules are analyzed.

Table 2 For each type profile, each agent's best alternative and feasible outcome of a voting by quota mechanism

(b_3^m, i)	(b_2^m, i)		(b_2^m, g)		(b_2^u, i)		(b_2^u, g)	
(b_1^m, i)	AAA	A	AAA	A	AAA	A	A(AC)A	A
(b_1^m, g)	AAA	A	CCC	C	AAA	A	CCC	C
(b_1^u, i)	AAA	A	AAA	A	AAA	A	A(AC)A	A
(b_1^u, g)	(AC)AA	A	CCC	C	(AC)AA	A	CCC	C
(b_3^u, i)	(b_2^m, i)		(b_2^m, g)		(b_2^u, i)		(b_2^u, g)	
(b_1^m, i)	AAA	A	AAA	A	AAA	A	A(AC)A	A
(b_1^m, g)	AAA	A	CC(AC)	A/C	AAA	A	CC(AC)	A/C
(b_1^u, i)	AAA	A	AAA	A	AAA	A	A(AC)A	A
(b_1^u, g)	(AC)AA	A	CC(AC)	A/C	(AC)AA	A	CC(AC)	A/C
(b_3^m, g)	(b_2^m, i)		(b_2^m, g)		(b_2^u, i)		(b_2^u, g)	
(b_1^m, i)	AAA	A	CCC	C	AAA	A	CCC	C
(b_1^m, g)	CCC	C	CCC	C	C(AC)C	A/C	CCC	C
(b_1^u, i)	AAA	A	(AC)CC	A/C	AAA	A	(AC)CC	A/C
(b_1^u, g)	CCC	C	CCC	C	C(AC)C	A/C	CCC	C
(b_3^u, g)	(b_2^m, i)		(b_2^m, g)		(b_2^u, i)		(b_2^u, g)	
(b_1^m, i)	AA(AC)	A	CCC	C	AA(AC)	A	CCC	C
(b_1^m, g)	CCC	C	CCC	C	C(AC)C	A/C	CCC	C
(b_1^u, i)	AA(AC)	A	(AC)CC	A/C	AA(AC)	A	(AC)CC	A/C
(b_1^u, g)	CCC	C	CCC	C	C(AC)C	A/C	CCC	C

4 Concluding Remarks

We have argued that the design of mechanisms that operate in societies that involve interdependent types is important and nontrivial, even when only two possible social outcomes are at stake. We have proven that the possibility to design satisfactory mechanisms in that context crucially depends on the domains over which they must be defined. In a context that admits the possibility of agents to be indifferent between the two potential alternatives, we have found characterization of those domains admitting nonconstant, ex post incentive compatible, and respectful mechanisms. We have also described a family of quota rules that satisfy these conditions in an example involving deliberative committees. These mechanisms, in addition, satisfy further properties, like anonymity and neutrality, which make them particularly attractive and hint that in the absence of knitness there is a wide range of mechanisms to choose from, some of which are far from trivial. While all quota rules have much in common, two of the quotas (one and two) lead to efficient outcomes, quota three rules may be inefficient. Further analysis of the compatibility between efficiency and good incentives deserves additional attention in further work.

Ackowledgements Salvador Barberà acknowledges financial support from the Spanish Ministry of Economy and Competitiveness, through the Severo Ochoa Programme for Centers of Excellence in R&D (SEV-2015-0563) and grant ECO2017-83534-P and feder, and from the Generalitat de Catalunya, through grant 2017SGR-0711. D. Berga and B. Moreno acknowledge the support from the Spanish Ministry of Economy, Industry and Competitiveness through grants ECO2016-76255-P and ECO2017-86245-P, respectively, and thank the MOMA network.

References

Austen-Smith, D., & Banks, J. S. (1999). *Positive political theory I. Collective preference*. Ann Arbor: The University of Michigan Press.

Austen-Smith, D., & Feddersen, T. J. (2006). Deliberation, preference uncertainty, and voting rules. *American Political Science Review, 100*, 209–217.

Barberà, S., & Jackson, M. O. (2004). Choosing how to choose: Self-stable majority rules and constitutions. *Quarterly Journal of Economics, 119*(3), 1011–1048.

Barberà, S., Berga, D., & Moreno, B. (2018). *Restricted environments and incentive compatibility in interdependent values models*. SSRN WP/BGSE WP 1024.

Bergemann, D., & Morris, S. (2005). Robust mechanism design. *Econometrica, 73*, 1771–1813.

Fishburn, P. C. (1973). *The theory of social choice*. Princeton, NJ: Princeton University Press.

Holmstrom, B., & Myerson, R. B. (1983). Efficient and durable decision rules with incomplete information. *Econometrica, 51*, 1799–1819.

May, K. O. (1952). A set of independent necessary and sufficient conditions for simple majority decisions. *Econometrica, 20*, 680–684.

Mechanisms in a Digitalized World

Gabrielle Demange

1 Introduction

Social institutions were the primary interests of Leonid Hurwicz. He developed a theory of how to analyze institutions and economic systems in terms of their incentives and enforcement properties in Hurwicz (1960) and Hurwicz (1973). In putting the emphasis on the crucial role of information for allocating resources efficiently, he formalized ideas from Hayek and Mises on the market as an aggregator of dispersed information. In doing so, as argued by Myerson (2009), he shed light on an old debate about socialism and central planning. The tools he introduced, relying on an analytical modeling of incentives, have a fundamental influence on current economics, both theoretical and applied. They paved the way to mechanism design, which considers how a designer (planner, institution, firm) who aims at achieving certain goals should choose the rules applied to individuals who act strategically. Mechanism design plays a critical role in the development of new market allocation procedures. To name a few, the allocation of students to schools, kidney exchanges, or auctions are all determined by mechanisms. As

This is written for the volume "Social Design: Essays in Memory of Leonid Hurwicz" edited by Walter Trockel. I thank him for his encouragement. This work is partially based on a talk I gave in the stimulating workshop "Social Responsibility of Algorithms" organized by Alexis Tsoukias at Paris-Dauphine University, December 2017.

G. Demange (✉)
Paris School of Economics-EHESS, Paris, France
e-mail: demange@pse.ens.fr

W. Trockel (ed.), *Social Design*, Studies in Economic Design,
https://doi.org/10.1007/978-3-319-93809-7_19

such, Leonid Hurwicz can be considered as a precursor of the new area of market design. Furthermore, the formal procedures computed by algorithms for solving public, economic, or social problems share common features with mechanisms as defined by economists.[1] As these procedures are developing fast due to computing and communication facilities, the impact of Hurwicz's work now extends to an even broader area than expected.

After pioneering the mathematic modeling of incentives and introducing mechanism design, Hurwicz became interested in the "human side" of mechanisms. The article "But who will guard the guardians?" (2008) refers to at least two human sides. The first one is alluded to in the title of the article, which is a question raised by Juvenal, described by Hurwicz as follows:

> In posing the famous question, the Roman author, Juvenal, was suggesting that wives cannot be trusted, and keeping them under guard is not a solution because the guards cannot be trusted either.

In the mechanism context, Juvenal's question could be rephrased as: Who are the guardians of the institutions? Who watches whether the announced constitution (or regulation) is correctly applied? In other words, Hurwicz raises the incentives' issue on the mechanism's designer rather than on the individuals on which the mechanism applies. Apart from the designer's incentives, one may also add: Who checks whether the mechanism is correctly computed? Who checks that errors in the mechanism do not generate large risks? Such issues are especially important for the complex mechanisms that are now computed through algorithms by computers.

The second human side raised by Hurwicz (2008) pertains to the individuals and their "illegal" strategies. This aspect is in line with his primary objective of studying the functioning of economic systems, impossible to describe fully by an analytical approach. While the analytical apparatus of mechanism design is suitable for studying institutions that must be precisely defined, such as electoral rules, it is too constrained to describe most institutions because of the numerous individuals' possibilities of action (the illegal or secret strategies).[2] Research has not much considered illegal strategies, but rather developed in studying well-defined settings ranging from the implementation literature to market design, in which case the issue of illegal strategies is irrelevant.

A third human consideration differs from the ones referred to by Hurwicz. A reluctance to mechanisms has been revealed now that they are implemented in the real world at a large scale through computers and algorithms. Though, the reluctance is only partly explained by the use of these computerized tools. The very genuine feature of a mechanism is to be mechanical, and this feature *per se* might be perceived as non-human, whether the mechanism is computed by hand or a computer.

---

[1] Naturally, the relationships do not apply to all algorithms, in particular to those computed in artificial intelligence.

[2] Such argument has the same flavor as the one saying that contracts are typically incomplete.

The plan is as follows. Section 2 describes rules and mechanisms, introduces some basic insights from the theoretical literature, and presents two examples of mechanisms run by a governmental agency -spectrum auctions in the US and assignments of students to universities in France. When the designer is a governmental agency, there is a legitimate demand for explanation. I will argue that the approach called "axiomatization" developed by social choice theory may help policymakers in providing such an explanation. Section 3 discusses algorithms (in the broad meaning used currently) in the economic and social areas and their links with rules and mechanisms. It presents mechanisms used by private firms and ends with the economic risks generated by their computing power.

2 Rules and Mechanisms

Ii is useful to consider rules before introducing mechanisms. A typical rule is a voting procedure for the election of a president among several candidates. The rule assigns the winner to the votes. No specific assumption is made on how voters vote. A mechanism instead—in the precise sense of the mechanism design literature following Hurwicz—assumes that the votes are cast strategically and considers the rule that results of these strategic votes. As the term mechanism is more broadly used in practice, I will refer in some places to a mechanism even when no specific strategic assumption is being made.

2.1 Rules

The rules considered here aim at solving some social issues between a group of "units" such as consumers, workers, or citizens. Specifically, a rule solves the issue by choosing an outcome, such as an allocation of resources, an assignment of tasks or a president. Crucially, the outcome is based on a profile of data, which specifies data for each unit, representing, for example, the unit's preferences, skills, resources. In formal terms, a rule assigns an outcome to each set of data. The rule thus starts with units' data without making assumptions on how they have been gathered. There are many rules in a variety of contexts, aiming at answering questions such as:

1. How to select a candidate? A voting procedure is a rule that assigns the selected candidate (the outcome) to the expressed votes over the candidates (the data). For some rules, a vote is a single name, while for others it is a list of admissible candidates, or a full ranking of the candidates.

2. How to rank a set of alternatives? A rule here assigns a full ranking of the alternatives, as considered by Arrow (1950):

   > By a "social welfare function" will be meant a process or rule which, for each set of individual orderings $R_1, \ldots, R_n$, for alternative social states (one ordering for each individual), states a corresponding social ordering of alternative social states, R.

   A social welfare function may be used to select a single alternative, the one at the top of the social ordering. As an illustration, Arrow considers a community that has to repeatedly choose between three alternative modes of social action, e.g., disarmament, cold war, or hot war. In that case a rule assigns a ranking of the alternatives as a function of individuals' preferences at the time of the decision.
3. How to rank a set of Websites?[3] Consider the search engine PageRank of Google, described to "bring order on the Web" by Page et al. (1999). PageRank rates the Websites corresponding to a query: the units are the Websites and the outcome is their rating, which determines in which order Websites are displayed on the screen. As described in 1999, PageRank is a method for rating Web pages *objectively and mechanically*, mainly based on the hyperlink structure: PageRank is a rule that determines the rating of the Websites as a function of data, where the data for each Website is composed of the list of Websites that point to it.
4. How to assign students to schools, to universities? An assignment procedure is a rule that assigns the students to universities based on students' preferences and universities' priorities, as discussed in Sect. 2.2. Similar assignment problems arise in other contexts, such as the allocation of social housing.
5. How to assign kidneys between receivers and donors? Here a rule defines an ordering and a matching between receivers and donors based on observable individuals' characteristics (compatibility, age, health status, etc.) in the waiting list (Roth et al., 2004).
6. How to allocate a painting? Here a rule determines who wins the painting, the price paid by the winner and the compensations to others (if any), as a function of the valuations of the potential buyers (their data). An auction procedure is such a rule. More complex auctions for multiple goods are now run, as seen in the example presented in Sect. 2.2.

The above questions can all be answered in a discretionary way. Instead, a rule specifies the answer for all possible data *prior* to the knowledge of the current data. A voting procedure, for example, is written in a constitution before knowing the citizens' votes.

The designer of a rule may be a state, a public agency, or a private firm. In the latter case, a firm is not required to explain its rule, which might not be transparent. For example, the procedure used by Google to rank the Websites (question 3 above) has evolved and made less transparent than the original PageRank. Section 3 will discuss mechanisms used by private firms. When the designer is a state or a public agency, as is the case in our two next examples, there is a legitimate demand for

[3]This question is very much related to the previous one, as noted by Dwork et al. (2001).

explanation. As said in the introduction, the axiomatization approach developed by social choice theory may help policymakers in providing such an explanation (Sect. 2.3).

2.2 *Two Examples*

I describe here in more detail two examples, a successful and a failed one.

Allocating Spectrum Rights The sale of spectrum licenses over the USA illustrates a successful procedure organized by the Federal Communications Commission (FCC). The procedure dramatically changed in 1993 (see McMillan 1994). Prior to 1993, the sale of spectrum licenses over the USA was an administrative decision, based on hearings and lotteries. There were obvious inefficiencies: some licenses were left unassigned and it happened that a winner of a license re-sold it quickly at a much higher price than the acquisition one. After 1993, thousands of licenses were sold through auctions. The FCC goals were multiple: to avoid monopoly, facilitate contiguous territories, favor access to certain minorities, avoid collusion. The auctions were carefully designed with the help of game theorists to satisfy these goals.[4] They turned out to be a big success on various grounds. They raised significantly higher revenues than previously and the absence of immediate resale or bilateral exchanges witnessed the efficiency of the allocation. Such types of large scale multi-item auctions are now conducted in many areas, like the millions of Internet ad-auctions. The interaction between game theorists and policy-makers to set up new allocation procedures thus proves to be fruitful.

Admission Post-Bac (APB) in France The procedure called APB put in place in 2009 for assigning students at their entrance to the French universities turned out to be a failure, resulting in its replacement in 2018. The higher education system in France is mostly public and any student who gets the second-degree diploma called baccalaureat is entitled to a seat. As a result, the admission system has to deal with a high number of candidates (in 2016, 335,696 entered). The APB procedure is based on the centralized deferred-acceptance algorithm introduced by Gale and Shapley (1962). The procedure works as follows. Students state rank the slots (a slot specifies the education curriculum and the university) and the universities rank the students. Then the algorithm computes an assignment, based on a virtual process of successive applications-rejections defined by these rankings. The algorithm is the building block of many successful matching mechanisms dealing with school choice or labor market clearing (for a presentation, see Roth (2008)).

A main property of the deferred-acceptance algorithm is that it produces a *stable* assignment, which was the main purpose of Gale and Shapley (1962). They defined

[4]As described in McMillan (1994), options were multiple, for example simultaneous versus sequential auctions, open versus sealed bids, or royalties versus reserve prices.

an assignment to be unstable if there are two applicants α and β who are assigned to colleges A and B, respectively, although β prefers A to B and A prefers β to α. As argued by Gale and Shapley, if this situation did occur, applicant β could indicate to college A that he would like to transfer to it, and A could respond by admitting β, letting α go to remain within its quota. The original assignment is therefore "unstable" in the sense that it can be upset by a college and applicant acting together in a manner which benefits both.

A second property of the deferred-acceptance algorithm has been shown by Dubins and Freedman (1981):

> Suppose a student, called Machiavelli, lies, that is, does not apply to the universities in the order of true preference. Can this help Machiavelli? The answer is no, not if the others continue to tell the truth. Similarly for coalitions of student liars.

Such a property, according to which students have no incentives to lie, is now called *strategy-proofness*. Accounting for incentives is one of the main concerns of Hurwicz, as discussed in Sect. 2.4. People may be doubtful about Dubins and Freedman's claim, especially when the method is applied to a large population, as in the APB mechanism, because there are many opportunities to lie. The argument is indeed not trivial, although it does not rely on any knowledge in mathematics.[5]

The APB procedure nevertheless revealed to be a fiasco in the last years. But it modified the deferred-acceptance algorithm in an important way, which might explain the failure. To cope with the required "no selection" principle, according to which any student with the baccalaureat is entitled to a seat in any field, no priority was set for the universities. When the number of applicants to some slots largely exceeded the number of seats, students were allocated at random to satisfy the no selection principle. The result was that some students lacking the background for succeeding in a field and almost certain to fail got a seat while some others, much better qualified, did not. The problem was exacerbated by the huge increase in the number of applicants to universities due to the 2000 baby-boom and the policy of proposing many new variants of the baccalaureat. But at the same time, neither the number of seats nor new curriculums appropriate to the background of the new students' population followed the trend. The absurdity of the system led to its rejection in 2017 after many debates and careful examinations, which culminated in a meeting organized by the Field medal C. Villani for French policymakers and deputies.

The blame has been put on the "non-human" aspect of the procedure, in particular to the random draws, due to the fact that it was implemented by an algorithm. As said previously, this non-human aspect is not the one referred to by Hurwicz (2008), but is related to the mechanical aspect of a mechanism/algorithm.[6] The new system

[5]The APB procedure did not satisfy this property because it modified the deferred-acceptance algorithm so that universities' choices depended on the students rankings. Though, students' misreporting their preferences does not seem to have been a big issue.

[6]Another example of reluctance to automatic systems is Centrelink put in place in Australia to recover social security overpayments. Though the system, dubbed "Robodebt"

that replaces APB starting 2018 instead is quite opaque with unclear specification of the universities' objectives, and the overall process might last a very long time.[7]

The result of APB's failure is a clear defiance towards "algorithms" from the French population. In my view, the failure is due to the absence of consistency and transparency in the policy, not in the way it is computed. Taking the viewpoint of social choice theory described in the next section would have been beneficial: explain the desirable properties the government wants to achieve and make explicit the constraints. It would have made clear that the random draw was resulting from the absence of selection and the space constraint. But this was not politically admissible.

2.3 *Justifying a Rule: Axiomatization*

The huge benefit of thinking of solving problems through rules is to specify the desirable properties—called "axioms" following Arrow (1950)—one would like a rule to satisfy. Ideally, these properties can be stated in words. This was basically the approach for designing the spectrum rights auction (along incentives issues). The "axiomatization" approach compares the rules on the basis of the properties they satisfy.

One may distinguish two types of properties: those bearing on the outcome specified by the rule for a given data profile and those bearing on the behavior of the rule when data varies, i.e. how the outcome varies with the data.

Here are a few representative examples of properties that bear on the outcome for a given data profile: the outcome should be efficient, envy-free (for example, an allocation of tasks is envy-free if each person prefers his/her bundle of tasks and compensation to that of anyone else), anonymous (neutrality with respect to labeling), stable in a suitable sense, as the assignments reached by the deferred-acceptance algorithm.

Here are properties on the behavior of a rule when data varies. Some reflect a monotonicity with respect to data. For example, in a representative election, a party who sees its number of votes to increase at the expense of another party should obtain at least as many seats. In the assignment problem, no student is worse off if more seats are available at universities. Other properties reflect a consistency "principle" (also called uniformity), which underlies studies in fair

by users, had some flaws initially—people being unable to complain or reach the service—the main objection relied on the automatic nature of the procedure. Information can be found in https://www.humanservices.gov.au/organisations/about-us/publications-and-resources/government-response-community-affairs-references-committee-report.

[7]Each student will first select a list of applications without order; each application in the list is sent to the corresponding university. Then there will be a succession of rejection or acceptance by universities and students (on their list). The number of applications in a list is a priori constrained to 10, but each application may regroup up to 20 slots.

division, bankruptcy problems (Young, 1987), apportionment problems (Balinski and Demange, 1989). In fair division for example, the principle says that if an allocation among a group of individuals is fair, then it should be perceived as fair when restricted to each subgroup of individuals (for a general presentation, see Thomson (1990)).

To sum up, the axiomatization approach is as follows: define desirable properties on the rule associated to the context under consideration and characterize the rules that satisfy them. Why is it a relevant and delicate question? In most settings, no rule enjoys all properties that sound desirable, as stated by the impossibility theorems, following Arrow (1950, 1951). A rule has to make a choice between properties.

2.4 Mechanism: Introducing Incentives

A rule needs units' data such as preferences, characteristics to compute its outcome. How to learn them? When people provide their preferences, are they truthful? We saw that students have no incentives to lie when the deferred-acceptance algorithm is used: it is strategy-proof. But many rules are not. Consider, for example, plurality voting with more than 2 candidates: a voter might be better off by not voting for her preferred candidate.

A major vision of Hurwicz has been to account systematically for the incentives of individuals to *provide* their data, possibly *strategically*. Formally, this is studied through a game in which individuals' strategies are the (non-verifiable) announcement of their data. The outcome due to the strategic players may result in a very different outcome than the one prescribed by the rule. There are however difficulties to address this issue if the rule is not strategy-proof: How do people behave? What do they know about others' behavior? Do they need to anticipate others' behavior? I give here a simple example that will be useful to illustrate the role of information in data collection.

Simple Buyers-Seller Games Let us consider the exchange of a painting between a seller S and two potential buyers B_1 and B_2. Each attaches a value to the painting. Let S's valuation be 70 (meaning that S benefits from selling at a price larger than 70), B_1's and B_2's be, respectively, 100 and 80 (meaning that they benefit from buying at a price, respectively, lower than 100 and 80).

Consider two rules, known as first price and second price auctions, assuming the above valuations known. In each rule, B_1 obtains the painting, but B_1 pays the highest valuation, here 100, in the first price auction and the second highest valuation, here 80, in the second price auction. These two rules prescribe an efficient outcome since the painting is acquired by B_1, the person whose valuation is the highest, but they produce a different share of the surplus.

Let us now assume that an auctioneer asks the buyers to announce their valuations. The buyers' incentives to announce their true valuations dramatically differ in the two auction rules. In the second price auction, B_1 does not benefit from lying about her valuation because as long as she wins, she will pay 80 (if B_2 does not lie). B_2 does not benefit from lying either because he can win the object only by bidding more than 100 (if B_1 does not lie), in which case the price becomes 100, higher than 80, his valuation. This argument holds more generally whatever the valuations: the second price auction is strategy-proof for the buyers (Vickrey, 1961). In the first price auction instead, B_1 benefits from lying and bidding just above B_2's bid, which is surely less than 80. Though, this strategy assumes that B_1 knows B_2's bid; if this is not the case, finding which amount B_1 should bid starts to be quite complex as it depends on B_1's expectation on B_2's bid, and vice versa.[8]

Let us now consider the seller. S also may have incentives to lie. In the second price auction for example, S benefits in posting a reserve price larger than 80, so as to increase the price paid by B_1, at the risk of not selling the painting and missing a benefit opportunity if S does not perfectly know B_1's valuation. Note, however, that *under perfect knowledge* of the bidders' valuations, the seller extracts all the surplus whatever the auction.

Neither auction thus elicits all three players to reveal their valuations. Furthermore, strategic behavior may result in inefficiency when valuations are unknown due to foregone opportunities to trade. Such analysis and results on auctions extend to situations with more buyers and general valuations or to multiple sellers and buyers (Demange and Gale, 1985). Currently, auctions are being applied at a huge scale on the Web, say for selling the ads and their positions on a Webpage (Varian, 2007). The inefficiency in auctions illustrated by the above example is a robust phenomena: Inefficiency is unavoidable when valuations are unknown, as first shown by Myerson and Satterthwaite (1983) in the case of one seller and one buyer: no exchange mechanism is efficient because strategic play induces foregone opportunities of exchange. Though no one is efficient, some mechanisms are better than others. Finding them is the main issue of mechanism design. Finally, players' information is crucial in determining the outcome and a player may benefit from his knowledge of others' data.

2.5 Developments

Due to the facilities offered by Internet and computers to run mechanisms, a tremendous amount of developments is being conducted by researchers in computer sciences. Spiddit, for example, is a not-for-profit Website that provides algorithms to compute the fair division of goods, credits, or tasks (see the description in Goldman and Procaccia (2015)). A line of research, at the boundary of game theory and computing science, is referred to as algorithmic mechanism design (Nisan and Ronen, 2001). A main objective is to adapt the classic analysis by

[8] An equilibrium as defined by Nash solves this feedback. Wilson (1967) provides a first analysis of the bidders' strategies in an equilibrium context, further developed by Milgrom and Weber (1982).

considering complexity issues (in the computational sense) or by analyzing complex settings. Strategy-proofness is weakened by accounting for the cost and complexity in lying. Dynamical aspects are introduced. Complex "combinatorial" auctions are considered, in which buyers bid for a combination of items (a package) such as a package made of take-off and landing slots at airports (see the book edited by Cramton et al. (2006)). Allowing bidders to bid for a package helps reaching efficiency when the goods in a package are "complements," meaning that a buyer values two goods more than the sum of each. For example, complementarities arise in the FCC auction (Sect. 2.2) when a mobile phone operator values licenses in two adjacent cities more than the sum of the individual license values, due to roaming between the cities.

3 Algorithms

In its original scientific sense, an algorithm is a computational tool solving a well-defined class of problems, such as computing a solution to a linear optimization problem or finding a stable assignment as the deferred-acceptance algorithm does (Sect. 2.2). It is composed of a list of instructions[9] that can be applied to different data sets, for example different students' preferences and universities' priorities. In everyday life, the usage of the word "algorithm" has spread and includes any computerized tool that relies on data in order to provide an outcome in social life, such as an assignment, recommendations to consumers, fiscal controls, or facial recognition. In the settings involving choices and interactions between users, an algorithm can often be viewed as a way for computing the outcome of a rule (as defined as in Sect. 2.1). As such, it is useful to distinguish the rule from the tool for computing it. Abstracting from the computational aspect, lessons can be drawn from the theoretical and strategic analysis of rules roughly described in the previous section.

3.1 *Algorithm as Computing a Rule*

Considering the rule computed by an algorithm, I retain three main lessons from the previous analysis.

Firstly, in most contexts, an algorithm has to operate a selection between fundamental desirable properties that cannot be satisfied simultaneously. Speaking in terms of a rule—prior to the algorithm used to compute it—helps making the selection explicit and thinking in terms of the priority goals one wants to achieve.

[9]This is not true for new algorithmic methods such as machine learning with evolving data, neuronal networks and all the techniques referred to as "artificial intelligence."

Secondly, exhibiting the properties satisfied (or not) by the rule underlying an algorithm helps explaining its rationale.

These two points are related to transparency. They are especially relevant when the algorithm is designed by a public agency, but less so when the designer is a private firm. To illustrate, consider PageRank, the search engine of Google (Sect. 2.1). The engine's computation of the ranking no longer follows the principles described in Page et al. (1999) and is largely unknown. According to the European Commission, Google has abused its market dominance as a search engine by giving an illegal advantage to another Google product, its comparison shopping service. As a result, the European Commission has fined Google 2.42 billion of euros for breaching EU antitrust rules. The judgement is based on statistics showing the bias in favor of Google's product, not on the fact that the firm does not explain how its ranking is computed.[10] In other words, a firm is not required to provide detailed information on its algorithms. When the environment is complex, the rule that is computed by the algorithm might be difficult to decipher.

Thirdly, information is crucial. When individuals provide their data voluntarily, the design of an algorithm should account for their incentives to distort them. When an algorithm instead extracts data for its designer, possibly without the individuals' consent, new issues arise. While data collection and processing bring social benefits in many areas, they also raise privacy concerns. Both public and private designers can use sophisticated techniques to extract data but often for different purposes. I examine in the next section the case of private firms.

3.2 *The Power of Algorithms*

This section discusses some economic aspects that pertain to the power of algorithms/mechanisms used by firms.

Collecting Personal Data: Extracting the New Oil? As we have seen on a simple example between a seller and buyers (Sect. 2.4), the information the participants have on each other's valuations dramatically affect the exchanges. In particular, a seller informed on the buyers' valuations can extract the surplus of the exchanges, provided there is no competing seller. When Internet users search or visit Websites, they provide accurate information to search engines or companies on their intent to purchase, thereby allowing for discriminate pricing and surplus extraction. This is part of the extraction of the "new oil."[11]

This point resonates with current "privacy" concerns. The EU General Data Protection Regulation (GDPR), to be enforced in May 2018, regulates the use of

---

[10]More information can be found at http://europa.eu/rapid/press-release_IP-17-1784_en.htm.

[11]Clive Humby, a data scientist, claimed in 2006 "Data is the new oil," now a popular maxim.

data.[12] A primary goal is to improve the users' knowledge about how their data are processed, not about the economic implication of data collection. Article 12, which deals with "Transparent information, communication and modalities for the exercise of the rights of the data subject" mainly specifies what the collecting entity should reveal to the subject whose data is collected. Though useful in some contexts, this information comes too late in others. Once an Internet user's intent to buy a product is known, a firm can charge him/her a higher price in a world where prices can be changed very quickly and be personalized. This is a major economic problem, complex to solve. So far, one could argue that firms and users are engaged in a kind of tacit barter agreement where firms deliver a valuable service (search engine, targeted proposals for products, targeted ads, access to friends, maps) in exchange for the uploading of users' data.[13]

Organizing Market Exchanges and Algorithmic Pricing Internet platforms (AirBnb, Uber, Amazon, etc.) organize exchanges through computerized tools. To illustrate, consider the Amazon platform Marketplace, which allows sellers (including Amazon) to display and sell their products. A main tool of Marketplace is the "Buy Box," which displays the price and the name of a particular seller corresponding to a search, as shown in Fig. 1 (see Chen et al. (2016) and the references therein). Being the Buy Box winner yields a significant advantage

Fig. 1 An example of the Buy Box on Amazon Marketplace from Chen et al. (2016)

[12]https://www.eugdpr.org/.

[13]Balinski and Demange (2018) study whether tax instruments are useful in reducing excessive level of data collection.

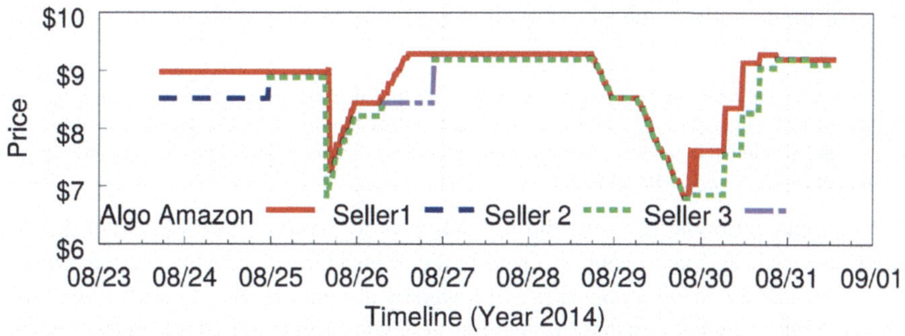

Fig. 2 Pricing strategies of Amazon and 3 sellers from Chen et al. (2016)

since 82% of the sales go through the Buy Box. The Buy Box winner, chosen by a mechanism/algorithm designed by Amazon, is not necessarily the seller with the cheapest product. The mechanism has unknown features, resulting in an informational gap between Amazon and the other sellers. Though, as a private firm, Amazon cannot be enforced to reveal its mechanism. Other sellers have two options: decipher Amazon's strategy or leave Marketplace.

Another gap exists between the sellers, specifically between those who set their prices using computer algorithms, known as *algorithmic pricing*, and those who do not. Sellers can adjust their prices at any time and have access to all the prices posted on Marketplace. Algorithmic pricing thus allows sellers to react very quickly to changes in the prices set by other sellers. For example, they can "track" others' prices, as illustrated in Fig. 2 between Amazon and seller 3 (resp. the red and green lines). There are periods of sharp decrease in the posted prices—each seller trying to be the cheapest but at the minimum rebate—and, as soon as the competitor raises its price, the other follows the increase. This can explain why there is no evidence that algorithmic pricing pushes prices down, on average: As first shown by Stigler (1964), dynamic pricing together with the price observation facilitates collusion.

Note finally that the sellers who do not rely on algorithmic pricing are unable to exploit all the available information on time. One may guess that they will tend to be eliminated.

Algorithmic Trading Algorithmic trading techniques, among which is high-frequency trading, are becoming prevalent on financial markets. In 2016, high-frequency trading is estimated roughly at 55% of the volume in the US equity markets and 80% of the volume in the foreign exchange futures (Millera and Shorter, 2016).

Each algorithmic trading here is a mechanism, which takes at each instant of time the observable market data, in particular the participants' orders, to generate orders, post prices, and execute trades. These automatic mechanisms interact between themselves and may result in snowball effects difficult to control, thereby calling for regulation. Pointing to the potential large and negative externalities generated by algorithmic trading, the EU Markets in Financial Instruments Directive (MiFID) prescribes:

> An investment firm that engages in algorithmic trading shall have the effective systems and risk controls suitable to the business it operates to ensure that its trading systems are resilient [...] and prevent the sending of erroneous orders [...] that may contribute to a disorderly market (Article 17(1) MiFID II).

The main rationale here for the regulation is to avoid a systemic event due to algorithm trading. Such event occurred in the "Flash Crash" of May 6 2010, which was generated by an erroneous order followed by a sequence of automatic reactions. The possible drawback of automatization first appeared in the stock market crash of October 19 1987, with the S&P 500 stock market index falling about 20%. At that time, a new technique, called portfolio insurance, was introduced, generating automatic sales when the market was falling, and some argued that it had a large role in the amplitude of the fall. But, according to the designers, the amplitude was due to a misunderstanding of portfolio insurance by the market's participants: Interpreting the initial automatic selling orders as bad news, the market's participants started to sell, triggering further automatic sales and creating a downward spiral (see Carlson 2007).

Currently, automatic trading has developed at a much larger scale, so algorithms and their interactions might create extreme disruptions difficult to control. There is a debate about the benefits (increased liquidity) and the costs of high frequency trading. The costs involve not only short-term price disruptions but also unfair competition because techniques such as (extremely fast) order cancellation allow high frequency traders to get advanced information on other traders' intentions. This led some to call for regulating the markets, by introducing a "Tobin" tax or by changing the price mechanism.[14]

4 Conclusion

The approach pioneered by Hurwicz and few other researchers in economics and game theory has a profound and growing impact in a variety of economic and social decisions. Such impact has been multiplied by the huge development in communications and computing facilities. Though, the use of heavily technical tools and the scale at which they are applied raise new challenging issues. Keeping Hurwicz's viewpoint, a new issue is now "But who will guard the algorithms"?

[14] I am grateful to Carmine Ventre for calling my attention on this point. For more information, see e.g. the book edited by Easley et al. (2014), in particular Chapter 4 by Golub, Dupuis and Olsen and Chapter 10 by Linton, O'Hara and Zigrand.

References

Arrow, K. J. (1950). A difficulty in the concept of social welfare. *Journal of Political Economy, 58*(4), 328–346.

Arrow, K. J. (1951) . *Social choice and individual values* (Vol. 12). London: Yale University Press (3rd ed. 2012).

Balinski, M.L., & Demange, G. (1989). An axiomatic approach to proportionality between matrices. *Mathematics of Operations Research, 14*(4), 700–719.

Bloch, F., & Demange, G. (2018). Taxation and privacy protection on Internet platforms. *Journal of Public Economic Theory, 20*(1), 52–66.

Carlson, M. A. (2007). A brief history of the 1987 stock market crash with a discussion of the federal reserve response. Downlable at https://www.federalreserve.gov/pubs/feds/2007/.../200713pap.pdf.

Chen, L., Mislove, A., & Wilson, C. (2016). An empirical analysis of algorithmic pricing on amazon marketplace. In *Proceedings of the 25th International Conference on World Wide Web* (pp. 1339–1349).

Cramton P., Shoham Y., & Steinberg, R. (Eds.). (2006). *Combinatorial auctions*. Cambridge, London: MIT Press.

Demange, G., & Gale, D. (1985). The strategy structure of two-sided matching markets. *Econometrica, 53*, 873–888.

Dubins, L. E., & Freedman, D. A. (1981). Machiavelli and the Gale-Shapley algorithm. *The American Mathematical Monthly, 88*(7), 485–494.

Dwork, C., Kumar, R., Naor, M., & Sivakumar, D. (2001). Rank aggregation methods for the web. In *Proceedings of the 10th International Conference on World Wide Web* (pp. 613–622). New York: ACM.

Easley, D., Prado, M. L. D., & O'Hara, M. (2014). *High-frequency trading: New realities for traders, markets and regulators*. London: Risk Books.

Gale, D., & Shapley, L. S. (1962). College admissions and the stability of marriage. *The American Mathematical Monthly, 69*(1), 9–15.

Goldman, J., & Procaccia, A. D. (2015). Spliddit: Unleashing fair division algorithms. *ACM SIGecom Exchanges, 13*(2), 41–46.

Hurwicz, L. (1960). *Optimality and informational efficiency in resource allocation processes* (pp. 27–46). Stanford, CA: Stanford University Press.

Hurwicz, L. (1973). The design of mechanisms for resource allocation. *The American Economic Review, 63*(2), 1–30.

Hurwicz, L. (2008). But who will guard the guardians? *The American Economic Review, 98*(3), 577–585.

McMillan, J. (1994). Selling spectrum rights. *Journal of Economic Perspectives, 8*(3), 145–162.

Milgrom, P. R., & Weber, R. J. (1982). A theory of auctions and competitive bidding. *Econometrica, 50*, 1089–1122.

Miller, R. S., & Shorter, G. (2016). *High-frequency trading: Overview of recent developments*. CRS Report, 44443.

Myerson, R. B. (2009). Fundamental theory of institutions: a lecture in honor of Leo Hurwicz. *Review of Economic Design, 13*(1–2), 59.

Myerson, R. B., & Satterthwaite, M. A. (1983). Efficient mechanisms for bilateral trading. *Journal of Economic Theory, 29*(2), 265–281.

Nisan, N., & Ronen, A. (2001). Algorithmic mechanism design. *Games and Economic Behavior, 35*(1–2), 166–196.

Page, L., Brin, S., Motwani, R., & Winograd, T. (1999). *The PageRank citation ranking: Bringing order to the web*. Venue: Stanford InfoLab.

Roth, A. E. (2008). Deferred acceptance algorithms: History, theory, practice, and open questions. *International Journal of Game Theory, 36*(3–4), 537–569.

Roth, A. E., Sönmez, T., & Ünver, M. U. (2004). Kidney exchange. *The Quarterly Journal of Economics, 119*(2), 457–488.
Stigler, G. J. (1964). A theory of oligopoly. *Journal of Political Economy, 72*(1), 44–61.
Thomson, W. (1990). The consistency principle. Game *Theory and Applications, 187*, 215.
Varian, H. R. (2007). Position auctions. *International Journal of Industrial Organization, 25*(6), 1163–1178.
Vickrey, W. (1961). Counterspeculation, auctions, and competitive sealed tenders. *The Journal of Finance, 16*(1), 8–37.
Wilson, R. B. (1967). Competitive bidding with asymmetric information. *Management Science, 13*(11), 816–820.
Young, H. P., (1987). On dividing an amount according to individual claims or liabilities. *Mathematics of Operations Research, 12*(3), 398–414.

Incentive Compatibility on the Blockchain

Jonathan Chiu and Thorsten Koeppl

1 Blockchain as a Distributed Ledger

A blockchain is a decentralized ledger that digitally records the ownerships of assets and the transfer thereof. Owing to its digital nature, an ownership record is simply a string of bits which can easily be copied and re-used repeatedly, leading to a *double-spending problem*. In a centralized system such as PayPal, this problem can be solved by relying on a trusted third-party to manage the ledger. This trusted central authority validates and enforces all transactions, preventing users from tampering with the ledger.

Blockchain systems aim to maintain a digital ledger without the need for a designated party to keep records and enforce the transfer of ownership. Instead, transactions are verified and processed by a network of potentially anonymous validators. In this system, a *block* is simply a set of transactions that transfer ownership between users in the ledger. From individual blocks a chain is formed by *time-stamping*. The blocks are linked together in a sequence where each block depends on the previous block in time. This creates a full, historical record of transactions where no past block can be changed without also changing all the subsequent blocks.

The ownership of assets is protected by the use of basic cryptographic principles. An owner holds a private–public key pair where the private key is kept secret and controls the entry in the blockchain, while the public key proves ownership to any

J. Chiu
Bank of Canada, Ottawa, ON, Canada
e-mail: jchiu@bankofcanada.ca

T. Koeppl (✉)
Queen's University, Kingston, ON, Canada
e-mail: thor@econ.queensu.ca

W. Trockel (ed.), *Social Design*, Studies in Economic Design,
https://doi.org/10.1007/978-3-319-93809-7_20

other party. A transaction is conducted by transferring an entry in the blockchain to a different private–public key pair.

In a truly decentralized blockchain such as Bitcoin, anyone can access the public ledger, verify its information, and even serve as a validator who updates the chain with a new block. The blockchain itself lives on a distributed network where users interact peer-to-peer and each keeps a copy of the ledger. This extreme redundancy introduces resiliency into the system. As long as one of the peers is live on the network the ledger can be accessed and transactions can be conducted.

The key challenge is to design the rules for updating the blockchain so that it is hard to tamper with and, thus, that users trust the information contained in the ledger. The security of a blockchain is based on three elements: (1) a consensus protocol, (2) confirmation lags, and (3) a reward scheme.

First, in a decentralized network, the system needs to ensure that, when new transactions are incorporated into the blockchain, a common consensus about the new update is agreed upon among all users. To reach such a consensus in a decentralized setting, validators are asked to compete for the right to append a new block to the chain. This competition can take various forms. In the most common consensus protocol, Proof-of-Work (PoW), this process is called mining. Validators, also called *miners*, need to solve a computationally difficult problem. The winner of this mining competition has the right to update the chain with a new block. In addition, the consensus protocol prescribes that the "longest" chain proposed by the network will be accepted as the trusted public record. A chain is considered to be the longest one if it has incorporated the most "work" by miners or, equivalently, has burnt the most resources. This ensures that there is agreement within the peers of the network what constitutes the true history of all past transactions.

The second element is a confirmation lag. This lag helps prevent users from altering the history of transactions through double spending. After having transferred ownership of an asset, a user can attempt to convince other users to accept an alternative history in which the transaction has not been conducted. For example, in the case of Bitcoin, a dishonest user can try to revoke a payment in Bitcoin after he has received goods from a seller. To do so, he needs to create an alternative history of transactions by winning the mining competition against all honest miners. If such an attack succeeds, the dishonest user effectively steals the goods from the seller, as he gets the goods without paying the seller. The seller can protect himself against double spending by delaying the delivery of goods until multiple confirmations of the trade have been recorded in the blockchain. This is the case when several new blocks have been stacked onto the block that contains the original record of the transaction. Double spending would then require waiting for the delivery of the good and then replacing all these other blocks and the original block containing the transaction.

The final aspect is that mining needs to be properly incentivized. Under a PoW protocol, the probability of winning the right to update a block is proportional to the fraction of computational power owned by a miner. Hence, sufficient overall mining activities help discourage dishonest behaviour. Since mining is costly, mining has to be induced by offering rewards. These rewards can either be financed by seignorage—issuing cryptocurrency or tokens—or by collecting transaction fees from traders. Importantly, increasing rewards will increase the effort and, hence, the computational investment by miners making it harder to double spend.

The key innovation of blockchain technology is to put the users in the system in charge of guarding the system itself. Nakamoto (2008) formalized a solution to the problem of having to trust a third-party as the guardian of a payment system. Beyond the original Bitcoin proposal, it has become clear that his somewhat brilliant idea applies more broadly. Ironically, it arrived at about the same time as Leo Hurwicz delivered his Nobel lecture that posed the problem of "But Who Will Guard the Guardians?" once again (see Hurwicz, 2008). Our attempt here is to go full circle by linking the idea of a blockchain back to an economic mechanism design problem.

In what follows, we first express the PoW protocol as a simple Cournot game of mining. We then formalize the double spending problem and show that it can be summarized as a simple incentive compatibility constraint given the Cournot game of mining. We then briefly describe how this constraint manifests itself in two examples. The first example concerns a securities settlement system where both the asset transfer and the payment are recorded on a blockchain. The second one is the classic example of using a cryptocurrency for payments as intended in the original Bitcoin proposal. We close out our contribution by briefly discussing some further issues.

2 Modelling the Proof-of-Work Protocol

We first set up a simple game form[1] that captures the basic idea behind the PoW protocol where consensus on the blockchain is reached by competition called mining. Time is continuous and there are M miners. The *protocol* specifies a computational problem and sets a difficulty D for the problem. It also specifies a reward R for the first miner to solve the problem and decrees that the first one to solve the problem is allowed to add a new block to the blockchain.

At the start of time, each miner invests a quantity q_i, $i = 1, \ldots, M$, into computing power to solve the computationally difficult problem. We denote the price per unit of computing power by α. The probability that miner i with computing power q_i solves the computational problem in a given time t is assumed to follow an exponential distribution with parameter q_i/D. Hence, a miner can solve the problem before t with probability

$$F(t) = 1 - e^{-\frac{q_i}{D}t}. \tag{1}$$

The expected time for a solution by miner i is then given by D/q_i.

[1]We use this terminology in the spirit of Hurwicz (2008) since the protocol specifies the strategy domain for the individual actors and the outcome function that maps strategies into outcomes.

The first solution (or proof-of-work (PoW)) among all M miners is then also an exponentially distributed random variable with parameter $\frac{1}{D}\sum_{i=1}^{M} q_i$. Given $\{q_i\}_{i=1}^{M}$, the expected time needed to complete the PoW is

$$\frac{D}{\sum_{i=1}^{M} q_i}, \tag{2}$$

with miner i having the probability

$$\rho_i \equiv \frac{q_i}{\sum_{i=1}^{M} q_i}. \tag{3}$$

of being the first one to solve the problem. By adjusting the parameter D, the protocol can thus ensure that—on average—a solution is found in a particular time interval.

The PoW protocol essentially formalizes an exponential race where miners decide how much to invest into the race to win a reward R. This implements a simple Cournot game where each miner maximizes his payoff

$$\max_{q_i} \rho_i R - \alpha q_i \tag{4}$$

taking as given the investments of all other miners $-i$. The symmetric Nash equilibrium to the Cournot game is given by

$$q_i = Q \equiv \frac{M-1}{\alpha M^2} R \tag{5}$$

which leads to the following result.

Lemma 1 *The total mining cost is given by*

$$C \equiv \alpha Q M = \frac{M-1}{M} R$$

and the expected time to solve a block (i.e. block time) is

$$T \equiv \frac{\alpha M}{M-1} \frac{D}{R}.$$

This allows us to immediately derive some comparative statics for the PoW protocol. Of interest are the role of (1) computing costs, (2) rewards, and (3) total mining capacity. Most importantly, as the number of miners increases, the total mining cost converges to R. Competition dissipates all rents from mining, with miners earning zero expected profits when $M \to \infty$. These results are summarized below.

Proposition 1 *The block time D decreases when mining rewards R increase, the number of miners M increases or when the cost of computing power α falls.*
Total mining costs are unaffected by the cost of computing α, but increase proportionally with rewards R.
Total mining costs also increase with the number of miners M and mining profits converge to 0 *as* $M \to \infty$.

The mining game we have formalized takes place for each block and does not directly depend on block time and the size of the block which determines how frequently blocks are added and how many transactions can be included into a single block. It does, however, crucially depend on the reward R for solving a block. This reward can be financed either through the issuance of tokens on the blockchain or through transaction fees posted by users for including their transactions into a block.

For tokens to serve as rewards, they need to have some real value. This value derives traditionally from their use as a payments instrument or, in other words, *cryptocurrency*. Users exchange their tokens against real goods. More recently, *initial coin offerings* have used tokens akin to shares in crowd investments where future tokens can be seen as additional shares being offered. Transaction fees arise when block size and block time are used to make settlement a scarce resource. Restricting block size and lengthening block time create *congestion* on the blockchain and, thus, exploit users willingness to pay for fast settlement.

In general, what matters is the reward per block. The difficulty D controls the reward that is available over a fixed amount of time. Note that D cannot be arbitrarily short. In a distributed network, due to network latency, it takes time for data to get from one designated point to another. Hence, some time delay is necessary to communicate updates of the blockchain and to ensure that all miners and users work with the same information on the blockchain. A lower difficulty speeds up settlement, but reduces user willingness to post transaction fees, and thus the reward. Similarly, any reward from newly created tokens needs to be split across blocks. Keeping block rewards fixed over a time interval, a faster block time will lower the rewards available per block.

Nakamoto (2008) was first to introduce the idea of using a PoW protocol for achieving consensus on a blockchain with his Bitcoin proposal. The PoW problem to be solved is to find a particular output to the SHA256 algorithm. This algorithm takes an input of any bit size and hashes it to produce a random, but unique and not invertible 256 bit output. Bitcoin's protocol requires a miner to produce a hash with a certain number of leading zeros using the transaction data in the block, a summary of the previous block, the blockheader, plus an additional random number. The difficulty is given by how many leading zeros the hash has to have. It is adjusted every 2016 blocks so that the average time it takes miners to solve a block is about 10 min.

The reward that the Bitcoin protocol offers for miners comes from two sources. First, each block includes a special transaction creating a certain number of new Bitcoins. In other words, winning a block creates seignorage for the winning miner. Second, users pledge transaction fees so that their transaction are included quickly into a block. The winner of a block also wins these transaction fees. As pointed out in Proposition 1, if the value of Bitcoin relative to the dollar costs of computing power increases, competition between miners will go up. The protocol then has to raise the difficulty for solving the problem to ensure that the target for a block time of 10 min is achieved on average.

3 Ruling Out Incentives to Cheat

3.1 Double Spending Problem

In a *permissionless* blockchain, any user can act as a miner to validate and process transactions. As pointed out before, cryptography ensures that only private key holders can transfer assets recorded on the blockchain. Therefore, a dishonest user cannot simply steal assets without stealing someone else's private keys. However, a user can still remove transactions that have been initiated by himself or by other users. To do so, he needs to alter the blockchain and have all other users believe in the altered blockchain. This refers to what we call a *double-spending problem.*

We will first describe the general problem of a user double spending and then consider two specific examples of the problem. Consider a blockchain based on a PoW protocol with mining reward R. Suppose that double spending requires to win the competition N times and gives an *additional payoff* denoted by Δ. The payoff for a user trying to double spend is given by

$$\Pi(N, R) = \max_{\tilde{\mathbf{q}}} \mathscr{P}[\tilde{\mathbf{q}}; N, Q(R)](NR + \Delta) - c(\tilde{\mathbf{q}}). \tag{6}$$

Here, the user chooses a vector of computing power $\tilde{\mathbf{q}}$ for the N blocks to maximize the net expected return, where the probability of success, $\mathscr{P}(\tilde{\mathbf{q}}; N, Q)$ increases with $\tilde{\mathbf{q}}$ and decreases with N and Q. When the user successfully wins the mining game $N + 1$ times, he gains Δ from altering the blockchain. But he also receives all the block rewards from winning the mining game $N + 1$ times. Finally, the user needs to incur a mining cost $c(\tilde{\mathbf{q}})$.

A user has an incentive to double spend whenever $\Pi(N, R) > 0$. Hence, in order to rule out that users tamper with the blockchain, we require that

$$\Pi(N, R) \leq 0 \tag{7}$$

which we call a *no double spending constraint* (NoDS). This is akin to an incentive constraint where the design of the blockchain—in particular rewards and the confirmation lag—plays a crucial role in ruling out incentives for users to engage in double spending.

3.2 Example 1: Blockchain for Securities Settlement

We first study an example in which a blockchain is used for settling asset trades (Chiu and Koeppl, 2018). In any security settlement system, it is important to avoid settlement failures where the seller of a security fails to deliver the security while receiving payment, or the buyer of a security fails to deliver payment while receiving the security. A delivery versus payment (DvP) mechanism typically ensures that the security and the cash are exchanged simultaneously in order to avoid such a settlement failure. When both the cash and the asset are recorded on the same blockchain, DvP can be enforced by a self-enforcing, autonomous program often called a *smart contract*. In particular, the two legs involving the transfer of security and the cash payment are executed either in their entirety or not at all. In database systems, this is referred to as an *atomic transaction* which is an indivisible and irreducible series of database operations such that either all or none of them occurs.

With settlement on a blockchain, when two counterparties agree to exchange a payment for a security, they jointly broadcast a transaction message about the terms of trade to the network so that the miners will validate the transaction and update the blockchain accordingly. Since the transfer of the asset and the payment are linked in an atomic transaction, it is infeasible for one side to undo the joint transfers unilaterally. However, any counterparty can eliminate the entire transaction by mining a block that changes ownership of one of the legs (cryptocurrency or security) so that the original transaction is invalid. If such double spending is successful before the original transaction has been included in the blockchain, the transaction has effectively never occurred.

The buyer in a security trade has an incentive to double spend if the agreed price is higher than the current value of the security, while a seller has an incentive to do so when the current value of the security is higher than the price received in the trade. For example, suppose the two counterparties originally have agreed to trade the security at a price p. After the arrival of new information, the buyer's valuation of the security becomes v_b while the seller's valuation becomes v_s. In this example, the buyer has an incentive to revoke the trade—and effectively default—if $p - v_b > 0$. The seller wants to cancel the trade when $v_s - p > 0$. Hence the maximum incentive for one of the two sides to double spend is

$$V = \max\{p - v_b, v_s - p\}. \tag{8}$$

As pointed out, an investor can revoke a transaction by simply including a message into a block that changes the ownership of the security or the cryptocurrency used for payment. Hence, a dishonest investor needs to win the mining game against honest miners just once. The probability of a successful double spend is therefore

$$\mathscr{P}(\tilde{q}; N = 1, Q(R)) = \frac{\tilde{q}}{\tilde{q} + QM} \tag{9}$$

where $\tilde{q}$ is the computing power invested by the dishonest investor who attempts to revoke the transaction. In general, a dishonest user might be subject to a higher mining cost than a regular miner. In addition, a dishonest investor may suffer a reputational damage or penalty if cheating is detected. We can capture this by assuming that the mining cost per block is given by

$$c(\tilde{q}) = \Gamma + \hat{\alpha}\tilde{q} \tag{10}$$

with $\Gamma \geq 0$ and $\hat{\alpha} \geq 1$.

The dishonest user therefore solves

$$\Pi(N = 1, R) = \max_{\tilde{q}} \frac{\tilde{q}}{\tilde{q} + QM}(R + V) - \Gamma - \hat{\alpha}\tilde{q}. \tag{11}$$

For an interior solution we obtain

$$\tilde{q} = QM\left(\sqrt{\frac{V + R}{\hat{\alpha}QM}} - 1\right) \tag{12}$$

with the gain from double spending given by

$$\Pi(N = 1, R) = \frac{\sqrt{\frac{V+R}{\hat{\alpha}QM}} - 1}{\sqrt{\frac{V+R}{\hat{\alpha}QM}}}(V + R) - \Gamma - \hat{\alpha}QM\left(\sqrt{\frac{V + R}{\hat{\alpha}QM}} - 1\right). \tag{13}$$

Proposition 2 *Suppose the cost function for double spending is given by* $c(\tilde{q}) = \Gamma + \hat{\alpha}\tilde{q}$ *while the cost function for honest mining is* $c(q) = q$.
As $M \to \infty$, *the NoDS constraint for users is given by*

$$V \leq \Gamma + 2\sqrt{R\hat{\alpha}\Gamma} + R(\hat{\alpha} - 1).$$

The proposition implies directly that—in order to discourage users from cheating—the fixed cost Γ or the marginal cost $\hat{\alpha}$ have to be sufficiently high. In particular, if double spending has no cost disadvantage over honest mining, one cannot rule out double spending in a securities settlement system based on a blockchain built on a PoW protocol.

3.3 Example 2: Blockchain for Cryptocurrency in Goods Transactions

We now consider a different example where a blockchain is used to record cryptocurrency transfers when purchasing real goods (Chiu and Koeppl, 2017). DvP in this context is not automatic anymore as the ownership of goods is not recorded digitally on the blockchain. Consider a spot trade where a buyer agrees to pay p units of cryptocurrency to a seller for a certain amount of goods. The buyer can cheat by mining a block where the transfer of cryptocurrency is not included, but instead the cryptocurrency is spent back to the buyer. If the attempt fails, the buyer pays the seller, but still gets the goods. If the attempt succeeds, the buyer gets the goods without paying the seller at all. Hence, the double-spending payoff for the buyer is the price of the goods, p, which effectively means he steals the good.

Since there is no DvP, a seller can make double-spending more difficult by introducing a confirmation lag. The goods are to be delivered only after the transaction has been confirmed sufficiently many times—say $N - 1$ times—in the blockchain. This forces the buyer to win the mining game N times in a row, keeping it a secret from the rest of the miners each time he finds a solution in the first $N - 1$ computational problems. Hence, we call a double spending attempt *secret mining*. This way the seller is fooled to deliver the good after $N - 1$ confirmations, only to lose the payment in the following block.

Suppose now for simplicity that the cost of secret mining for a dishonest buyer is the same as that of a regular miner

$$c(\tilde{q}) = \tilde{q} \tag{14}$$

where we have normalized $\alpha = 1$. If there are no confirmation lags (i.e. $N = 0$), then Proposition 2 from the Sect. 3.2 above immediately implies that buyers have no incentives to double spend if and only if $p < 0$. Hence, confirmation lags are necessary for preventing double spending.

To see the effect of a confirmation lag, suppose $N = 2$ so that the buyer needs to win the mining game twice in order to reclaim the payment p. If he succeeds, he earns $p + 2R$, while if he fails, he earns 0. The dishonest buyer chooses his investment in computing power for the first and second blocks $(\tilde{q}_1, \tilde{q}_2)$ sequentially. The probability of a successful double spending attempt is given by

$$\mathscr{P}(\tilde{q}_1, \tilde{q}_2; N = 2, Q) = \frac{\tilde{q}_1}{\tilde{q}_1 + QM} \frac{\tilde{q}_2}{\tilde{q}_2 + QM} \tag{15}$$

as the buyer needs to win both blocks in order to revoke the payment.[2] We can solve the problem backward starting from the second block. Conditional on having solved the first block, the optimal secret mining investment $\tilde{q}_2$ for the second block is a solution to the following problem

$$\max_{\tilde{q}_2} \frac{\tilde{q}_2}{\tilde{q}_2 + QM}(p + 2R) - \tilde{q}_2. \tag{16}$$

[2]This is only an approximation to the decision problem of double spending. See Sect. 4 for a discussion of the issue.

Hence, the optimal investment is positive and given by[3]

$$\tilde{q}_2 = QM\left(\sqrt{\frac{p+2R}{QM}} - 1\right). \tag{17}$$

The expected payoff from secret mining is also positive and given by

$$\Pi_2 = QM\left(\sqrt{\frac{p+2R}{QM}} - 1\right)^2. \tag{18}$$

Given this solution, the optimal investment in secret mining for the first block solves

$$\max_{\tilde{q}_1} \frac{\tilde{q}_1}{\tilde{q}_1 + QM}\Pi_2 - \tilde{q}_1. \tag{19}$$

Hence, the optimal investment is given by

$$\tilde{q}_1 = \max\left\{QM\left(\sqrt{\frac{p+2R}{QM}} - 2\right), 0\right\} \tag{20}$$

and the expected payoff from mining secretly is

$$\Pi(N=2, R) = \max\left\{QM\left(\sqrt{\frac{p+2R}{QM}} - 2\right)^2, 0\right\}. \tag{21}$$

It follows immediately that $\tilde{q}_1 < \tilde{q}_2$ because the chance of successful double spending has gone up once the buyer has successfully mined the first block. The following proposition derives a constraint on the buyer to rule out double spending.

Proposition 3 *Suppose there is one confirmation lag so that $N = 2$ and suppose the mining costs for double spending are $c(\tilde{q}) = \tilde{q}$.*
As $M \to \infty$, the NoDS constraint for users is given by

$$p < 2R.$$

[3]Note also that the buyer—having started to mine secretly—has no incentive to announce that he has found a block. He would immediately get the block reward R but void the transaction which is worth at least p. Hence, conditional on winning the first block with secret mining, the buyer has an incentive to keep on mining secretly independent of $p > 0$. For details, see Chiu and Koeppl (2017).

Therefore, requiring confirmation in at least one block—which means that a double spending has to solve $N = 2$ blocks to be successful—the transaction is double-spending proof if the payment size is small relative to the mining rewards. Chiu and Koeppl (2017) show that, for any given N, the NoDS constraint is given by

$$p < RN(N - 1). \tag{22}$$

Hence, larger transaction sizes require longer confirmation lags or higher rewards for mining to ensure that there are no incentives for buyers to double spend.

4 Discussion

4.1 The Costs of Cryptocurrencies

Keeping records on a blockchain is not a free lunch. It is necessary to offer rewards to rule out double spending which directly or indirectly increase the cost of maintaining a distributed ledger. In case of a cryptocurrency, rewards can be offered by seignorage. Such seignorage causes inflation which levies indirect costs in form of an inflation tax on users. However, there are also direct costs that arise from investment into computational power (mainly energy) that uses up most of the revenue from seignorage. A traditional currency does not waste seignorage, but raises revenue for the issuer. In the case of a modern central bank, this generates profits above operational costs that can be used to offset other distortionary taxes used by the government. A quantitative assessment has shown that a low-inflation currency regime dominates any cryptocurrency (see Chiu and Koeppl (2017)).

Still cryptocurrencies can exploit a trade-off between settlement speed and rewards to deter double spending. As shown in Sect. 3.3, for any given transaction increasing the confirmation lag reduces the reward necessary for a tamper proof blockchain. Notwithstanding, increasing confirmation lags has a time cost as the delivery of goods is delayed.

This points to settling transactions in cryptocurrencies being a public good. Settling one transaction does not preclude settling more transactions at any given time. Interestingly, double spending is driven by transactions that have the largest incentives to do so. Hence, all other transactions with lower incentives can free-ride once double spending has been ruled out. In contrast, however, a single transaction with very large incentives to double spend requires very large rewards for mining blocks. These costs are then indirectly borne by all other users which can make using a cryptocurrency unnecessarily expensive. This implies that a cryptocurrency works best for a fairly homogeneous group of transactions with small incentives to double spend such as retail payments (see again Chiu and Koeppl, 2017).[4]

4.2 Double Spending as a Poisson Race

For simplicity, we have modelled secret mining as an exponential race for each update against a group of honest miners. In order to double spend, a user had to win the race N times in a row, but kept his result secret. This is not entirely accurate when looking at actual PoW protocols employed for blockchain technology. These users can catch up and only need to generate at least N blocks faster than all the other miners. Hence, secret mining is really a Poisson race against a fringe of honest miners that play a sequence of simple exponential races to find one block in each race.

This is related to the so-called “51% attack” problem. If a miner controls more than half the computational power among all miners, confirmation lags in theory lose their power in controlling double spending incentives. The dishonest miner creates an arrival rate that is larger than those of the other honest miners combined. This implies that he eventually will outrun other miners for sure in generating a longer chain and, thus, can always cheat by double spending (see, for example, Rosenfeld (2014)). However, from an economic point of view, this requires that a dishonest miner has deep pockets and is risk neutral. These assumptions tend to be unrealistic and in practice users have little economic incentives to launch such an attack especially when the computational investment by other miners is large.

This is reflected in our approach where larger confirmation lags reduce the probability of successful double spending. More generally, rather than ruling out double spending altogether it could be sufficient to ensure that double spending only occurs with a sufficiently small probability. Interestingly, there could then even be competition for double spending where there are multiple dishonest users. If coordination of such behaviour is difficult, then double spending from the perspective of an individual transaction is small.

4.3 Other Consensus Protocols

We have focussed exclusively on PoW protocols to achieve consensus. Many other protocols have been discussed that try to save on the costs associated with running a blockchain. Protocols based on *Proof-of-Stake* (PoS) allocate the right to update the

[4]In other applications, like securities settlement systems, it may be necessary to counteract the public good character of settling on a blockchain. One has to create some congestion so that users have an incentive to post transaction fees. Settlement on a blockchain then becomes a club good.

blockchain randomly across users. The chance of any user to win the right is linked to his stake in the system; for example, the number of units of cryptocurrency the user owns. However, these alternative systems usually do not possess a key feature of PoW: one needs to spend a large amount of resources to be successful in cheating and being unsuccessful means that one has incurred a large, irretrievable sunk cost.

Another alternative is a voting type arrangement where a majority or super-majority of users are needed to agree on a new block. The classic protocol in this area is *Practical Byzantine Fault Tolerance* (PBFT) where for an update 2/3 of the users in a network need to agree that 2/3 of the users have agreed on a new block.[5] However, any blockchain with too many nodes cannot implement such a protocol as it introduces too much latency due to extreme communication requirements. Consequently, such protocols have been explored mainly in "closed" or "permissioned" blockchains where a small group of known validators are charged with updating the blockchain.

References

Chiu, J., & Koeppl, T. (2017). The Economics of Cryptocurrency – Bitcoin and Beyond. Queen's University, Working Paper 1389

Chiu, J., & Koeppl, T. (2018). Blockchain-based Settlement for Asset Trading. Queen's University, Working Paper 1397

Hurwicz, L. (2008). But who will guard the guardians? *American Economic Review, 98*, 577–585

Lamport, L., Shostak, R., & Pease, M. (1982). The Byzantine generals problem. *ACM Transactions on Programming Languages and Systems, 4*, 382–401

Nakamoto, S. (2008). Bitcoin: A peer-to-peer electronic cash system. White Paper

Rosenfeld, M. (2014). Analysis of hashrate-based double spending. arXiv:1402.2009

[5]The consensus protocol is built to tolerate failures up to 33% of nodes in a distributed system. This is the theoretical limit of failures such a system can sustain when having some form of syncronization (see Lamport et al., 1982).

Contextual Mechanism Design

Pierfrancesco La Mura

1 Introduction

We argue that the context in which a mechanism is meant to be implemented can have a significant influence on its performance, and hence should be taken into explicit account in its design phase. Such influences can arise not only from the physical and information-theoretic context in which the participants operate, but also from the subjective context in which they make their decisions.

To exemplify the first type of contextual influences consider the following decentralized market scenario. A bidding team, comprised of Alice and Bob, is active on separate auction markets for complements and substitutes. Alice and Bob would like to coordinate their actions (namely, whether to buy or not) depending on local information (say, which good is on sale on that day on their respective market), but once they reach the two markets they have no opportunity to communicate. What is the optimal performance they can achieve as a distributed bidding team? As we shall see, an answer to this question cannot be given without a specification of the physical and information-theoretic context in which the team is assumed to operate, and in particular of which classes of signals from their shared environment may be available to the team members.

To exemplify the second type of contextual influences consider the two following auction scenarios. In the first scenario a seller has a single object for sale, and two bidders (A and B) have a value of 1 for the object. The seller proposes the following decreasing-price mechanism: at time $t \in 1, 2, ..$ the price of the object is γ^t, with $\gamma \in (0, 1)$. Bidder A gets a chance to buy (at the current price) whenever t is odd, while bidder B can only buy when t is even. As soon as the object is purchased by

P. La Mura (✉)
HHL Leipzig Graduate School of Management, Leipzig, Germany
e-mail: plamura@hhl.de

W. Trockel (ed.), *Social Design*, Studies in Economic Design,
https://doi.org/10.1007/978-3-319-93809-7_21

either bidder at time t, the two bidders must bargain on how to split the resulting surplus $1 - \gamma^t$. They bargain with alternating offers, starting with the bidder who purchased the object. Both bidders have a time discount factor given by $\delta \in (0, 1)$, that only applies to the bargaining stage, with $\delta < \gamma$.

The above mechanism is a combination of a Dutch auction (with alternating bidders) and infinite-horizon bilateral bargaining. The reader will also recognize in it a structure similar to that of the Centipede game (Rosenthal, 1981), and indeed—just as in the Centipede—the only prediction compatible with common knowledge of rationality (in the sense of Aumann, 1995) is that bidders should purchase at the current price as soon as they are able to, then bargain over the resulting surplus as first movers.

Should the seller really expect, when offering an object for sale through a similar mechanism, that it performs as expected on theoretical grounds? The empirical analysis of the Centipede suggests that standard game-theoretic predictions here could easily be misleading, and that actual decisions would depend much on context, even to the point that two indistinguishable scenarios, from the perspective of a hypothetical observer/experimenter/designer, could lead to dramatically different predictions.

As a second exemplary scenario consider a sequential, first or second-price auction with two identical objects for sale, namely, 1 and 2 (auctioned in the same order), and three bidders, namely, A, B, and C, with independent private values uniformly distributed in [0, 1]. Weber's Martingale Theorem (Weber, 1983; Weber and Milgrom, 2000) predicts that, if the price paid for the first object is P_1, this should also be the expected price for the second object. Yet, there are well-documented order effects (e.g., the "afternoon effect" of decreasing prices in art or wine auctions) that deviate from this theoretical prediction (Ashenfelter, 1989; Deltas and Kosmopoulou, 2004; Andersson and Andersson, 2017). How to explain and predict such effects in a principled way?

2 Decentralized Bidding in Markets for Substitutes and Complements

A first source of contextual effects that are potentially relevant to the mechanism designer is given by the type of physical and information-theoretic environment in which the mechanism is meant to operate. In order to demonstrate the importance of such influences on the performance of a mechanism let us consider the decentralized auction setting mentioned in the introduction.

A team comprised of Alice and Bob is active on two separate markets: Alice on market 1, and Bob on market 2. Once they reach their respective markets they have no opportunity to communicate with each other. On each market, with equal probability and independently of what happens on the other market, exactly one of three types of object is put on auction, say: a table, a sofa or a bed. The expected price of each item on either market, when available, is always P.

Let us assume that Alice and Bob strive to maximize the expected value from their joint market activity, net of the total cost for procuring the objects. Specifically, let us assume the following team payoffs from procuring different combinations of the three items. If they jointly procure a single object, say, one table, their team payoff is equal to $s = v - P > 0$, namely, the (positive) surplus from winning a single object at a price of P. If they procure two identical objects (say, two tables) at the same unit price of P, then the second one brings no additional value, and the team payoff is given by $s - P$. If they procure two different objects, say, a table and a bed, then their payoff is given by $2s + \epsilon$, with $\epsilon > 0$, reflecting the fact that for Alice and Bob the two objects are partial complements. If they buy no objects, then the team payoff is zero.

The two conditional payoff tables below summarize what Alice and Bob obtain by buying (B) or not buying (N), respectively, depending on whether the two markets offer the same type of object for sale (Same) or different types of object (Different).

Same	B	N
B	$s - P$	s
N	s	0

Different	B	N
B	$2s + \epsilon$	s
N	s	0

Alice and Bob do not know (and, in the absence of communication, have no way to find out) whether the two markets have the same type of object for sale, and hence the team is playing with the payoffs in the left-hand side table, or different types, in which case the payoffs are those on the right-hand side. Finally, let us assume that the expected price P is large with respect to the potential surplus s. What is the best performance that Alice and Bob can achieve as a decentralized bidding team? What strategy should they follow? The answer, perhaps surprisingly, turns out to depend on the physical and information-theoretic context in which they operate. Specifically, if Alice and Bob are able to share an entangled quantum state before reaching their respective markets, and to make their actions after accessing the local information contingent on the outcome of measurement from their part of the quantum state, then they can win just as often—but obtain a larger surplus—than they would in case they only had access to classical signals when implementing their decentralized strategy.

To see how this can be the case, observe that a good team strategy must avoid whenever possible the large negative payoff $s - P$, and in fact if the participants operate in a classical (non-quantum) environment one can easily see that, for P sufficiently large, all efficient strategies must be of the following type. The three goods are partitioned into two nonempty subsets, which are assigned to Alice and Bob, respectively. Then Alice and Bob only buy if the object on sale on their respective market belongs to the subset assigned to them.

For instance, Alice could be appointed to buy if, and only if, the object on sale on her market is a table or a bed, while Bob if and only if the object on sale on his market is a sofa. With probability $1/3$ the two markets will offer the same type of object for sale, in which case according to the above strategy Alice and Bob will procure exactly one item, obtaining a payoff of s. With probability $2/3$ Alice and Bob will bid for different objects, in which case: with probability $(1/3) * 1 = 1/3$ both will buy, obtaining a team payoff of $2s + \epsilon$; with probability $1/3 + 0 = 1/3$ only one of them will buy, obtaining a team payoff of s; and with the remaining probability neither will buy, and the team payoff will be zero. Hence, the expected payoff under the above strategy is given by

$$(1/3)s + (2/3)((1/3)(2s + \epsilon) + (1/3)s) = s + (2/9)\epsilon$$

This game belongs to a class of team decision problems (namely, those representable via Kuhn trees) in which correlated equilibrium payoffs coincide with the convex hull of Nash equilibrium payoffs (La Mura, 2005; Brandenburger and La Mura, 2016). Hence, the above is also the efficient team payoff in case Alice and Bob are able to correlate their strategies through classical (non-quantum) signals from their shared environment.

Let us compare the above scenario with one in which Alice and Bob, in addition to classical signals, can also make use of quantum signals. Specifically, let us assume that Alice and Bob, before they separate to go to their respective markets, are able to share and preserve an entangled pair of quantum bits (qubits), on which they can later perform one of several alternative measurements. Then Alice and Bob could make their actions (after they learn which good is on sale on their respective market) contingent on the outcome of measurement of their own qubit. The following conditional probability tables, for the two cases in which the measurements operated on the two qubits are the same (left) or different (right), can be realized quantum-mechanically with a suitable preparation of the entangled quantum bit pair and choice of three alternative measurements on each qubit (say, M_1, M_2, and M_3).

Same	Yes	No
Yes	0	1/2
No	1/2	0

Different	Yes	No
Yes	3/8	1/8
No	1/8	3/8

If Alice and Bob operate measurement M_1, M_2, or M_3 when the good on sale on their market is a table, sofa or bed, respectively, and then buy if, and only if, the outcome of their measurement is a Yes, then the expected payoff under such quantum-correlated strategy is given by

$$(1/3)s + (2/3)((3/8)(2s + \epsilon) + (1/4)s) = s + (1/4)\epsilon.$$

This is strictly larger than the optimal team payoff calculated above. Hence, we conclude that a decentralized team of bidders active on separate markets in the presence of both complementary and substitutable items can procure a higher expected surplus if a suitably prepared quantum state is shared in advance among team members. In turn, this suggests that the availability of quantum information-theoretic resources (or the designer's ability to prevent participants from accessing them) may become an important factor in the design and performance of high-speed, high-frequency electronic marketplaces where communication among market participants would be too slow to be useful.

3 Decisions with Contextual Preferences

Turning to the second family of contextual effects mentioned in the introduction, and motivated by the two auction scenarios already discussed there, we would like to provide an extension of the ordinary mechanism design formulation that accounts for an element of context in the participants' preferences. Specifically, we would like to take the view that preferences are only well defined with respect to acts mapping states of Nature into lotteries over a set of subjective consequences, which may or may not coincide with the objective outcomes understood by the mechanism designer. Those subjective consequences belong in the "large world" (in the sense of Savage, 1954) in which the individual participant formulates his or her preferences. We postulate that, for all practical purposes, such "large world" is generally not accessible to the designer, who can only identify and set lotteries in the "small world" of objective outcomes (e.g., price-allocation vector pairs). Hence, we submit, the work of the designer should be carried on behind a "veil of ignorance" about the broader context of decision, represented by the individual, "large worlds" of subjective consequences of the mechanism participants.

As we would like to represent the same decision in two distinct frames of reference, namely, the objective world of the designer and the subjective one of the participant, we seek a representation in which the two perspectives can be conveniently related. For this reason we shall identify lotteries with unit vectors in L^2, rather than in L^1 as it is customarily done. This is because Euclidean space is the only L^p space that is also Hilbert, and Hilbert spaces have the unique property that the set of unit vectors is invariant with respect to basis changes. Specifically, we introduce the following setup and notation.

S is a finite set of states of Nature.

$\langle . | . \rangle$ denotes the usual inner product in Euclidean space.

Ω is the natural basis in $\mathbb{R}^n$, identified with a finite set $\{\omega^1, \dots, \omega^n\}$ of lottery outcomes (prizes).

Z is an orthonormal basis in $\mathbb{R}^m$, with $m \geq n$, identified with a finite set of subjective consequences $\{z^1, \dots, z^m\}$. V is an arbitrary $(m \times n)$ matrix chosen so that, for all ω^i in Ω, $V\omega^i$ is a unit vector in $\mathbb{R}^m$. Observe that V is always well defined as long as $m \geq n$. When $m = n$, we conventionally set $V \equiv I$, where I is the $n \times n$ identity matrix.

Lotteries correspond to L^2 unit vectors $x \in \mathbb{R}^n_+$; X is the set of all lotteries.

Since Ω is the natural basis, $\langle \omega^i | x \rangle^2 = x_i^2$; this quantity is interpreted as $p(\omega^i | x)$.

The quantity $\langle z^j | V x \rangle^2$ is interpreted as $p(z^j | x)$, the conditional probability of subjective consequence z^j given lottery x. In particular, $\langle z^j | V \omega^i \rangle^2$ is interpreted as $p(z^j | \omega^i)$, the conditional probability of subjective consequence z^j given the degenerate lottery which returns objective outcome ω^i for sure. Once the subjective consequences z^j are specified, for any lottery x one can readily compute $p(\omega^i | x) = x_i^2$ and $p(z^j | x) = \langle z^j | V x \rangle^2$. Moreover, given the latter probabilistic constraints, one can readily identify a lottery x and an orthonormal basis Z which jointly satisfy them. Hence, in the above construction lotteries are identified with respect to two different frames of reference: objective lottery outcomes, and subjective consequences.

An act is identified with a function $f : \mathcal{S} \to X$. H is the set of all acts.

$\Delta(X)$ is the (nonempty, closed, and convex) set of all probability functions on Z induced by lotteries in X. M is the set of all vectors $(p_s)_{s \in \mathcal{S}}$, with $p_s \in \Delta(X)$.

For each $f \in H$ a corresponding risk profile $p^f \in M$ is defined, for all $s \in \mathcal{S}$ and all $z^j \in Z$, by $p_s^f(z^j) := \langle z^j | V f_s \rangle^2$.

As customary, we assume that the decision-maker's preferences are characterized by a rational (*i.e.*, complete and transitive) preference ordering $\succsim$ on acts. Next, we proceed with the following assumptions, which mirror those in Anscombe and Aumann (1963).

Axiom 3.1 (Projective) *There exists a finite orthonormal basis* $Z := \{z_1, \ldots, z_m\}$, *with* $m \geq n$, *such that any two acts* $f, g \in H$ *are indifferent if* $p^f = p^g$.

In Anscombe and Aumann's setting, the above axiom is implicitly assumed to hold with $Z \equiv \Omega$. Because of Axiom 3.1, preferences on acts can be equivalently expressed as preferences on risk profiles. For all $p^f, p^g \in M$, we stipulate that $p^f \succsim p^g$ if and only if $f \succsim g$.

Axiom 3.2 (Archimedean) *If* $p^f, p^g, p^h \in M$ *are such that* $p^f \succ p^g \succ p^h$, *then there exist* $a, b \in (0, 1)$ *such that* $ap^f + (1-a)p^h \succ p^g \succ bp^f + (1-b)p^h$.

Axiom 3.3 (Independence) *For all* $p^f, p^g, p^h \in M$, *and for all* $a \in (0, 1]$, $p^f \succ p^g$ *if and only if* $ap^f + (1-a)p^h \succ ap^g + (1-a)p^h$.

Axiom 3.4 (Non-degeneracy) *There exist* $p^f, p^g \in M$ *such that* $p^f \succ p^g$.

Axiom 3.5 (State Independence) *Let* $s, t \in \mathcal{S}$ *be non-null states, and let* $p, q \in \Delta(X)$. *Then, for any* $p^f \in M$, $(p_1^f, \ldots, p_{s-1}^f, p, p_{s+1}^f, \ldots, p_n^f) \succ (p_1^f, \ldots, p_{s-1}^f, q, p_{s+1}^f, \ldots, p_n^f)$ *if, and only if,* $(p_1^f, \ldots, p_{t-1}^f, p, p_{t+1}^f, \ldots, p_n^f) \succ (p_1^f, \ldots, p_{t-1}^f, q, p_{t+1}^f, \ldots, p_n^f)$.

Theorem 3.1 (Anscombe and Aumann) *The preference relation* $\succsim$ *fulfills Axioms* $1-5$ *if and only if there is a unique probability measure* π *on* $\mathcal{S}$ *and a non-constant*

function $u : Z \to \mathbb{R}$ *(unique up to positive affine rescaling) such that, for any* $f, g \in H$, $f \succsim g$ *if, and only if,*
$\sum_{s \in \mathcal{S}} \pi(s) \sum_{z^i \in Z} p_s^f(z^i) u(z^i) \geq \sum_{s \in \mathcal{S}} \pi(s) \sum_{z^i \in Z} p_s^g(z^i) u(z^i)$.

The following theorem, proven in La Mura (2009), relates the Anscombe and Aumann result to the small world of objective outcomes.

Theorem 3.2 *The preference relation* $\succsim$ *fulfills Axioms 1–5 if and only if there is a unique probability measure* π *on* $\mathcal{S}$ *and a symmetric* $(n \times n)$ *matrix* U *with distinct eigenvalues such that, for any* $f, g \in H$, $f \succsim g$ *if, and only if,*
$\sum_{s \in \mathcal{S}} \pi(s) f_s^{\mathsf{T}} U f_s \geq \sum_{s \in \mathcal{S}} \pi(s) g_s^{\mathsf{T}} U g_s$.

Let us call the above a projective expected utility (PEU) representation. Observe that any two objective outcomes a and b identify a sub-matrix of the general payoff matrix U that we can always write in the form below.

$$\begin{bmatrix} u(a) & \epsilon_{a,b}||u(a) - u(b)|| \\ \epsilon_{a,b}||u(a) - u(b)|| & u(b) \end{bmatrix}$$

Observe that the PEU from an equal objective mixture of a and b is given by

$$(\sqrt{1/2}, \sqrt{1/2})U_{a,b}(\sqrt{1/2}, \sqrt{1/2})^{\mathsf{T}} = u(a)/2 + u(b)/2 + \epsilon_{a,b}||u(a) - u(b)||.$$

Hence, if $\epsilon_{a,b} > 0$, the objective lottery involving an equal probability of a and b has a higher utility than a situation involving equal probabilities but purely subjective uncertainty (in which case, the PEU and EU values coincide and are given by $u(a)/2 + u(b)/2$). If we further assume that a and b are monetary amounts, then a linear (resp. concave, convex) specification of u captures risk neutrality (resp. risk aversion, risk loving), while a zero (resp. positive, negative) $\epsilon_{a,b}$ captures ambiguity neutrality (resp. aversion, loving).

4 Mechanisms with PEU Agents

It is shown in La Mura (2009) that every finite game with PEU-maximizing players has an equilibrium, possibly involving a combination of objective randomization and subjective uncertainty. Backward induction reasoning can be performed in games with PEU agents, but the solution may or may not coincide with the one obtained assuming EU-maximizing agents, depending on contextual effects. This is due to the fact that lotteries involving strictly dominated strategies may not be dominated. For instance, when in the last round of a Centipede game player 2 chooses with what probability p to quit, the PEU payoff is given by

$$(\sqrt{p}, \sqrt{1-p})U(\sqrt{p}, \sqrt{1-p})^{\mathsf{T}} = u(a)p + u(b)(1-p) + 2\epsilon||u(a) - u(b)||\sqrt{p}\sqrt{1-p}$$

where $u(a), u(b)$ are the payoffs from quitting and staying, respectively, and $u(a) > u(b)$. The first order condition with respect to p is given by

$$u(a) - u(b) + \epsilon||u(a) - u(b)||\frac{1-2p}{\sqrt{p}\sqrt{1-p}} = 0.$$

Observe that, for an ambiguity-averse decision-maker, the FOC is never satisfied at $p = 0$ or $p = 1$. In particular, when ϵ is positive and small the optimal probability of quitting is very close to, but strictly less than, one.

Why would an ambiguity-averse decision-maker prefer to commit to a random variable giving a probability of success that is close, but not equal to one? One can imagine cases in which what appeared to the decision-maker as a negative outcome at the time of decision turns out against all odds to be a positive one (a "Frog Prince"), or vice versa. If there is any ambiguity on the actual payoffs associated to each outcome, the only possibility for the decision-maker to reduce it would be to remain open, even with very low probability, to those outcomes which appear dominated at the moment of decision. If both players are ambiguity-averse, this guarantees a positive probability of continuation at all rounds of the Centipede. Yet, unless ϵ is sufficiently large the payoff from continuing remains lower than that from quitting, and hence the backward-induction solution still involves quitting, with probability very close to one, at every opportunity independently of the number of rounds.

Let us call a mechanism whose participants maximize PEU a PEU mechanism. Which general results from the theory of economic mechanisms, and in particular auction theory, carry over to the case of PEU mechanisms?

Let us consider mechanisms for PEU agents with private values, and preferences that are quasi-linear with respect to money, where possible contextual effects only take the form of varying degree of ambiguity aversion on surplus, controlled by a single parameter ϵ. Specifically, let us assume that for any two mechanism outcomes (a, x) and (b, y), where x and y are monetary expenditures, the corresponding utility matrix is given by

$$\begin{bmatrix} v(a) - x & \epsilon||(v(a) - x) - (v(b) - y)|| \\ \epsilon||(v(a) - x) - (v(b) - y)|| & v(b) - y \end{bmatrix}$$

Some general foundations, such as the Revelation Principle (Myerson, 1981), do not depend on whether participants are assumed to maximize EU or PEU.

The Revenue Equivalence theorem for single-object auctions still holds in case all agents have diagonal payoff matrices (in which case PEU reduces to EU), but fails with general payoff assignments. This will become apparent in the next section, where we analyze bidders' behavior in first and second price auctions with varying attitudes towards ambiguity on surplus.

The Vickrey-Clarke-Groves mechanism (VCG) with PEU participants is still individually rational and truthful whenever $\epsilon \leq 0$. By contrast, for positive and small values of ϵ (and hence, for strictly ambiguity-averse agents) the VCG mechanism is never truthful. Yet, it can be modified in a simple way to obtain a truthful mechanism that is still interim efficient, but may fail to be ex-post efficient. This will also become apparent in the next section, when we discuss truthfulness in second-price auctions.

5 First and Second Price Auctions with Ambiguity-Sensitive Bidders

Consider a second-price auction for a single item with PEU-maximizing bidders. Each bidder has quasi-linear utility, with payoff matrix

$$\begin{bmatrix} v_i - P & \epsilon||v_i - P|| \\ \epsilon||v_i - P|| & 0 \end{bmatrix}$$

so that $v_i - P$ is the surplus for i from winning the object at price P.

For general values of ϵ, if p is the probability of winning the object, then the PEU payoff is given by

$$(\sqrt{p}, \sqrt{1-p})U(\sqrt{p}, \sqrt{1-p})^{\top} = (v_i - P)p + 2\epsilon||v_i - P||\sqrt{p}\sqrt{1-p}).$$

Taking the first order condition with respect to p, and multiplying both sides by $\sqrt{p}\sqrt{1-p}$, one obtains

$$(v_i - P)(\sqrt{p}\sqrt{1-p}) + \epsilon||v_i - P||(1-2p) = 0.$$

Solving for p, we find that $p^* = 1/2 \pm (1/2)\sqrt{1-4\epsilon^2}$. Hence, the first order condition implies that, if ϵ is zero, the bidder will want to set p to either zero or one in case the surplus at price P is negative or positive, respectively. Evaluating the FOC near $p = 0$ and $p = 1$ one finds that, for negative values of ϵ, the bidder would still want to set the probability of winning to zero or one depending on whether the surplus is positive or negative, just as above. By contrast, for small and positive values of ϵ the bidder will want to set a probability very close to (but strictly less than) one when the surplus is positive, and a probability very close to (but strictly more than) zero when the surplus is negative. This cannot be obtained in case the bid is always set to the true value. Hence, we conclude that second-price auctions cannot be truthful in case bidders are averse to ambiguity. Yet, in such scenarios a simple modification of the second-price auction, namely, one in which a small amount of noise is added to each bid on behalf of the players, could be made into a truthful mechanism that would still be interim (but not ex-post) efficient. In particular, consider the very simple scenario of a second-price auction with a single participant, who wins the object at a price of zero if, and only if, her bid is strictly positive. Then a modification of the format, in which the accepted bid is set by a suitable random variable to be either the amount reported by the participant (with probability p^*) or zero (with probability $1 - p^*$), would be truthful because it would exactly replicate the optimal strategy on the participant's behalf.

Let us now consider first-price auctions, in a simple scenario with two bidders, with values that are i.i.d. and uniform in [0, 1]. If bidder $j \in 1, 2$ is bidding $y = v_j/2$, then bidder i $(i \neq j)$ by offering $x \in [0, 1]$ wins with probability $2x$, and receives a payoff of $v_1 - x$, otherwise gets nothing and pays nothing. The expected payoff for bidder i is $(\sqrt{2x}, \sqrt{1-2x})U_i(\sqrt{2x}, \sqrt{1-2x})^{\mathsf{T}}$, where U_i is the payoff matrix above with $P = x$.

Assuming $x \leq v_i$, the PEU payoff is given by

$$(\sqrt{2x}, \sqrt{1-2x})U_i(\sqrt{2x}, \sqrt{1-2x})^{\mathsf{T}} = (v_i - x)(2x + 2\epsilon\sqrt{2x(1-2x)}).$$

Taking the first order condition, and rearranging, yields

$$(v_i - 2x)\sqrt{2x(1-2x)} + \epsilon((v_i - x)(1-4x) - 2x(1-2x)) = 0.$$

Observe that, when ϵ is zero, the unique solution is $x = v_i/2$. This identifies the unique symmetric equilibrium with EU bidders, in which both bid half of their true value.

Evaluating the left-hand side of the first-order condition at $x = v_i/2$ with general ϵ one obtains $-\epsilon v_i/2$, which is never zero except for the zero-value type. When ϵ is positive the above expression is negative, and hence in that case the bidder would want to bid less than half of the true value. By contrast, for negative ϵ the bidder would want to bid strictly more than that amount.

Recalling that a positive ϵ in the PEU payoff matrix captures ambiguity aversion, we can interpret the above result along the following lines: bidders who are averse (resp., propense) to ambiguity will tend to bid less (resp., more) aggressively in a first-price auction than ambiguity-neutral ones. This is consistent with experimental results on first and second price auctions, which find that bids in first price auctions are lower in the presence of ambiguity (Chen et al., 2007).

The pattern of lower bids in first price auctions by ambiguity-averse bidders could also be used, in principle, to explain afternoon effects in sequences of first price auctions for identical objects. For instance, consider a scenario with $n - 1$ objects for n bidders, where values are i.i.d. and uniform in [0, 1]. Then with EU bidders the following is an equilibrium. In each auction bidder $i \in I$ bids $x_i = v_i/2$, which is i's expected price, and also expected surplus, in case of winning. In equilibrium bidder i expects a surplus of $v_i/2$ with probability $1 - (1 - v_i)^{n-1}$, and zero otherwise.

If bidder i is ambiguity averse, the PEU payoff at the EU equilibrium is given by

$$(v_i/2)(1 - (1 - v_i)^{n-1}) + 2\epsilon(v_i/2)\sqrt{1 - (1 - v_i)^{n-1}}\sqrt{(1 - v_i)^{n-1}},$$

which for n large is approximately the same as the EU payoff. As the number of remaining objects decreases the probability of winning an object also decreases for each of the remaining bidders. Observe that the difference between EU and PEU payoff, and hence the distortion, becomes highest when the probability is close to $1/2$, while in case the probability of receiving an object is close to one or zero the distortion becomes negligible. When n is sufficiently large bidder i initially expects to win one of the $n-1$ objects with probability close to one, in which case distortion and underbidding are both negligible. For each remaining bidder the probability of winning an object decreases with each new auction, but as long as it remains above 1/2 an ambiguity-averse bidder will underbid more and more with each new auction. By contrast, in scenarios when only few objects remain on sale among many bidders then the probability of winning an object will eventually become lower than 1/2 for most bidders, in which case the distortion and the underbidding again decrease. This suggests that sequences of first price auctions with ambiguity averse bidders will tend to produce afternoon effects when the probability of winning for most bidders remains sufficiently high (few bidders per object), and reverse afternoon effects in case that probability becomes sufficiently low (many bidders per object).

Acknowledgements We are grateful to Adam Brandenburger and Lukasz Swiatczak for valuable comments. Financial support from the German Bundesbank is gratefully acknowledged.

References

Andersson, O., & Andersson, T. (2017). Timing and presentation effects in sequential auctions. *Journal of Mechanism and Institution Design, 2*(1), 39–55. Society for the Promotion of Mechanism and Institution Design, University of York.

Anscombe, F. J., & Aumann, R. J (1963). A definition of subjective probability. *The Annals of Mathematical Statistics, 34*, 199–205.

Ashenfelter, O. (1989). How auctions work for wine and art. *Journal of Economic Perspectives, 3*(3), 23–36.

Aumann, R. J. (1995). Backward induction and common knowledge of rationality. *Games and Economic Behavior, 8*(1), 6–19.

Brandenburger, A., & La Mura, P. (2016). Team decision problems with classical and quantum signals. *Philosophical Transactions. Series A, Mathematical, Physical, and Engineering Sciences, 374*(2058), 20150096.

Chen, Y., Katuscak, P., & Ozdenoren, E. (2007). Sealed bid auctions with ambiguity: Theory and experiments. *Journal of Economic Theory,136*(1), 513–535.

Deltas, G., & Kosmopoulou, G. (2004). 'Catalogue' vs 'Order-of-sale' effects in sequential auctions: Theory and evidence from a rare book sale. *The Economic Journal, 114*(492), 28–54.

La Mura, P. (2005). Correlated equilibria of classical strategic games with quantum signals. *International Journal of Quantum Information, 03*, 183.

La Mura, P. (2009). Projective expected utility. *Journal of Mathematical Psychology, 53*(5), 408–414.

Myerson, R. B. (1981). Optimal auction design. *Mathematics of Operations Research, 6*(1), 58–73.

Rosenthal, R. (1981). Games of perfect information, predatory pricing, and the chain store paradox. *Journal of Economic Theory, 25*, 92–100.

Savage, L. J. (1954). *The foundations of statistics*. New York: Wiley.

Weber, R. J. (1983). Multiple-object auctions. In R. Engelbrecht-Wiggans, M. Shubik, & R. M. Stark (Eds.), *Auctions, bidding, and contracting: Uses and theory*. New York: New York University Press.

Weber, R. J., & Milgrom, P. R. (2000). A theory of auctions and competitive bidding II. In P. Klemperer (Ed.), *The economic theory of auctions*. Cheltenham: Edward Elgar Publishing Ltd.

The manufacturer's authorised representative in the EU is Springer Nature Customer Service Centre GmbH, Europaplatz 3, 69115 Heidelberg, Germany. If you have any concerns regarding our products, please contact ProductSafety@springernature.com

Printed and bound by CPI Group (UK) Ltd, Croydon, CR0 4YY
30/09/2026
02225651-0012